EXTRAORDINARY ANIMALS

EXTRAORDINARY ANIMALS

An Encyclopedia of
Curious and Unusual Animals

Ross Piper

Illustrations by Mike Shanahan

GREENWOOD PRESS

Westport, Connecticut • London

Library of Congress Cataloging-in-Publication Data

Piper, Ross.
  Extraordinary animals : an encyclopedia of curious and unusual
animals / by Ross Piper ; Illustrations by Mike Shanahan.
      p. cm.
  ISBN-13: 978–0–313–33922–6 (alk. paper)
  ISBN-10: 0–313–33922–8 (alk. paper)
  1. Animals—Encyclopedias.   I. Title.
  QL7.P57   2007
  590—dc22   2007018270

British Library Cataloguing in Publication Data is available.

Library of Congress Catalog Card Number: 2007018270
ISBN-13: 978–0–313–33922–6
ISBN-10: 0–313–33922–8

First published in 2007

Greenwood Press, 88 Post Road West, Westport, CT 06881
An imprint of Greenwood Publishing Group, Inc.
www.greenwood.com

Printed in the United States of America

The paper used in this book complies with the
Permanent Paper Standard issued by the National
Information Standards Organization (Z39.48–1984).

10 9 8 7 6 5 4 3 2 1

In memory of my Dad

CONTENTS

x   CONTENTS

PREFACE

Extraordinary Animals is an exploration of the animal kingdom, a cherry-picking of these fantastically diverse organisms whose ways and characteristics are astounding and often stranger than fiction. The book covers a wide variety of animal life, including many obscure but exceptionally interesting creatures, the likes of which can only be discovered in the confines of specialized, very inaccessible textbooks. Not only is the diversity of the subject matter unique, but the content has been thoroughly researched for scientific accuracy and is written in a way that it is clear, engaging, and enthusiastic.

The audience for *Extraordinary Animals* is basically anyone with an interest in nature—the sort of people who buy books from the natural history section of a bookstore or who enjoy nature documentaries. The main purpose of *Extraordinary Animals* is to highlight just how remarkable animals are in a way that just about anyone can read and understand. Textbooks are full of fascinating information, but all too often, they are inaccessible to general audiences. This book provides a bridge to those resources for anyone who has even the slightest interest in the natural world.

In this book, you will find 120 animals separated into one of eight categories. You can dip into the book wherever you want to as it is not laid out so that you have to read it from cover to cover. Each piece contains information on how the animal is classified, what it looks like, how big it is, and where it lives. The main body of the piece is devoted to the extraordinary natural history or characteristics of the animal. A number of bulleted facts give some extra, interesting information on the animal. Some of the animals in the book can quite easily be found in a backyard or in places that are not that exotic, and in these cases, there is a "Go Look!" section that gives tips on how and where to find them, how to watch them, and how to look after them in captivity for short periods of time.

It was the initial intention to include a list of Web sites to which the reader could go to find additional information on these animals; however, the content of these sites can never be guaranteed, and with the constant reshuffling of pages on the Web, links can rapidly become inactive or useless. For those readers keen to trawl the Web for extra information, the best way is to type the Latin name, or perhaps the common name, into an Internet search engine. The amount of information on the Web today is such that there will be numerous pages on most of the animals in this book, but only those sites ending in .gov or .edu will carry information that has been thoroughly researched and edited. At the end of many entries, there is a list of resources for further reading. These lists, as well as the selected bibliography at the end of the book, include textbooks and journal articles that can be found in any decent library. Some of these books have an asterisk (*) appearing next to them—it is these resources that I heartily recommend you buy as they are a treasure trove of information for anyone interested in the natural world.

Wherever possible, I have tried not to use jargon. There is a whole dictionary of specialized zoological terms, which can sometimes be confusing or difficult to say. I have tried to write in more general terms without using this specialized language. However, there is a glossary at the end of the book to explain any jargon that was unavoidable.

ACKNOWLEDGMENTS

I would like to thank the following individuals for their comments and suggestions on earlier versions of the manuscript: John Alcock, Christian Bordereau, Tom Buckelew, Jason Chapman, Steve Compton, Paul Cziko, Ian Denholm, Stephanie Dloniak, Jack Dumbacher, Mark Eberhard, Howard Frank, Megan Frederickson, Douglas Fudge, Ram Gal, Mike Howell, Robert Jackson, Jeff Jeffords, David Julian, Uwe Kils, Alex Kupfer, Jim Macguire, Andrew Mason, David Merritt, William Miller, Claudia Mills, Sarah Munks, Phil Myers, Dick Neves, Arne Nilssen, Jerome Orivel, Robert Presser, Galen Rathbun, Thomas Roedl, Ernest Seamark, Andrei Sourakov, Erhard Strohm, Paul Sunnucks, Laurie Vitt, Ashley Ward, Marius Wasbauer, and Philip Weinstein. I would also like to thank Mike Shanahan for his brilliant illustrations, Lucy Siveter of Image Quest Marine for sourcing images, Adam Simmons for reviewing an early draft of the manuscript, and Roger Key. I would like to thank Bart Hazes and Mike Howell for going out of their way to help me. Special thanks go to Kevin Downing for giving me the chance to write this book.

INTRODUCTION

The earth, from a purely celestial point of view, is unremarkable. It is a small planet in a solar system orbiting a medium-sized star in a smallish galaxy, the Milky Way, which contains billions of solar systems. The Milky Way is but one of billions upon billions of galaxies in the incomprehensible vastness of the universe. Yet, in one respect, the earth is special beyond compare. It is the only place we know of on which there is life.

Life is such a small, seemingly insignificant word, yet it encompasses a fantastic diversity of living forms. The exact time and nature of life's appearance on the earth has divided scientists for decades, and it will continue to do so because the time spans with which we are dealing are huge, almost impossible for us to grasp, and the evidence is fragmentary and hard to come by.

What we do know is that the earth is very, very old—4.6 billion years old to be exact—but for the vast majority of this time, it was a lifeless globe cooling from the fires of its creation, circling the sun in the young solar system, while being heavily bombarded by asteroids. Over hundreds of millions of years, the earth changed and the asteroid impacts became less frequent. Oceans formed and our planet became slightly more hospitable, but conditions on this primordial Earth were still very different from the comparatively balmy conditions we enjoy today. And then, more than 3 billion years ago, the first life evolved. Where and how are questions we can only make good educated guesses at, but an experiment conducted in the 1950s by scientists in the United States showed that lightning bolts discharged through an atmosphere, the likes of which could have shrouded the young Earth, could have produced biological molecules—the precursors of the first simple cells. Although these experiments have since been called into question, as more recent findings suggest that the mix of gases used by the scientists to mimic the atmosphere of the young earth was probably inaccurate, they do give us an idea of what may have happened all those millions of years ago. The complexity of these first biological molecules increased over the eons, eventually forming the first self-contained biological systems, which in turn gave rise to the first proper cells—the first life.

This first life was no more than simple, single-celled organisms, and these organisms had the earth to themselves for a long, long time. In the atmosphere that shrouded the earth at this time, oxygen was as good as absent, but it is thought these first life-forms created oxygen as a waste gas. Over more immense stretches of time, the levels of oxygen in the atmosphere steadily grew until oxygen became quite abundant. Then, around 700 million years ago, this simple life gave rise to increasingly complex forms. From that point onward, the diversity of life on earth exploded. Lots of different life-forms and body plans appeared, some of which were successful, spawning long, unbroken lines of descent, while others disappeared into prehistory. The life-forms interacted with and adapted to each other, becoming ever-more entwined. The extraordinary diversity of life on earth today reflects the relationship between organisms and their environment—an intricate web of interactions with the continual processes of adaptation and change, fine-tuning every species over time to its environment. Life-forms have become so attuned to their environment that scarcely a niche is vacant; in almost every conceivable habitat on earth, animals can be found. In the deepest parts of the ocean, more alien to us than the surface of Mars, creatures thrive. Even in the coldest and highest places on earth, you would be hard pressed not to see some form of animal life. To us, perhaps the most bizarre place to live is inside another animal, yet many creatures have taken to this parasitic way of life and have become very good at it.

Traditionally, scientists have classified life on earth into five different kingdoms based primarily on shared characteristics. This system has undergone many modifications, but it is straightforward and I use it in this book to describe how animal life is categorized. This system of classification is hierarchical, starting with the kindgom level, followed by phylum, class, order, family, genus, and species. Organisms are named by a two-word system, the genus followed by the specific epithet, which together give the species' scientific name. For example, the European honeybee's scientific name is *Apis* (genus) *mellifera* (specific epithet). In this classification scheme, the first and most primitive kingdom is that of bacteria. Next are the plants, familiar to us as they dominate terrestrial ecosystems. The fungi represent the third kingdom. The fourth kingdom is the one to which we belong—the animals. The fifth kingdom, a sort of taxonomic dustbin, includes the protists, organisms that do not really fit in to any of the other four kingdoms.

The topic of this book is the diversity of the animal kingdom, but you will not find an inexhaustible list of all of the animals on earth between the covers of this book—far from it. Such a book would take years to read, and it would hardly be the sort of thing you could easily keep on your bookshelf. No, this is essentially a cherry-picking of the animal kingdom. It only covers living animals because although long extinct animals are fascinating, their lives are the stuff of guesstimation. Bones and impressions of long dead bodies can only tell so much. This book is a selection of those animals whose fantastic habits and lives really hammer home the message of how remarkable our planet is. All the animals you will read about are real. Some are found outside your back door; others dwell in habitats where humans rarely venture. Some are miniscule, barely visible to naked eye, and some are massive, thousands of times larger than a fully grown human. Many of them are rarely seen, and there is still a great deal to learn about their lives.

The scientists who study animals are known as zoologists, and it is these people who unravel the mysteries of the animal kingdom. Scientists, by their very nature, have an urge to categorize and order the things they study, and zoologists are no exception. All animal life may be divided into 38 different categories, or phyla. Each phylum contains animals that in one way or another are very similar, and they may be grouped by shared physical characteristics or genetic similarities. Animals are continually being shifted within and between these phyla as scientists understand more about DNA and genetics. It has to be remembered that these phyla are a construct of the human mind and are merely an abstraction that allows us to make sense of the natural world. The number of species within these phyla is a huge bone of contention. Estimates range from 1.5 million to 100 million, but we may never know the true number.

Regardless of the human need to categorize and identify things, it goes without saying that animals are a source of intense interest for a large percentage of the population. Perhaps this stems from the more primitive days of the human race when we lived in much greater harmony with nature. Before the days of agriculture and even civilization, our forebears would not have lasted long without a thorough understanding of the animals that shared their environment— which species they could use for food and which species they should avoid. Today, you can still see this impressive level of understanding in the ways of the tribes that survive in the more remote reaches of our planet. For thousands of years, aboriginal people have lived in the same way thanks to their intimate knowledge of the world about them.

Today, most people's interest in animals begins in childhood with the creatures found in a typical backyard, inevitably, the ones lovingly described as creepy-crawlies—the insects and their relatives. These animals are easy to find and easy to keep in glass jars or old margarine tubs. This fascination with bugs grows, and before long, you find yourself reading about other animals, some of which you will probably never see but whose origins, diversity, and astonishing lives amaze you. This book is an encapsulation of this path of curiosity, and it draws on things I have read and things I have seen. I hope that whoever reads this will find the animals contained herein as interesting as I do.

STRENGTH IN NUMBERS: ANIMAL COLLECTIVES

ACACIA ANT

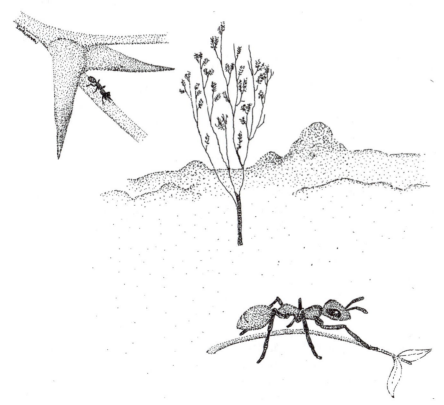

Acacia Ant—The ant on the acacia tree on which it lives. (Mike Shanahan)

Scientific name: *Pseudomyrmex ferruginea*
Scientific classification:
  Phylum: Arthropoda
  Class: Insecta
  Order: Hymenoptera
  Family: Formicidae
What does it look like? This ant is a slim-bodied species, with the workers measuring around 3 mm in length. They are orangey brown and have very large eyes.
Where does it live? This is an arboreal ant, and it is to be found on or around a certain species of acacia tree that is found throughout Central America. The ants nest inside the large thorns of these acacias.

A Relationship among the Thorns

Acacia trees with their succulent little leaves are relished by a large number of herbivorous animals from tiny insects to huge mammals. To protect their foliage from these hungry mouths, they have evolved a number of ways to keep the leaf eaters at bay. Many acacias have vicious-looking spines, while others have leaves packed with repellent and noxious chemicals. Some acacias, however, have gone even further and actually depend on other animals for protection. One such acacia, the bull's horn acacia, has its own species of dedicated ant bodyguard. The relationship begins when a young queen ant, newly mated, lands on the acacia looking for a place to start a nest. The thorns on this acacia are great little ant havens as they are bulbous at their base and hollow. The queen, convinced by the odor of the tree that she is in the right place, starts to nibble a hole in the tip of one of the thorns, eventually breaking through to the cavity within. In the safety of her new nest she lays 15–20 eggs, and soon enough, these give rise to the first generation of workers. The embryonic colony grows, and as it does, it expands into more of the bulbous thorns. When the colony has exceeded around 400 individuals, the repayment to the acacia for lodgings can begin, and the ants assume their plant-guarding role. The ants become aggressive and do not take kindly to any creature they find trying to surreptitiously munch the acacia's leaves, regardless of whether it is a cricket or a goat. It doesn't take much to set them off. Even the whiff of an unfamiliar odor sees the ants swarming from their thorns and toward a potential threat. Herbivorous insects are killed or chased away, and browsing mammals are stung in an around their mouth, which quickly persuades them to look elsewhere for less well-defended fodder. Apart from these active defending duties, the ants also have gardening to tend to—they leave the tree and scout around its base looking for any seedlings that would eventually compete with their acacia for light, nutrients, and water. If they do find any, they destroy them, and the ants even go so far as to prune the leaves of nearby trees so that their host is not shaded out.

Not only does the tree supply the ants with nesting sites, but special glands at the base of the tree's leaves produce a nectar rich in sugars and amino acids that the ants lap up. The tips of the leaves also sprout small, nutritious packets of oils and proteins (Beltian bodies), which the ants snip off and carry away to feed to their grubs. The grubs even have a little pouch at their head end which the Beltian body can be tucked into while they feast on it.

This charming relationship between the ants and the acacias is as not as wholesome as it first appears. The ants will repel most herbivorous insects, but they turn a blind compound eye to the feeding antics of scale insects, which suck the sap of the host acacia, thus weakening it and providing entry for disease. The ants tolerate and even protect the scale insects because they produce sweet honeydew, which the ants relish.

- This is not the only example of a symbiotic relationship between a tree and an ant species. There are at least 100 species of ants that live in a close partnership with a plant. In return for the services provided by the ants, the plants furnish them with accommodation. The diversity of the relationships encompasses all the conceivable parts of a plant. Some plants have modified swellings on their branches and twigs, while others have cavities within their stems and trunks or modified roots that house their insect guests.

- Some of these nests can be very small, only a few centimeters across, while others can be large and elaborate.

- In South America, there is a tree, *Duroia hirsute*, which is sometimes found to dominate small areas of the rain forest, forming areas that the local people call "devil's gardens." Within special cavities on the tree's trunk there are special cavities in which the ant, *Myrmelachista schumanni*, makes its nest. Any saplings sprouting near the host tree are attacked by the ants and stung with formic acid, which kills them, thus removing competition for their host's resources.

- In some of these relationships, the cost of the ant's protection can be quite expensive. *Cordia* trees in the Amazonian rain forest have a kind of partnership with *Allomerus* ants, which make their nests in modified leaves. To increase the amount of living space available to them, the ants will destroy the tree's flower buds. The flowers die and leaves develop instead, providing the ants with more dwellings. Another type of *Allomerus* ant lives with the *Hirtella* tree in the same forests, but in this relationship, the tree has turned the tables on the greedy ants. When the tree is ready to produce flowers, the ant abodes on certain branches begin to wither and shrink, forcing the occupants to flee and leaving the tree's flowers to develop free from attack by the ants.

- In the mangrove forests of Southeast Asia and New Guinea there lives an odd plant called *Myrmecodia*. This green, spiky, small football-sized plant clings to the branch or trunk of a mangrove tree. Scuttling over its surface are numerous ants, and the odd plant is their home. Open it and an elaborate system of tunnels and chambers will be revealed. Some of these chambers are the nest's rubbish tips, and the waste therein is used by the plant as a fertilizer, allowing it to flourish even though its roots will never come into contact with the soil.

- As you can see, ants have struck up some amazingly complex relationships with plants. On the whole, the two very different organisms help each other, but occasionally, there are freeloaders. These stories exemplify the degree to which insects and flowering plants have become inextricably linked over millions of years. Ever since the flowering plants first appeared on earth many millions of years ago, the insects have gravitated toward them and evolved with them, resulting in the complex relationships we see today.

Further Reading: Frederickson, M., Greene, M. J., and Gordon, D. M. "Devil's gardens" bedevilled by ants. *Nature* 437, (2005) 495–96.

ANTARCTIC KRILL

Antarctic Krill—A whale moving in to engulf a swarm of Antarctic krill. (Mike Shanahan)

Antarctic Krill—An adult krill clearly showing the feeding basket formed by its forelimbs. (Uwe Kils)

Scientific name: *Euphausia superba*
Scientific classification:
    Phylum: Arthropoda
    Class: Malacostrata
    Order: Euphausiacea
    Family: Euphausiidae

What does it look like? The Antarctic krill can be about 6 cm long when fully grown, and it is more or less transparent, with a pair of big black eyes. The antennae are long and sprout from the very front of the head. The thorax bears several pairs of specialized limbs that form a basketlike structure. The abdomen has several pairs of swimming limbs and ends in a paddle called the *telson*.

Where does it live? This crustacean is found in the southern waters surrounding the continent of Antarctica. Their preferred habitat varies depending on how old they are. As youngsters, they dwell at great depths, but young adults and adults spend their time in surface waters.

Swarming Crustaceans

There can be few animals whose importance in the planet's ecosystems is as great as the Antarctic krill. Singly, they are not that impressive. They look like a myriad of other shrimplike animals, but what they lack in appearance they more than make up for in sheer abundance. The life of an Antarctic krill starts as a fertilized egg, about the size of a period, sinking into the abyss. As the egg descends, its cells divide and differentiate to form the young larva. At a depth of between 2,000 and 3,000 m, the baby krill hatches and begins to ascend, developing and growing as it makes slow but steady progress through the icy waters, sustained by the remaining yolk from its egg.

In the surface waters, the young krill that have made their way successfully from the depths begin to form huge groups, known as swarms. The individuals in these groups continue growing, and it can take between two and three years from the time they hatch for them to reach maturity. The swarms are composed of adults and young adults and can be huge. They can stretch over an area of ocean equivalent to several city blocks and can be as much as 5 m thick, with as many as 60,000 krill in 1 cu. m of water. From the air, one of these swarms has been likened to a gigantic amoeba as it moves slowly through the water, changing shape as it goes. The preferred food of krill are the tiny, single-celled plantlike organisms called diatoms. These diatoms rely on the sun's rays for energy, using the power of sunlight to convert carbon dioxide and water into simple sugars—the all-important process of photosynthesis. As they are sun worshippers, these diatoms are only found in surface waters. They are found in such huge numbers that they form a kind of soup along with other minute organisms. These diatoms not only float freely in the water but also coat the underside of the pack ice, forming verdant lawns. The krill graze these upside-down pastures like tiny, multilimbed cows and also swim through the plankton using their front limbs like a straining basket to separate the edible cells from the water. They scrape these appendages clean with their mouthparts and swallow the green paste. They are messy eaters, and a lot of the green globules miss the crustaceans' mouths and sink slowly to the seafloor. The digestive system of krill is also far from efficient, and quite a large proportion of what they take in is egested without being broken down. These strings of krill waste follow the feeding debris to the sea bottom. All of this accumulating waste is known as a so-called biological pump, as an incomprehensible amount of atmospheric carbon dioxide, utilized by the diatoms, is locked away on the seafloor for around 1,000 years. This process is massively important in the regulation of the earth's climate. Today we are seeing the consequences of too much carbon dioxide in the atmosphere. Just suppose that for some unknown reason these delicate little animals were to disappear from the face of the earth tomorrow. If they were inexplicably whisked away, we would not only see the collapse of marine ecosystems everywhere, as they are eaten by so many other animals, but also the full and relentless fury of runaway global warming.

- There are around 86 species of what can be described as krill. Regardless of the species, they are all considered *keystone* species in marine ecosystems. They occur in such huge numbers that many animals depend solely on them for food. The huge cetaceans, like the blue whale, are a good example. Their diet consists of krill and whatever else happens to be swimming among the swarm.
- The total mass of Antarctic krill in the ocean during the peak of the season is estimated to be on the order of 125–725 million tonnes, making this species probably the most successful animal on the planet, in terms of biomass at any rate.
- For reasons that are not fully understood, krill numbers go through cyclical peaks and troughs that are thought to be linked to the abundance of pack ice surrounding Antarctica. In years where there is lots of pack ice, it provides numerous little nooks, crannies, and caverns in which the young krill can shelter from their many predators. They appear to suffer when there is little pack ice. In these lean years, they are replaced as the dominant plankton animal by jelly-bodied creatures called *salps*.
- In some areas of the Southern Ocean there are unusual places rich in nutrients, but where the diatoms and other photosynthesizing, single-celled organisms are

surprisingly rare. As there is no food for them, the krill are absent from these areas. It turns out that these odd tracts of ocean lack iron. Injecting iron gives the plantlike organisms what they need, and before long they bloom, attracting the attention of the gigantic swarms of krill. It has been suggested that ships could circle the Southern Ocean and inject iron into the water. This would stimulate the diatom populations and, in turn, the krill, providing a way of engineering the environment to increase the amount of carbon dioxide that is locked away in the deep ocean.

- Around 100,000 tonnes of this Antarctic krill species, *Euphausia superba*, is taken every year for animal and human consumption. In Japan, processed krill is known as *okiami*.

- The Antarctic krill, like all its relatives, sheds its skin very regularly. It is peculiar not only for the speed with which it can do this but also for its ability to grow smaller at each successive molt if food is scarce. The Antarctic krill can quite literally jump out of its old skin, leaving the skin floating in the water, where it may act as decoy to confuse predators.

Further Reading: Everson, I. *Krill: Biology, Ecology and Fisheries*. Blackwell Science, Oxford 2000; Hamner, W. M., Hamner, P. P., Strand, S. W., and Gilmer, R. W. Behavior of Antarctic krill, *Euphausia superba*: chemoreception, feeding, schooling and molting. *Science* 220, (1983) 433–35; Loeb, V., Siegel, V., Holm-Hansen, O., Hewitt, R., and Fraser, W. Effects of sea-ice extent and krill or salp dominance on the Antarctic food web. *Nature* 387, (1997) 897–900; Ross, R. M., and Quetin, L. B. How productive are Antarctic krill? *BioScience* 36, (1986) 264–69.

APHIDS

Scientific name: *Aphids*
Scientific classification:
> Phylum: Arthropoda
> Class: Insecta
> Order: Homoptera
> Family: Aphididae

What do they look like? Aphids are small insects, varying in size from 1 to 10 mm. They have soft bodies with long, spindly legs. There is normally a pair of thin turrets projecting from the animal's back end, which secrete a waxy substance. In each species there are winged and nonwinged forms. The mouthparts are formed into a long, thin structure called the *rostrum*, which is held under the body when the animal is not feeding. The eyes are small and relatively simple compared to other insects.

Where do they live? Aphids are found worldwide, but they are more common in temperate regions. They are found on a quarter of all plant species.

Aphids, Aphids Everywhere

Aphids are not held in high regard. The damage they do to plants has made them enemies of gardeners and farmers the world over. From a purely zoological point of view, however, they are a fascinating and very successful group of animals. One of the most remarkable things about aphids is their reproductive ability. In a short amount of time, a plant free from aphids can be swarming with them. For much of the year, many species of aphids reproduce without mating.

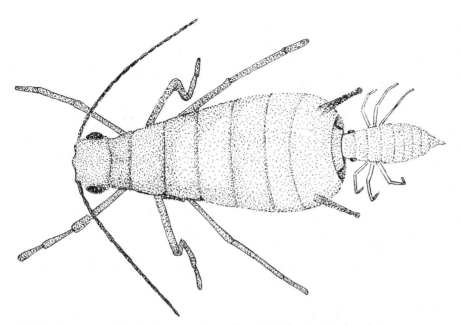

Aphids—A female aphid giving birth to a clone of herself. (Mike Shanahan)

This cycle begins with a female that hatches from an egg laid in a suitably secluded spot, such as the deep fissures in tree bark, during the previous year. This founding female had a mother and a father, but due to the odd make up of the aphid's chromosomes, a mating between a male and female can only ever produce daughters. These daughters survive the winter, and within them, they carry the seed of the new population. The founding female is already carrying a daughter, and within this embryo, another embryo develops—three generations in the body of one small animal, all thanks to the phenomenon of parthenogenesis, which enables animals to reproduce without sex. These daughters are born as miniature replicas of their mother, and they, too, give birth to further replicas, until there are huge numbers of aphids—all originating from the original female that survived the winter as an egg. The reproductive capacity of aphids is astounding. Theoretically, if all of the offspring from a single cabbage aphid managed to survive, there would be 1.5 billion, billion, billion aphids by the end of the season.

During the autumn, the aphid colony will start producing males and females whose function it is to mate and produce the founding females for the following year. In certain species of aphids, some of the clones, although genetically identically to the original female, will look slightly different and perform certain tasks, such as guarding the colony. These castes are commonly soldiers with enlarged front legs and a spiky head, which is jabbed at animals threatening the colony.

During the feeding season, the aphid colony may become too big, resulting in overcrowding that may kill the host plant. In these situations, the aphids start giving birth to winged individuals. These *alates*, as they are known, will leave the colony to search for new food plants.

> ◆ There are more than 4,000 species of aphids, and they are believed to have appeared more than 280 million years ago when there were far fewer plant species than there are today. Around 100 million years ago, there was an explosion in the variety of flowering plants, and the aphids diversified to exploit this new abundance of plant

Go Look!

Aphids can be found on all types of plants, including houseplants, garden plants, trees, and crops. You will see a whole array of interesting behaviors if you watch a colony of aphids. Typically, a colony will feed on the plant with their mouthparts inserted into the plant, greedily sucking the plant's sap. If you look carefully, you may see some of the females giving birth to miniature replicas of themselves. If the colony is really crowded, winged aphids will be testing their wings, hoping to fly away in search of new host plants. On the same plant, ants could be tending the aphids, waiting patiently at the animals' rear ends for drops of sweet honeydew to appear. To encourage the aphids to produce some honeydew, the ants stroke the aphids with their antennae. Stalking the aphids may be a range of predatory insects, including adult ladybirds and their active larvae, and lacewings and their fearsome looking larvae with huge sickle-shaped mandibles. Small maggots also hunt the aphids. These are the larval stage of hoverflies, a common sight on flowers. You may notice small flying insects buzzing around the aphids; these are female parasitic wasps. Watch a wasp carefully, and you will see it approach an aphid, touching it all over with her long antenna to "smell" whether it has already been parasitized before carefully injecting an egg into the sap sucker's soft body.

species. Over the eons, the aphids, as with many of the insects, have evolved hand in hand with the flowering plants, resulting in some amazingly complex relationships.

+ Aphids live in colonies, and in some species, certain individuals have a specific task to fulfill, such as guarding the colony. The only other insects to live in colonies where all of the individuals are very closely related to one another and where the colony members are divided into castes are the ants, wasps, bees, and termites.

+ All aphids use their piercing, strawlike mouthparts to penetrate the phloem vessels of their food plant. These vessels transport nutrients, as sap, to all parts of the plant, and the aphids gain access to these tubes through the leaves, stem, or roots. As the aphid inserts its feeding straw into the plant, the tip produces a fluid that hardens, forming a tube. To enter the phloem, the aphid must slowly rupture the plant cells, and to prevent the hole from healing, the aphid produces special proteins that fool the plant's defense mechanisms. It can take a long time for the aphid to get its first sip of the plant's fluids—anywhere between 25 minutes and 24 hours. To help digest the plant fluids, aphids enlist the services of bacteria or yeast. These microorganisms are present in the gut of the insect and feed on the sap, producing nutrients vital to the aphid. This is an example of symbiosis.

+ Much of what the aphid sucks from the plant is little more than sugary water, and so to stop from inflating itself with liquid, the aphid must get rid of the excess fluid as it is feeding. Droplets of sugary fluid, a substance that is commonly known as honeydew, emerge from the aphid's back end. Many animals have a special taste for this sugary treat, none more so than ants. Ants like honeydew so much that they treat the aphids like cows: herding them, protecting them, and milking them. This mutually beneficial arrangement is another reason why aphids are so successful. Many of the animals that would feed on them are deterred by the presence of their ant guardians.

+ Aphids don't get everything their own way. Although they have ants as minders, paid with honeydew, there are many animals, especially other insects, that feed on aphids or parasitize them. The ladybirds are one such example, as both the larval and adult stages of these beetles eat huge numbers of aphids. Many parasitic wasps hunt aphids, injecting eggs into their soft little bodies. These eggs hatch into small grubs and feed on the aphids' internal organs, eventually killing them.

- Plants lose many vital nutrients to aphids, which may explain the evolution of various defense mechanisms to discourage the aphids from feeding, such as small spines, hairs, scales, and secretions on all parts of the plant.
- Aphids are not strong fliers, but they utilize lofty air currents to travel many kilometers, often flying up to reach these currents in the calm conditions of a summer evening.

GIANT JAPANESE HORNET

Giant Japanese Hornet—Japanese honeybee workers forming a defensive ball around a giant Japanese hornet. (Mike Shanahan)

Giant Japanese Hornet—A worker perched on an index finger to give an idea of scale. (Takehiko Kusama)

Scientific name: *Vespa mandarinia japonica*
Scientific classification:
  Phylum: Arthropoda
  Class: Insecta
  Order: Hymenoptera
  Family: Vespidae
What does it look like? The giant Japanese hornet is a large insect. The adult can be more than 4 cm long with a wingspan of greater than 6 cm. It has a large yellow head with large eyes, a dark brown thorax, and an abdomen banded in brown and yellow. Three small simple eyes on the top of the head can be easily seen between the large compound eyes.
Where does it live? This subspecies of the Asian hornet is found on the Japanese islands. They prefer forested areas where they make their nests in tree holes.

Marauding, Hive-Raiding Hornets

Japanese beekeepers, in an attempt to increase productivity, try to keep European honeybees in Japan as they produce more honey than the indigenous Japanese honeybees. However, the giant Japanese hornet often thwarts this enterprise. This hornet is a formidable brute of an insect, which is in fact one of the largest living wasps. When a hornet locates a hive of European honeybees, it leaves a pheromone marker all around the nest, and before long, its nest mates pick up the scent and converge on the beehive. The hornets fly into the beehive and begin a systematic massacre. The European honeybee is no match for the hornet as it is one-fifth the size. A single hornet can kill 40 European honeybees in one minute, and a group of 30 hornets can kill a whole hive, something on the order of 30,000 bees, in a little over three hours. The defenseless residents of the hive aren't just killed but are horribly dismembered. After one of these attacks, the hive is littered with disembodied heads and limbs as the hornets carry the thoraxes of the bees back to their own nest to feed their ravenous larvae. Before they leave, the hornets also gorge themselves on the bees' store of honey.

This amazing natural phenomenon begs the question, well what about the native Japanese honeybees? Do they get attacked? The answer is no, and the reason is particularly neat. The hornet will approach the hive of the Japanese honeybee and attempt to leave a pheromone marker. The Japanese honeybees sense this and emerge from their hive in an angry cloud. The worker bees form a tight ball, which may contain 500 individuals, around the marauding hornet. This defensive ball, with the hornet at its center, gets hot, aided by not only the bees vibrating their wing muscles but also by a chemical they produce. The hornet, unlike the bees, cannot tolerate the high temperature, and before long, it dies and the location of the Japanese honeybees' nest dies with it.

- Aside from its large size and fearsome appearance, the giant Japanese hornet also has very potent venom, which is injected through a 6.25 mm stinger. The venom attacks the nervous system and the tissues of its victim, resulting in localized tissue damage where the flesh is actually broken down. A sting from this insect requires hospital treatment, and on average, 40 people are killed every year after being stung by giant hornets, due to anaphylactic shock. Typically, the hornets are not aggressive animals, but when threatened, they will attack. An attack initially involves one individual, but the release of alarm pheromones will quickly attract its sisters. Not only is the venom dangerous, but the sting is also very painful. A Japanese entomologist said of the sting: "It was like a hot nail through my leg."
- Hornet workers continually forage to feed their siblings developing in the nest. They will take a range of insects, including crop pests, and for this reason, they are considered beneficial. The insects they catch are dismembered, and typically, the most nutritious parts, such as the flight muscles, are taken back to the nest where they are chewed into a paste before being given to a larva. The larva returns the favor by producing a fluid that the worker eagerly drinks.
- The fluid produced by hornet larvae has aroused interest recently as it is the only sustenance the adult worker imbibes during its life. This substance somehow makes prodigious feats of stamina possible, as worker hornets fly for at least 100 km a day at speeds of up to 40 km/h. The substance produced by the hornet larvae, known as vespa amino acid mixture (VAAM), somehow enables intense muscular activity over extended periods, perhaps by allowing the increased metabolism of fats. A company has started producing VAAM commercially, and it has apparently improved the performance of many athletes.
- Not only is VAAM popular amongst the Japanese athletic community, but the fully grown larvae of the hornet are considered something of a delicacy and are eaten in mountain villages, either deep-fried or as hornet sashimi.
- The defensive-ball strategy of the Japanese honeybee works because the temperature inside the ball rises to 47°C, and the lethal temperature for the giant hornet is 44°C–46°C, whereas the lethal temperature for the bee is 48°C–50°C.

+ Recent studies have shown that the brains of the Japanese honeybee workers responsible for attacking a scout giant hornet are infected with viruslike particles. It is thought that the infection triggers aggressive behavior in worker bees. More experiments are being carried out in an attempt to understand this interaction.

Further Reading: Demura, S. Effect of amino acid mixture intake on physiological responses and rating of perceived exertion during cycling exercise. *Perception and Motor Skills* 96, (2003) 883–95; Ono, M., Igarashi, T., Ohno, E., and Sasaki, M. Unusual thermal defence by a honeybee against mass attack by hornets. *Nature* 377, (1995) 334–36; Tsuchita, H. Effects of a vespa amino acid mixture identical to hornet larval saliva on the blood biochemical indices of running rats. *Nutrition Research* 17, (1997) 999–1012.

LEAF-CUTTER ANTS

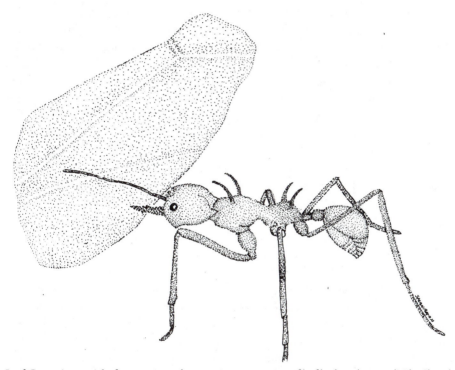

Leaf-Cutter Ants—A leaf-cutter ant worker carrying a cut section of leaf back to the nest. (Mike Shanahan)

Scientific name: *Atta* and *Acromyrmex* species
Scientific classification:
    Phylum: Arthropoda
    Class: Insecta
    Order: Hymenoptera
    Family: Formicidae
What do they look like? Within a single nest of leaf-cutter ants there are several types of workers, which vary in appearance. Generally, a leaf-cutter ant is brown, and its body supported on long, thin legs. The mandibles are well developed, and the eyes are small.

Where do they live? The leaf-cutter ants are found in Central and South America and in the southern United States. They like a warm, humid climate and throughout their range are found wherever there is sufficient vegetation to allow them to maintain a colony.

Six-Legged Farmers

Leaf-cutter ants form the largest and most complex animal societies on Earth. The workers in each nest are divided into several castes, all of which have a specific job to do. A colony of leaf-cutter ants is founded by a single female, the queen. This queen would have begun her life in another nest, tended and lavished with care by innumerable workers until the day she was ready to leave. The would-be queen crawls from the underground chambers of the nest into the open air. All around her there are other potential queens and males obeying with unerring synchronicity their reproductive instincts. The winged females and males take to the air for the first and only time in their lives where they meet each other in flight to mate. During this nuptial flight (the *revoada*), the female mates with perhaps eight different males to collect the 300 million sperm she needs to set up a colony. With her sperm bulbs full, she descends to the ground and rummages through the vegetation and leaf litter to find a suitable crevice or hole that she can call home. She has no further need of her wings, so they fall off, easing her underground activities. Safely beneath the ground, the queen excavates a small chamber and uses a small scrap of fungus carried in a pouch beneath her chin to grow a fungus garden. The fungus is integral to the success of the colony, and the queen took a small piece from her birth nest. She does not eat this nutritious fungus but relies instead on body fat and her now useless wing muscles for sustenance. She starts laying eggs, and the first young ants to hatch, her first daughters, will be gardener-nurses. Their responsibilities are to look after the fungus garden and lovingly tend further eggs produced by the queen. These first workers collect leaves from plants on the surface, on which the fungus grows. The queen, now relieved of menial tasks, can apply herself to the important job of enlarging the colony by pumping out eggs in huge numbers, producing different types of workers to look after different jobs within the nest. Some of the eggs will develop into nest generalists that will undertake all manner of miscellaneous tasks in the nest, while others will become foragers and excavators, collecting leaves for the fungus gardens and enlarging and maintaining the nest. Some eggs develop into heavy-headed workers with huge jaws whose job it is to defend the colony. After a few years, the nest, founded by a single female, has grown to monstrous proportions. It may descend 6 m underground, with a central nest mound more than 30 m across and smaller mounds extending out to a radius of 80 m. A single nest can take up 30–600 sq. m and contain 8 million individuals at any one time. One of these huge colonies operates like a superorganism, affecting the surrounding environment in profound ways. The ants can completely defoliate whole areas of forest, breaking the forest up into small glades that are important to many plants and animals. In the lifetime of one nest, 40 tonnes of soil can be turned over and aerated and fertilized by the huge amount of waste produced by the colony. This amazing collective of tiny insects working as one is all started by a single female that in her 10–15 year life span may produce 150 million daughters!

+ There are approximately 38 species of leaf-cutter ants.
+ During their nuptial flights, potential queens can fly more than 11 km from their birth nest.

+ The complex society of a leaf-cutter ant nest is held together by pheromones produced by the queen and built-in instincts. All the ants in a nest of leaf-cutters are sisters, and it is therefore in the interest of every individual to pull together for the sake of the colony. Everything that takes place in an ant nest is for the good of the colony, so that more would-be queens can be produced to carry the colony's genes into the future.

+ When a queen is spent and dies, the colony will lose its driving force and soon falter and collapse.

+ When the colony is young, the queen will produce mostly smaller workers, but as the colony matures, the size of the workers will increase. Scientists have found that it is possible to deceive the queen of a mature colony into perceiving that she is in charge of a young nest. This was done by removing ants from the nest, reducing its size.

+ Leaf-cutter ants are one of the only animals apart from humans to farm another organism. The crop in question is the fungus, and whatever its origins, it has developed in such a close relationship with the ants that it is found only in their nests.

+ When a cut section of leaf finds its way to the fungus chambers, the gardener ants will deposit a small drop of fluid from their abdomen on the leaf with a tiny piece of fungus. The ant's secretion acts like a form of fertilizer, allowing the fungus to proliferate through the leaf.

+ The fungus gardens are prone to a bacterial disease, which can spell disaster for the ant colony. To stop this bacterial infection in its tracks, the ants can produce a potent antibiotic, which is applied to the gardens in much the same way as farmers apply pesticides to their crops.

+ Starting a colony is a difficult business, and only 2.5 percent of potential queens will be successful. Many will be eaten during the nuptial flight or when they land, and many may successfully start a colony only to lose it to disease in the first three months.

+ In some parts of their range, leaf-cutter ants can be quite a nuisance to humans, defoliating crops and damaging roads and crops with their nest-making antics.

Further Reading: Fowler, H., and Robinson, S. Foraging by *Atta sexdens* (Formicidae: Attini): seasonal pattern, caste, and efficiency. *Ecological Entomology*, 4, (1979) 239–47; Wilson, E. Caste and division of labor in leaf cutter ants—I. The overall pattern in *A. Sexdens. Behavioral Ecology and Sociobiology* 7, (1979) 143–56; Wilson, E., and Holldobler, B. *The Ants.* Belknap Press of Harvard University Press, Cambridge, MA 1990; Wilson, E., and Holldobler, B. *Journey to the Ants.* Belknap Press of Harvard University Press, Cambridge, MA 1994; Wirth, R., Herz, H., Ryel, R., Beyschlag, W., and Holldobler, B. *Herbivory of Leaf-Cutting Ants.* Springer, New York 2003.

NAKED MOLE RAT

Scientific name: *Heterocephalus glaber*
Scientific classification:
  Phylum: Chordata
  Class: Mammalia
  Order: Rodentia
  Family: Bathyergidae

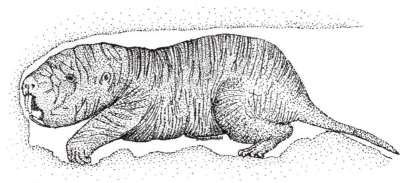

Naked Mole Rat—A naked mole rat excavating soil at the face of one of the colony tunnels. (Mike Shanahan)

What does it look like? Based on looks alone, the naked mole rat must surely rate as one of the most bizarre mammals. Except for a few sensory hairs and whiskers, it is completely bald, pink, and wrinkly. The head is large, but much of this bulk is taken up by the jaw muscles. Its eyes are minute and give it a squinting look. The lips of the animal close behind the huge, curved, ever-growing incisors. They have hairs between their toes, allowing them to use their feet like miniature brooms.

Where does it live? This rodent is a burrow-dwelling creature of arid areas and savannah in parts of Kenya, Somalia, and Ethiopia.

Where Females Rule

The naked mole rat is a fascinating little mammal. These creatures shun the sky and do everything they need to underground in extensive burrow systems, which may have 4 km of tunnels. Some of these tunnels are just below the surface, while others can be more than 2 m underground. One of these tunnel networks is occupied by one colony of mole rats, which can contain between 70 and 300 individuals. Perhaps the most bizarre aspect of the naked mole rat's life is the way that it reproduces. Like the bees, wasps, and ants, there is only one female in the mole rat colony that gives birth to young—the queen. No other mammal is known to reproduce in such a way. The queen may live for many years, producing as many as 900 pups in her lifetime, and it is very likely that all of these offspring will stay with the colony as dispersal in these rodents is very rare indeed. With migration and immigration being so low, inbreeding in the nest is high, so most of the individuals in the colony will be very closely related to one another, which is also the case for bee, wasp, and ant colonies. As the queen has a monopoly on breeding, all of the other individuals in the colony help to rear the young, which are produced prolifically. The queen can produce a litter of pups every 80 days and can have five litters a year. The average number of pups in each litter is 12, but the record is 27, so you can see, these rodents know how to breed. Not only do the other colony members help to rear the young, but they are also responsible for finding food, tunnel making, maintaining the tunnels, and defending the colony. The other individuals in the colony have the biological apparatus needed for breeding, but their natural urges are somehow curbed. The animals all share a latrine where they defecate and urinate, and chemicals in the queen's urine are responsible for suppressing the reproductive tendencies of the other females in the colony.

With their libido quashed by the queen's urine, the other animals in the colony can concentrate on more industrious tasks. Just after the rains, when the soil is softened, the naked mole rats begin some frantic team digging to enlarge and maintain the tunnel network. They form a chain gang, with a digger at the front putting its huge teeth and jaw muscles to good use to scrape away the tunnel face. Sweepers behind the digger brush the loose soil to their rear with their feet and shuffle backward, hugging the tunnel bottom until they reach the ejector mole rat that kicks the soil from the tunnel entrance, forming a molehill-like structure—the only clear, above ground evidence of these remarkable mammals.

+ There are nine species of mole rat, all of which are found in sub-Saharan Africa. As their name implies, they are ratlike rodents that have become brilliantly adapted to an underground lifestyle.

+ Although the mole rats have a very poor sense of sight, all of their other senses are very well developed. They have an exceptional sense of touch, feeling their way around their dark burrows with the long sensory hairs scattered all over their bodies.

+ Not only does the naked mole have a peculiar social structure, but due to the stable temperature within its burrow systems, it is the only mammal that cannot regulate its own body temperature. It is effectively cold-blooded. When naked mole rats are cold, they huddle together or bask for a while in the shallow tunnels. If they are too warm, they retreat to the deeper, cooler parts of the tunnel system.

+ Temperature regulation is energetically expensive and requires a lot of food, but as the naked mole rat has abandoned this way of life, its appetite is small. It also grows more slowly than similarly sized mammals. Its metabolic rate is much lower than other small mammals, so its longevity is considerably extended. In captivity, queens can live for more than 20 years, which is astonishing for such a small mammal.

+ Naked mole rats feed on roots and tubers located by burrowing through the soil. Some of these tubers are very large, and the mole rats make excavations into the nutritious interiors while leaving the skin intact, which allows the plant to survive and yield further food in the future. In some areas, the feeding activities of this animal can do a lot of damage to crops, especially sweet potatoes. Other mole rat species also chew through underground cables and undermine roadways when building their subterranean homes.

+ To increase the nutrition that can be extracted from its food, the naked mole rat has a high density of bacteria in its gut, and to make digestion as efficient as possible, it often eats its own feces. Feces are also fed to the young to inoculate their gut with the bacteria.

+ The skin of naked mole rats lacks a chemical called *substance P*. All other mammals have this substance, which enables them to detect pain and injury to their skin. As a result, the mole rat feels no pain, in its skin at least. Why this should be is a mystery, but it may be because the mole rat lives in such tightly packed communities. In-fighting in these communities would be very damaging to the colony as a whole; therefore, the lack of substance P could be a way of eliminating aggressive individuals from the population. Since this rodent cannot detect wounds, molerats with a propensity for fighting would get wounded, and as there is no sensation of pain, the injuries could easily be severe enough to cause the death of the animal from blood loss or infection. Another possibility for the lack of substance P is as a consequence of the high levels of carbon dioxide in the mole rat tunnels due to all the frantic activity. Carbon dioxide at high concentrations can be painful to the lips, nostrils, and eyes; but as mole rats have no substance P, they feel nothing.

Further Reading: Sherman, P. W., and Jarvis, J. U. *The Biology of the Naked Mole-Rat*. Princeton University Press, Princeton, NJ 1991.

NEW ZEALAND BAT-FLY

New Zealand Bat-Fly—The adult fly hitching a ride on the short-tailed bat. (Mike Shanahan)

Scientific name: *Mystacinobia zelandica*
Scientific classification:
  Phylum: Arthropoda
  Class: Insecta
  Order: Diptera
  Family: Mystacinobiidae
What does it look like? The bat-fly is a wingless insect with rudimentary eyes and elaborate claws. The abdomen of the sexually mature females is swollen with ovaries and eggs.
Where does it live? This fly is found only on the northern tip of New Zealand's North Island and only in association with the short-tailed bat, which lives in hollow trees.

A Fly that Does Not Fly

Bats the world over are coveted by a variety of freeloaders. One group of animals in particular has been very successful in exploiting these small nocturnal mammals. Numerous fly species, commonly known as bat-flies, live on bats, typically sucking their blood with piercing mouthparts. However, on the island of New Zealand, there exists a bat-fly that is very different from the norm and that has struck up a remarkable relationship with the short-tailed bat, which is also native to New Zealand.

Caves and other rocky nooks are in short supply on the northern tip of New Zealand's North Island, so the short-tailed bat has taken to living in the hollows inside large trees, such as the giant, primitive kauris. In these woody refuges, the bats live together in small roosts. Alongside the bats lives the New Zealand bat-fly, also in small colonies. The female flies in the colony lay their eggs together in a nursery, and when the young maggots hatch they, too, stay close together. Unlike other bat-flies, the New Zealand bat-fly does not suck the blood of the bats it lives with, instead it gorges on the flying mammal's droppings. This excrement accumulates in the bottom of the roost and is known as *guano*. On this nutritious diet the flies have time to engage in social activities. As the adult flies and their young live side by side in the bat roost, the females will often sidle over to the crèche to groom their offspring. The adults will also groom each other, cleaning and caressing their colony mates with their forelegs.

Not only is this social and maternal behavior very peculiar, but there is also what appears to be the beginnings of a caste system, similar to the different types of individual found in a colony of ants or termites. In the New Zealand bat-fly colony, some of the males live beyond their normal reproductive age. These elderly males take on the role of colony guards, and if a hungry bat approaches the flies too closely, the bat will be met with a cacophony of high-frequency buzzing produced by the guards.

Occasionally, a colony of bat-flies will become too big for the bat roost, so some will have to leave and found a new colony in another roost. To do this, they climb aboard one of the resting bats as it dangles from its perch and burrow into the bat's fur. There they wait for the bat to begin its nighttime sorties in the hope that they will be ferried to another roost. As many as 10 bat-flies can leave the nest in this way, clinging to the fur of one bat.

+ The New Zealand bat-fly is quite unlike the typical bat-flies found in other parts of the world. The New Zealand bat-fly is actually more closely related to the flies known as blue bottles and green bottles.
+ The species (*Mystacinobia zelandica*) was only discovered in 1973 by Beverly Holloway, a New Zealand entomologist, when a giant kauri tree in the Omahuta Kauri sanctuary fell over and an examination of its hollows revealed a roost of short-tailed bats and their fly cohabitants.
+ No other species of fly shows this level of social behavior and maternal care.
+ The flies have been very successful at taking advantage of birds and mammals. Several types, known as *keds*, live on a range of mammal species, including domestic species like sheep. Many of these highly modified flies move very quickly and with a sideways crablike gait. For people who work with sheep, the sheep ked is an unpleasant element of their job as its bite is particularly painful. Other species live on birds and have become very flattened, which enables them to slip beneath the feathers of their avian hosts. Like fleas, the parasitic flies thrive on those animals that build nests, which provides safety and food for their young.

+ The bat-fly and the short-tailed bat are two examples of the unique fauna of New Zealand that has developed because of geological processes. Many millions of years ago, New Zealand, Australia, Antarctica, and South America were fused into a huge landmass known as Gondwanaland. The animals and plants of this landmass evolved along a very different path from those of the Northern Hemisphere. When Gondwanaland spit into the land-masses we see today, the flora and fauna the landmasses carried evolved still further, but in isolation. Australia and South America have marsupials and an abundance of unique plant life. Antarctica was probably very similar, but its slow drift southward saw it fall into an icy grip. New Zealand, for some reason, had no land mammals apart from bats, and these may have colonized at a later date. In isolation, life flourished, free from the tooth and claw of predatory mammals. Birds took to living on the ground. Insects evolved to fill the niches occupied in other parts of the world by small mammals, and a host of primitive plants flourished in the cool, moist climate. Sadly, this paradise was not to last forever, and the ar-rival of humans, first Polynesians and then Europeans, spelled devastation and extinction.

+ A second species of New Zealand bat-fly lived in association with the greater short-tailed bat, but both became extinct when rats invaded Big South Cape Island in 1965.

Further Reading: Holloway, B. A. A new bat-fly family from New Zealand (Diptera Mystacinobndae). *New Zealand Journal of Zoology* 3, (1976) 279–301.

PORTUGUESE MAN-OF-WAR

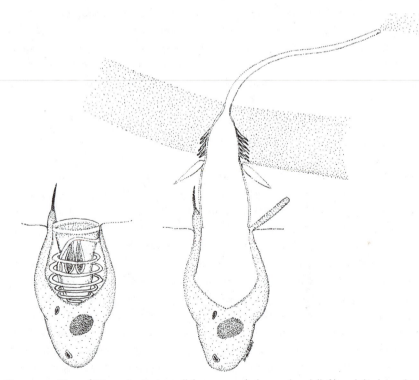

Portuguese Man-of-War—A stinging cell (nematocyst) shown closed (left) and discharged into the flesh of an animal (right). (Mike Shanahan)

Scientific name: *Physalia physalis*
Scientific classification:
   Phylum: Cnidaria
   Class: Hydrozoa
   Order: Siphonophora
   Family: Physaliidae

What does it look like? The Portuguese man-of-war is very bizarre. It has a large gas-filled bladder, tinged with blue and pink, which can be 30 cm long. Dangling in the water, below this bladder, are many tentacles.

Where does it live? This animal is found in many parts of the oceans and is frequently seen off the coast of Europe, North America, and Australia. It may prefer warm water, but ocean currents and storms will often push the man-of-war north and south.

Life on the Ocean Waves

The Portuguese man-of-war looks like some manner of jellyfish. You could be forgiven for thinking this, and you wouldn't be too far wrong. It is related to the jellyfish, but it is a quite distinct collection of creatures. It is not a single animal, but a close-knit colony of four different types of individual polyps, or zooids. The first of these is the bladder polyp, which is responsible for producing the gaseous bladder that the colony uses as a buoyancy aid. The bladder is filled with mostly carbon monoxide, and in times of danger, it can be rapidly deflated so that the colony can sink out of reach of a potential predator. Not only does the bladder keep the Portuguese man-of-war afloat but it also acts like the sail of a ship, catching the wind and carrying the colony around the seas. The second type of individual polyp in the colony is the feeder, whose job it is to catch food using tiny poison barbs, which can be shot into small fish and other sea animals. The feeders bear the long tentacles that hang below the Portuguese man-of-war, and when they catch some food, they contract, bringing the prey in reach of the tentacles of the third type of polyp—the gastrozooid, which is the stomach of colony that digests the food and provides nutrients for the rest of the polyps in the colony. The last type of polyp is the gonozooid, whose job it is to make more colonies, producing small larvae that grow to become the buoyancy bladder and the other three zooids of a complete colony.

The way in which the Portuguese man-of-war catches its prey is very interesting. The fishing tentacles of the colony, the dactylozooids, are armed with thousands upon thousands of stinging structures called *nematocysts*. These are beautifully adapted hunting tools. The stinger, produced by a special type of cell, looks like a small bulb, and inside it is a coiled thread. On the outside of the stinger, in the water, is a tiny hair trigger. An animal will brush past this hair and cause the bulb to fire, which it is does with astonishing ferocity. The coiled thread is shot from the bulb at a velocity of 2 m per second, and for something so small, this means its tip is accelerating at 40,000 Gs (by comparison, a race-car driver experiences 2–3 Gs speeding around a corner). The thread penetrates the prey and injects potent venom. A larger animal could tolerate the effect of one of these cells, but there is strength in numbers, and hundreds or thousands of nematocysts are fired at once. The venom is neurotoxic, and it rapidly paralyzes the prey so that it can be easily transferred to the gastrozooids.

   ◆   The phylum Cnidaria is a fascinating group of animals numbering more than 10,000 species. The Portuguese man-of-war is but one of these, and its relatives, such as the typical jellyfish and the anemones, are very familiar animals whose beauty can only be appreciated when they are seen in water. Their soft bodies, composed mostly of water,

Go Look!

If you live near the coast in the eastern United States, you may be lucky enough to see a Portuguese man-of-war colony washed up on the shore or still floating in a pool left by the retreating tide. Because of their bladders, they are at the mercy of the wind and the ocean currents, and several may be found on a single beach after a storm. You will notice the bladder with its pink/blue tinge and the tentacles hanging from its underside. The longer ones are the fishing tentacles. Be careful not to touch the tentacles, because even if the animal is dead or washed up on the shore, the stinging cells are still active and will be discharged at the slightest touch.

lose their shape when they are washed up on the shore, rendering them little more than featureless blobs.

- The cnidarians are an ancient group with the longest fossil history of any animal, extending back over 700 million years. Although anatomically simple animals, they successfully compete with organisms much more complex than themselves.

- The venom of cnidarians is very toxic. That of the Portuguese man-of-war can cause some very painful injuries if a swimmer accidentally brushes against the long fishing tentacles. The pain is extreme and immediate, and the sting can even be fatal to the young, elderly, and people with certain allergies. The victim should be taken from the water, and an ice-pack should be placed on the affected area, but in no circumstances must the sting be washed with vinegar (a common treatment for some other jellyfish stings).

- Although the sting of the Portuguese man-of-war should be treated with respect, it is by no means the most dangerous of the venomous cnidarians. This title belongs to the sea wasp, also known as the box jellyfish, which is frequently found in the waters off Australia and forces swimmers from the water. This animal has caused at least 64 deaths since 1884. Death, if it occurs, happens 3–20 minutes after stinging. The wounds caused by nonlethal stings can be severe and often take a long time to heal. The reason that these innocuous-looking blobs of jelly have such lethal toxins is that they do not have limbs, or anything for that matter, with which to grab and immobilize prey; therefore, they use fast-acting, paralyzing venom.

- The Portuguese man-of-war has developed an interesting relationship with several types of fish, including the shepherd fish, the clown fish, and the yellow jack, species which are rarely found elsewhere. The fish accompany the colony on its travels around the high seas. The clown fish can swim among the tentacles with impunity, very likely made possible thanks to skin mucus, which does not stimulate the hair triggers of the stingers. The shepherd fish seems to avoid the larger fishing tentacles and will feed on the smaller tentacles directly beneath the bladder. The presence of these fish may attract other animals on which the Portuguese man-of-war can feed.

- The name Portuguese man-of-war comes from the likeness of the gas bladder of the colony to the sail of the old Portuguese fighting ships.

SPONGES

Scientific name: Porifera
Scientific classification:
    Phylum: Porifera
    Class: Calcispongiae, Hyalospongiae, Demospongiae, Sclerospongiae

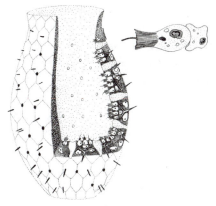

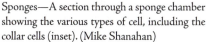

Sponges—A section through a sponge chamber showing the various types of cell, including the collar cells (inset). (Mike Shanahan)

Sponges—A stovepipe sponge photographed in the Caribbean. (Bart Hazes)

What do they look like? The sponges have a fibrous body wall, which can be a variety of bright colors. They can be encrusting or upright and branching, and a whole host of shapes in between. Their very odd body plan is built around an elaborate network of water canals.

Where do they live? Sponges are aquatic animals and can be found in both marine and freshwater habitats, although the vast majority are found in the former. They are normally found in shallow water the world over, although some species can be found in the cold, dark depths.

Successful Simplicity

In aquatic habitats throughout the world, sponges abound. To the uninitiated, sponges look like plants or small geological features. Some cling to rocks, while others extend out into the water, forming thin turrets or broad columns. Nothing in the outward appearance of these peculiar organisms gives any indication that they are in fact animals. Among the animals, the sponges are the most primitive group, and to see what makes them an animal, one look must look at them very closely. There are no organs within a sponge, just a network of cavities and canals. Lining the sponge's cavities there are so-called collar cells, which drive a current of water through the animal's body and latch onto and filter whatever edible particles may be suspended in the water. The water-pumping ability of even a small sponge is very impressive. Each day, a 10 cm long specimen is capable of channeling more than 22L of water through the network of collar-cell-lined chambers, which may number more than 2 million. Most of the particles ingested by the sponge are very tiny, even too small for a conventional microscope to see. These particles are absorbed by another type of cell that looks a lot like an amoeba. As well as playing a role in digestion, these cells also have the ability to turn into any one of the other cells that make a fully functioning sponge.

The outside of these animals is pocked with numerous pores and holes linking the interior of the animal to the surrounding water. Other cells in the sponge secrete an elaborate, glassy skeleton of silica, while others secrete a protein called *spongin* that fleshes out the frame.

In many ways, the sponge is simply an assemblage of different types of cells, and although there is a division of labor among these units, similar to that seen in other animals, the differentiation is nowhere near as complex. The simplicity of the sponge makes it a champion of regeneration. The tissue of a living sponge can be forced through a silk mesh, turning the animal into a so-called cell soup. Other animals would be very hard pressed to recover from such brutal treatment, but the sponge's separated cells quickly reorganize, forming themselves into several new sponges. In commercial sponge-growing enterprises, a large specimen may be cut into pieces, which are attached to cement blocks and then submerged. Each piece grows into a new sponge, which can be harvested after a few years.

These bizarre and simple creatures were probably among life's first attempts at an organism with numerous cells instead of just one, but they developed along a track that spawned no descendents. They are, in an evolutionary sense, a dead end.

- There are more than 5,000 species of sponge, and only around 150 of these can be found in freshwater. The rest are marine animals and can be found wherever there are suitable surfaces for them to attach to. They range in size from tiny species only a few millimeters long with an internal scaffold of calcium carbonate, to huge loggerhead sponges, which may be more than a meter long and wide.
- It was only in the midseventeenth century that naturalists first saw the circulation of water inside sponges, an observation that singled them out as animals. Up until this time, the lack of any obvious movement and their odd internal structure led naturalists to label sponges as plants.
- The earliest sponge fossils are at least 500 million years old, and there are claims of even older fossils. Regardless of when they appeared, sponges were at the peak of their success during the Cretaceous period.
- Along with their high silica content and the spiky nature of their skeletons, many species of sponges contain noxious substances to further deter potential predators. Even with this armory of defenses, many animals eat sponges, and some have specialized on a diet consisting solely of sponges. The sea turtles are a good example. The feces of certain turtles may contain more than 95 percent sponge fragments.
- A sponge larva is a free-swimming animal, albeit briefly. It swims out of its parent and floats in the water for a short time before settling and developing into the familiar creature.
- The complex internal arrangement of the sponges makes the larger species excellent refuges for a number of delicate aquatic animals. One large loggerhead sponge was found to contain more than 16,000 shrimps.
- In a very interesting relationship some species of freshwater sponge are preyed upon by small insects known as sponge flies, which are, in fact, close relatives of lacewings. The female sponge fly lays her eggs on vegetation overhanging water. The larvae drop into the water and seek out a sponge to feed on. The larvae of some species cling to the surface of the sponge, while others explore the sheltered confines of the sponge's internal cavities.
- Sponges have long been of commercial value to people all around the world. The larger, cylindrical forms are valued as bathroom accessories or trinkets, although decreasingly so in the former use due to the availability of synthetic sponge. The interesting biochemistry of sponges has also attracted interest, and many compounds extracted from these animals are being investigated as starting points for novel medicines.

STONY CORALS

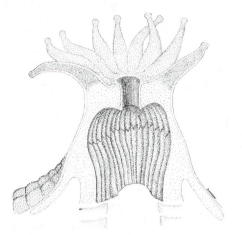

Stony Corals—A close up view of a tiny coral polyp showing the tentacles that identify it as a relative of jellyfish. (Mike Shanahan)

Stony Corals—A number of stony coral species can be seen here, showing the diversity of this type of colonial animal. (Bart Hazes)

Scientific name: Scleractinia
Scientific classification:
    Phylum: Cnidaria
    Class: Anthozoa
    Order: Scleratinia
    Family: several
What do they look like? Corals grow in a huge variety of forms. They can be thin and branching, moundlike, flat and spreading like a tabletop, and every intermediate form in between. They can be small and self-contained or huge rambling structures. The animal responsible for these structures, the coral polyp, is a tiny anemone-like creature, with a short, squat body crowned with numerous tentacles.
Where do they live? The stony corals are found throughout the world's oceans but are at their most impressive in the warm, shallow waters of the tropics and subtropics. The Indo-Pacific region, including the Red Sea, Indian Ocean, Southeast Asia, and the Pacific, accounts for the greatest abundance of stony corals.

Many Polyps Build Great Structures

The animal responsible for producing coral is one of the most inconspicuous creatures; yet its activities can fashion whole ecosystems and change the earth's climate. To the uninitiated, a stony coral looks like a rock, a dead, inanimate object; however, on closer inspection, the outer surface is actually a thin, living veneer. Pitting its surface are numerous small pores or cups, each containing a small polyp, 1–3 mm across. This polyp looks like a small sea anemone, nestled in a receptacle of calcium carbonate of it own creation. Only the animal's feeding tentacles project from this cup, and even these can be withdrawn and tucked out of the way of hungry sea animals. The polyps, like all jellyfish and their kin, are predatory. The tentacles of many species wave serenely in the current, waiting for prey ranging from microscopic floating animals to small fish to come within range of their batteries of explosive stinging cells. Other species secrete copious quantities of mucus to entrap tiny prey and edible particles. The prey, paralyzed by the powerful venom, is transferred to the polyp's equivalent of a mouth. Another source of nourishment comes from an interesting relationship many polyps have struck up with various forms of single-celled, algaelike organisms. These algal cells live within the cells lining the polyp's digestive cavity, and via the process of photosynthesis, they produce sugars and other nutrients, a proportion of which goes to the polyp. With a nourishing diet, the polyps grow quickly, constantly adding more calcium carbonate to their little cups. It is these small limestone cups that are the building blocks of a coral colony. Over decades, centuries, and even millennia, successive generations of polyps lay down layer after layer of calcium carbonate until the thin living layer surrounds a huge skeleton of what is essentially rock. The rate of colony growth achieved by the minute polyps, considering they are secreting rock, is extraordinary. Some of the branching corals can grow in height or length by as much as 10 cm per year (about the same rate at which human hair grows). Other corals, like the dome and plate species, are more bulky and may only grow by 0.3 to 2 cm per year. This may not seem like much, but it can be sustained for thousands of years, forming huge and complex reefs. The Great Barrier Reef, off the coast of Australia, is the largest reef complex in the world and is composed of a multitude of coral colonies. The structure we see today is thought to be 6,000–8,000 years old, although the modern structure has developed on a much older reef system, thought to be 500 million years old. This huge reef system stretches for over 2,000 km and can be seen from space. Reefs like the Great Barrier Reef support huge numbers of marine organisms, and by affecting the direction and flow of currents, they directly affect the earth's climate—all of this from the tireless deposition of rock by a minute sea creature.

- The stony corals are divided into reef-building species and solitary species. The former are the more familiar as they are found in shallow water and support huge assemblages of marine organisms. The latter are found throughout the world's oceans, even down to depths as great as 6,000 m.
- In a coral colony, all of the polyps are genetically identical. They are all connected by a network of channels allowing the sharing of nutrients and symbiotic algae.
- For much of the year, corals reproduce asexually. Either they spilt down the middle, producing two identical individuals each regenerating the missing half, or the crown of tentacles around the top of a polyp develops a small bud, which grows into a miniature polyp that is eventually released. At other times of the year, dictated by the phases of the moon, huge numbers of eggs and sperm are released by female and male polyps or hermaphrodite polyps with both sets of sex organs. The eggs are fertilized, and the resulting larvae drift through the water in the hope of eventually founding another colony in a suitable location.
- The ecosystems formed by stony coral reefs are some of the most diverse habitats on the planet. They are rich in nutrients and provide an intricate maze of hidey-holes for marine animals. For example, the Great Barrier Reef is composed of at least 400 species of stony and soft coral, forming a home for more than 5,000 species of mollusks and no fewer than 1,500 species of fish.

+ As the stony coral reefs form huge, natural barriers, ocean currents may be impeded or redirected, allowing the build up of nutrients in areas of ocean that may otherwise be quite nutrient poor. These ocean currents carry heat energy around the earth; therefore, by altering them, the large reef systems can affect the planet's climate.

+ Unfortunately, although the coral reefs look indestructible, they are, in fact, quite delicate structures and are particularly susceptible to pollution and other human activities. Many pristine areas of reefs have already been damaged, but it is hoped that with increased awareness the importance of these astonishing natural structures will be recognized.

Further Reading: Smith, S. V., and Buddemeier, R. W. Global change and coral reef ecosystems. *Annual Review of Ecology and Systematics* 23, (1992) 89–118; Veron, J. E. N. *Corals in Space and Time: The Biogeography and Evolution of the Scleractinia.* University of New South Wales Press, Sydney 1995.

TAR-BABY TERMITE

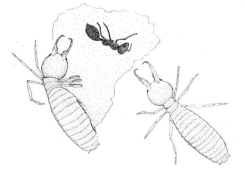

Tar-Baby Termite—A tar-baby termite worker rupturing its body to defend the nest against attacking ants. (Mike Shanahan)

Tar Baby Termite—A small number of worker termites around their nest. (Christian Bordereau)

Scientific name: *Globitermes sulphureus*
Scientific classification:
   Phylum: Arthropoda
   Class: Insecta
   Order: Isoptera
   Family: Termitidae
What does it look like? The soldiers of this termite species are small and eyeless. They are furnished with a fearsome pair of mandibles and have an abdomen that has a distinctive yellow hue.
Where does it live? The tar-baby termite is native to Southeast Asia where it builds its large nest in a wide variety of habitats. It builds its mound on the forest floor and anywhere else the soil is sufficiently soft for burrowing and nest making.

Ask Not What Your Colony Can Do for You, but What You Can Do for Your Colony

Along with the ants, bees, and wasps, termites are the only other insects that live in colonies of closely related individuals. The nests in which termites live are plundered by predators galore as they are essentially a larder of soft-bodied, walking snacks. Against larger marauders the termites have few if any defenses, but smaller animals such as ants can be repelled in a number of ways. One of the most amazing tactics for keeping these smaller aggressors at bay is demonstrated by the tar-baby termite. Should certain species of ant find a breach in the walls of this termite's nest, the occupants will first try to fix the breach. Although the workers are rapid builders, ants will invariably get inside the nest, where they seek quarry to take back to their own nest. The termite nest's second line of defense is its soldiers—a distinct type of worker whose responsibility is to protect the colony. Initially, the soldiers use their formidable mandibles to slash at the ants to persuade them to abandon their attack. The termite's mandibles are big and sharp, but in close-quarter fighting, termites will have little chance of stemming an ant invasion. An ant's body is very tough compared to the soft abdomen of the termite, and ants also have good eyes and potent stings. Fortunately, the soldier termites have a secret weapon. Occupying some of their thorax and much of their abdomen is the large frontal gland containing a yellowish fluid. The walls of this gland are very thin, and when the termite has repelled the invaders as much as it can with its mandibles, it ruptures its own body wall and lets the fluid ooze out. The termite soldiers can also rupture themselves when they sense danger. It used to be thought that these soldier termites actually exploded, spraying attacking ants with the contents of their defensive gland, hence the names "walking chemical bombs," "kamikaze soldiers," and so forth. To the contrary, the fluid comes out slowly, but the effect is no less dramatic. On contact with the air, the fluid becomes very sticky. Like a miniature tar pit, the fluid traps the ants until they become hopelessly entangled. They die in the grip of this gloop, as do some of the defending termites. Needless to say, the split in the termite's little body is severe and fatal. This suicidal gesture is the soldier termite laying down its life for the good of the colony. Although many soldiers may be lost defending the nest, their tactics must be effective as this species is very common in Southeast Asia where it is considered a pest. Not only can they use their powerful mandibles to pierce and slash ants, but as a last resort, they can commit suicide, taking an ant or two with them—a brilliant example of altruistic behavior.

+ Termites, like ants, are very successful animals. Although small and quite delicate insects, they possess strength in numbers. They are housed in nests that can often be large, very complex structures containing millions of individuals all produced by a single, massive queen or a group of queens at the heart of the colony. Unlike ant colonies, a termite colony also has kings.

+ There are at least 3,000 described termite species. Termites have been around for at least 100 million years, in which time they have evolved a huge range of defensive adaptations to protect their nests from predators.

+ The soldiers of some termite species have a large, cylindrical head, which is very tough. This can be used to plug the nest's entrances and to shore up the nest if the walls are breached in an attack. These soldiers also have short but powerful mandibles to deliver a potent nip.

+ In some termite soldiers, the mandibles evolved into beating weapons. The semiflexible mandibles are pushed against one another in the same way as you click your fingers. One mandible is released and is used to slap the attacker. This adaptation is even more finely tuned in some termites where asymmetrical mandibles can only be used to deliver powerful blows to the left, which are forceful enough to disable small attackers.

+ The soldiers of certain termites have almost completely lost their mandibles, but on their head they have a long nozzle that is used to squirt nasty chemicals, secreted by their frontal glands, at attackers. The nozzle looks like a long, thin nose, but the chemicals it emits are enough to kill small enemies or at least repel them.

+ Nozzles are also found in some soldiers with powerful piercing mandibles. The mouthparts inflict a puncture wound before the nozzle squirts some noxious chemicals.

+ Yet more bizarre are those termite soldiers with a head that extends into a long brushtipped snout. Their mandibles are small and are only used for carrying, but the brush at the tip of the snout is used like a paint brush to smear poisons onto their enemies. The chemicals daubed by this brush are very toxic to ants and other species of termites.

+ The tar-baby termite ruptures a gland to good effect, but other soldier termites can pop their gut at will. The resultant mess is far from attractive to predators. Along the same theme, some worker termites regurgitate the contents of their gut and rush out at attackers with the odorous liquid bubbling from their mouth. Other worker termites tense their abdominal muscles to the point where the whole hind gut pops out of the anus.

Further Reading: Bordereau, C., Robert, A., van Tuyen, V., and Peppuy, A. Suicidal defensive behavior by frontal gland dehiscence in *Globitermes sulphureus* Haviland soldiers (Isoptera). *Insectes Sociaux* 44, (1997) 289–97; Deligne, J., Quennedey, A., and Blum, M. S. "The enemies and defense mechanisms of termites." In Hermann, H. R. (ed.) *Social Insects Vol. II.* Academic Press, New York, 1982; Prestwich, G. D. Defense mechanisms in termites. *Annual Review of Entomology* 29, (1984) 201–32; Scheffrahn, R. H., Kreck, J., Su, N. Y., Roisen, Y., Chase, J. A., and Mangold, J. R. Extreme mandible alteration and cephalic phragmosis in a drywood termite soldier (Isoptera: Kalotermitidae: Cryptotermes) from Jamaica. *Florida Entomologist* 81, (1998) 238–40.

TRAP ANT

Scientific name: *Allomerus decemarticulatus*
Scientific classification:
  Phylum: Arthropoda
  Class: Insecta
  Order: Hymenoptera
  Family: Formicidae
  Subfamily: Myrmicinae
What does it look like? The trap ant is a small species. Adult workers are around 2 mm long. They are a golden amber color with a large head and powerful mandibles. The eyes are relatively small.
Where does it live? These ants are native to the rain forests of South America. It is likely that they are found throughout the Amazon basin, but they have only been recorded from the east coast of South America. They are an arboreal species, spending all of their time in their host trees.

Let the Food Come to You!

The trap ant of South America is a small, unassuming insect, staying out of sight much of the time among the branches of a tree that they have developed a very close relationship with. For much

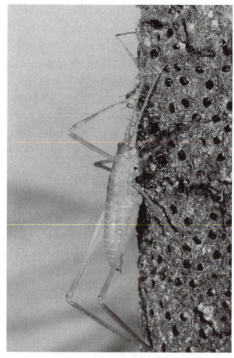

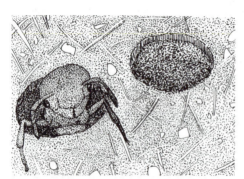

Trap Ant—Trap ant workers waiting in the holes of their trap for a victim. (Mike Shanahan)

Trap Ant—A cricket wanders on to the trap and is snagged by the worker ants hiding in the holes. (Alain Dejean)

of the time, these tiny animals are busy in their little nest pouches beneath the leaves of their host tree. Catching prey is difficult if you are an ant. Most animals that you would like to get your mandibles into are too big to subdue alone and may also have wings or other means of evading your snapping jaws. These problems are magnified for tree-dwelling ants. Even if some prey animal comes within range, all it has to do is leap from the trunk or branch to be carried away from the ant and the ant's nest mates. The trap ant has evolved an intriguing way of solving this problem of securing prey. Beneath the branches of its host tree, the trap ant workers construct small galleries, which look like pock-marked areas of dense grey webbing. More intriguing still is the way they actually build these galleries. Using hairs stripped from the surface of their host tree to act as a sort of fibrous matting and saliva and a specially cultivated fungus as resin, the ants construct what is essentially the insect equivalent of fiberglass. These galleries have been known about for a long time, but it was thought that they were simply a refuge for the ants while they were away from the nest foraging for food. As it turns out, the real purpose of these galleries is much more macabre. The worker ants mass in these galleries with their little heads just inside the holes, their powerful serrated jaws agape. Here they wait in ambush, and before long, a large, plump cricket ambles in to view, exploring the branch tentatively with its long limbs and sensitive antennae. The hairs that clothe the tree's outer surface are a deterrent to walking insects. The smooth surface of the gallery, on the other hand, will seem like a fine place to rest or have a snack. Sensing nothing out of the ordinary about the ant webbing, the cricket walks straight on to it. Two or three of its clawed feet may probe the holes for purchase, and it is then that the waiting ants strike. They grab the ends of the cricket's legs and heave, pinning the cricket to the surface of the gallery. Other workers rush out and begin dragging the remaining limbs and antennae into the holes until the prey is well and truly snared. Other workers then swarm all over the prey and sting it.

With the prey immobilized and at death's door, the ants can begin their gory work. Using their tenacious jaws, they begin butchering the carcass of the cricket, chopping out chunks of flesh to carry back to the nest pouches where the developing young can be found. The prey is a rich source of protein for the developing ant grubs. The amazing trapping strategy employed by these ants ensures that the nest has sufficient protein in an arboreal environment where it would otherwise be lacking in this type of nourishment.

- There are several species of *Allomerus* in the tropics. They are mostly small ants and all of them are specialist tree dwellers.
- *Allomerus decemarticulatus* is the only known species of ant that collectively constructs a trap to snare prey.
- When attacking their prey and delivering a sting, they must go for the soft parts of the body, such as the thin membrane between the body and limb segments. This arthrodial membrane as it is known can be easily penetrated by the little stinger of these ants.
- Not only have these ants evolved a unique way of catching animals much larger than an individual in the colony, but the whole setup is the product of a symbiotic relationship between the ant, its host tree (*Hirtella physophora*), and the fungus. The ant has a safe place to nest in the leaf pouches of the tree, a favor repaid by ridding the tree of herbivorous insects that can't wait to get their mandibles into the succulent leaves of the *Hirtella* tree. The fungus is used as an adhesive in the construction of the traps, but in being so used, it is cultivated by the ants and spread to areas where it might not reach by itself.
- There are a few spider species that build huge collective webs in which to snare prey. Should an insect become entangled in the silken threads, the small spiders run from all over the web to deliver bites to the unfortunate prey. Individually, one of these spiders would stand no hope of subduing such prey, but a multitude of fangs can quickly deliver a death blow to an insect massively larger than a single spider.

Further Reading: Dejean, A., Pascal, J. S., Ayroles, J., Corbara, B., and Orivel, J. Arboreal ants build traps to capture prey. *Nature* 434, (2005) 973.

2

THE WORLD IS A DANGEROUS PLACE: DEFENSIVE TACTICS

ARMORED SHREW

Armored Shrew—Cutaway of the armored shrew's body, revealing its massively modified spine. (Mike Shanahan)

Scientific name: *Scutisorex somereni*
Scientific classification:
    Phylum: Chordata
    Class: Mammalia
    Order: Insectivora
    Family: Soricidae

What does it look like? The armored shrew has a body length of 12–15 cm and a tail measuring 6.8–9.5 cm. It weighs between 30 and 115 g. The fur of the shrew is dense, coarse, and gray in color. The animal possesses the typical shrew features: short legs, slender snout, and tiny eyes.

Where does it live? This shrew is found in the forest belt of Africa in places such as southwestern Uganda, eastern Zaire, and northern Rwanda. Little if anything is known of its exact habitat preferences, but it is found in wet, waterlogged lowland forest with trees, palms, and dense undergrowth.

Carrying a Burden on Your Back

The armored shrew, also known as the hero shrew, is a small, unassuming insectivorous mammal, and for many years, it was regarded as an unremarkable, yet large African shrew. That was until its skeleton was examined. The spine of this animal has been called a "large bony buttress," and it is the most elaborate of any backboned animal. Normally, mammals have 5 lumbar vertebrae; yet the shrew has 11. The structure of the bones in the vertebral column, between the rib cage and hips, is also unique as they sport interlocking spines on their sides and lower surfaces. These spines mesh with the projections on the vertebrae behind and in front, creating an incredibly strong, yet flexible structure. The ribs of the shrew are also thicker than those of other small mammals. The spine is so strong that this tiny animal can take the full weight of a man on its back, around 70 kg. This is approximately 1,000 times greater than the shrew's body weight and is equivalent to a human bearing the weight of 10 elephants. Balancing with one foot on any other small mammal would kill it as the spine would simply collapse, crushing the internal organs, but the structure of the armored shrew's spine is so strong that it can bear a huge weight and protect the delicate organs. Although the spine is very strong, it is also very flexible. One of these shrews in a tube or burrow little wider than its body can somehow turn 180°.

The spine is so elaborate and modified that it accounts for 4 percent of the shrew's body weight. For other small mammals, the spine typically accounts for 0.5–1.6 percent of the overall mass. Ironically, other parts of the shrew's skeleton, such as the limbs, are not modified at all. They are not robust like the spine and ribs, but look like the limbs of any other shrew.

Along with the vertebral column, the spinal muscles are also greatly modified. The transverse muscles are reduced, while the muscles that extend and flex the spine are well developed, causing the shrew to walk in a rather snakelike fashion. In other animals, the transverse muscles minimize spinal twisting, but in the armored shrew, the greatly modified spine probably fulfills this role. In vertebrates, the backbone is generally the most conserved part of the skeleton, showing only minor modifications from one species to the next. Why, then, does this shrew have such a greatly modified vertebral column? No one knows. There is simply no plausible explanation for why the shrew has such a strong spine, but nature is never extravagant without reason, so there may be some aspect of the shrew's life, as yet unseen, in which the amazing spine is vitally important.

+ Like other shrews, the armored shrew is a secretive animal, scuttling around in the forest undergrowth. There are few people in these areas, so it is not surprising that this animal is so rarely seen and so poorly known by scientists.
+ In the wild, the armored shrew feeds on earthworms, insects, and small amphibians. In captivity, it will eat bird and mammal meat.
+ In its homeland, the shrew is revered by local tribes. They believe it to have magical powers because of its ability to support the weight of a grown man without being

squashed. They believe the animal will bestow upon them powers of bravery and make them invulnerable in battle. Any part of the animal, even a tiny amount of its ashes, is sacred, and warriors will go into battle with part of the armored shrew about their person.

♦ There are two types of hair on the shrew's back. Some hairs are sensory in function, while others are used to mark territory and can be actively ejected. This scent marking is something the armored shrew does almost continuously. If the animal is placed in a cage, it will frantically scent mark its new territory, which is done by contorting its body into unusual positions. The scent is produced in large quantities, staining the fur yellow. It is thought that the odor produced by the glands along the flanks of the shrew is repellent to others of its own species, demarcating its home territory where trespassers are not welcome.

Further Reading: Cullinane, D. M., and Aleper, D. The functional and biomechanical modifications of the spine of *Scutisorex somereni*, the hero shrew: spinal musculature. *Journal of the Zoological Society of London* 244, (1998) 453–58; Cullinane, D. M., Aleper, D., and Bertram, J. E. A. The functional and biomechanical modifications of the spine of *Scutisorex somereni*, the hero shrew: skeletal scaling relationships. *Journal of the Zoological Society of London* 244, (1998) 447–52; Kingdon, J. *East African Mammals*. Academic Press, London 1974.

BALLOON FISH

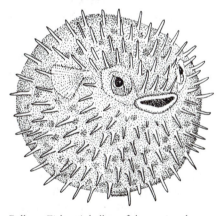

Balloon Fish—A balloon fish, sensing danger, inflates itself in defense. (Mike Shanahan)

Balloon Fish—An adult fish swimming around a reef in its normal, nonthreatened pose. (Robert Patzner)

Scientific name: *Diodon holocanthus*
Scientific classification:
   Phylum: Chordata
   Class: Actinopterygii
   Order: Tetradontiformes
   Family: Diodontidae
What does it look like? The balloon fish is a boxy-looking species with an angular head and large eyes. In its normal, nonthreatened pose, the long spines (modified scales) on its skin lie flat against its body. The pectoral and tail fins are small and delicate. The skin is pale, and there are dark blotches on its skin, giving it a mottled look. Fully grown, the balloon fish can reach lengths of 50 cm.

Where does it live? The balloon fish is associated with tropical coral reefs, mangroves, sea grass beds, and rocky seabeds. It can be found around the world. In the Atlantic, it is found from the Bahamas to western Brazil, and in the Pacific, it is distributed widely through the numerous island groups. It is a shallow water species, rarely venturing to depths of more than 100 m.

In Times of Danger Inflate Yourself

The balloon fish, with its spiky body and blotched coloration, is a handsome inhabitant of the warm, coastal waters of the tropics. It is a nocturnal hunter, preferring to spend the day secreted in crevices or small caves. When night falls, the fish leaves its daytime refuge to begin its prowling, closely hugging the coral or rock of its home and only rarely venturing into more open water. The underwater performance of the balloon fish is far from electrifying. The little fins and weak tail provide good stability but limited power. Luckily, the preferred prey of the balloon fish are not the sort of animals that can leave a hungry predator disorientated in a cloud of sediment. Its favored tidbits are snails, sea urchins, coral polyps, and hermit crabs, all of which can be rapidly dispatched with the balloon fish's bony mouth and strong jaw muscles. The reef at nighttime is not the safest place to be. Many predators have retired for the night, but under the cover of darkness, other, fierce animals lurk. The balloon fish is just the size to be on the menu of many self-respecting reef predators, yet it casually searches for its dinner with scarcely a care in the world. This nonchalance is not without good reason, because if the balloon fish is threatened by a predator, it has one of the most fantastic defensive tactics in the whole animal kingdom. Rapidly, the fish sucks water into its very elastic stomach. The stomach keeps swelling until it takes up most of the fish's body, and it keeps on distending until the whole fish starts to inflate. As the body grows, the spines on its surface become erect, until the fish looks like a huge, spiky ball. To the hopeful predator, what once appeared to be a tasty snack now looks pretty intimidating. Sometimes, a predator's lunge may be very rapid, and the balloon fish may not have time to inflate itself and ward the danger off. In these situations, the predator may swallow the balloon fish, which is more than enough to trigger its defensive ploy. The swollen, spiky fish sticks fast in the predator's throat, and it will take a whole lot more than mouthful of water or a pat on the back to dislodge it. Both the balloon fish and the predator will die in such an encounter, but for the most part, large reef hunters will know to give this unusual fish a wide berth as soon as it starts inflating.

+ The balloon fish is a type of porcupine fish, of which there are 19 species. They are closely related to puffer fish, which have a similar defensive ability, but not quite as pronounced.
+ The internal organs of these fish, such as the ovaries and liver, contain a potent toxin known as *tetrodotoxin*, which is at least 1,200 times more poisonous than cyanide. It acts on the nerves, disabling their ability to transmit electrical signals. Tetrodotoxin is produced by several types of bacteria dwelling inside the fish. Interestingly, the bacteria are somehow obtained from the fish's diet. Fish bred in captivity are not poisonous.
+ In the Far East, especially Japan, the flesh of the puffer fish and its relatives is a rare delicacy. The term *fugu* is applied to both the fish and its meat. As food, the

meat of these fish requires very careful preparation by very experienced chefs. The toxin is not damaged by washing or cooking, and an ill-judged filleting or misplaced slice can easily cause the accidental, fatal poisoning of a fugu lover.

+ A drug called *tectin* has been developed from tetrodotoxin, which shows great potential as a powerful pain reliever that is able to dull the discomfort of cancer and drug withdrawal.

+ The balloon fish and other types of porcupine fish are also fond of nibbling coral to extract the succulent polyps within. The calcium carbonate skeleton of the coral is swallowed, and some fish have been caught with over 500 g of the crushed material in their stomachs.

+ For reasons that are not yet understood, the puffer fish and its kin have the smallest genomes of any known vertebrate. Their complement of DNA appears to have little of the surplus carried by other species.

+ Even out of water these fish can inflate themselves quite happily using air, but due to the different properties of air and water, they have difficulty deflating themselves and can only do so if returned to the water.

+ In the Far East, the dried, inflated bodies of these fish are often made into curios or bizarre lanterns illuminated from the inside by an electric bulb. The Pacific islanders once used the dried, swollen bodies to make fetching ceremonial helmets.

BOMBARDIER BEETLES

Bombardier Beetles—A bombardier beetle releasing a noxious cloud of gases and liquids at an inquisitive ant. (Mike Shanahan)

Scientific name: *Brachinus* species
Scientific classification:

    Phylum: Arthropoda

    Class: Insecta

    Order: Coleoptera

    Family: Carabidae

What do they look like? The bombardier beetles are small beetles, normally a little over 1 cm in length. They usually have a metallic sheen. Each of the hardened forewings is modified into a tough shield called an *elytron*, which forms a type of shell that protects the soft abdomen beneath. The functional wings are below this shield, but in the bombardier beetles, they are tiny and completely useless.

Where do they live? The bombardier beetles are found throughout the temperate zone. They can be found in a variety of habitats, depending on the species. Some prefer the moist conditions of deciduous forest floors, while others may inhabit dryer habitats. All of them depend on some sort of shelter to spend the daylight hours, such as the space beneath a stone or log.

Arthropod Artillery

The animal kingdom is full of interesting defense mechanisms. Lizards can shed their tails at will, moths have large colorful eyespots, possums play dead, and so forth. Many of these tactics pale in insignificance when compared to the defensive ploys of the bombardier beetles. These beetles may be small, but they certainly pack quite a punch! On the outside, these beetles look like any other small ground beetle, but in their behind, they carry an impressive weapon. For much of the time, the beetle goes about its everyday business, looking for food and mates. However, when the bombardier beetle feels threatened by a potential predator, a unique chain of events is initiated.

Near the end of the animal's abdomen, there are two glands. One of these produces hydroquinones, while the other produces hydrogen peroxide—both of which are noxious chemicals. These substances flow into a reservoir, which opens into an explosion chamber via a one-way valve. The lining of this thick-walled reaction chamber is studded with cells that exude enzymes called catalases and peroxidases. These enzymes break down the hydroquinones and the peroxides into simpler molecules with the liberation of large quantities of oxygen. The chemical reactions also cause the cocktail of molecules to heat up beyond boiling point, triggering the vaporization of some of the mixture. The pressure in the reaction chamber builds up and up until it escapes from the animal through small openings at the tip of abdomen with ferocious speed in a series of around 70 fast explosions, the popping sounds of which are audible to the human ear. The pulsing effect is due to the build up of pressure in the reaction chamber, closing off the one-way valve into the reservoir. Only when the contents of the reaction chamber have reacted and shot out of the back end of the animal can more chemicals flow in via the valve. This whole sequence is over in a fraction of a second. The minute canon of some similar African bombardier beetles can be swiveled through 270° and thrust through the animal's legs so it can be discharged in all sorts of directions with considerable accuracy. The flexibility is made possible by a pair of tiny plates that deflect the stream of gas and liquid. The superheated cloud of noxious chemicals can inflict fatal wounds on other insects that irritate or threaten the bombardier beetle. The explosions can even burn the skin of a human who handles the insect without due care and consideration.

> ### 🔍 Go Look!
>
> There are several species of bombardier beetle native to the United States. They commonly look like the insect shown in the photograph. The elytra are normally dark and metallic, whereas the thorax and head are orangey. They are often found in moist floodplains of rivers and close to lakes, where temporary ponds are formed after storms. The easiest way to find them would be to search the ground or leave some pitfall traps in a potentially suitable habitat. Should you find what you think to be one of these beetles, handling it should provide the proof as it will give an explosive display in protest. Look for the puff of gases and fluids from the rear end of the beetle and the popping sound of the small explosions. There are some species of ground beetle that look very like the bombardier beetle, but which do not share its weaponry.

+ There are more than 500 species of beetle, distributed around the world, which can emit small explosions from their back ends. They are all ground beetles (carabids), but they are not all closely related to one another; therefore, this amazing defensive mechanism must have evolved independently in several different types of ground beetle. One possible driving force for the development of this adaptation is that bombardier beetles cannot fly, and taking to the air is a very handy way of evading those animals that would eat a tasty beetle. An explosive pulse of nasty chemicals shot from the behind is an excellent solution to the problem of defense in a flightless insect.

+ The bombardier beetles are predatory insects and will take a range of small ground-dwelling creatures.

+ Many insects and other terrestrial creepy-crawlies have glands that produce hydroquinones and hydrogen peroxides as a means of defense against various predators. Common examples are the small millipedes found beneath logs and stones. Handling one of these little creatures will make it rather angry, and it will begin to exude a cocktail of chemicals from pores along its body, including hydroquinones. These chemicals are toxic and have a distinctive smell that is more than enough to deter a hungry bird.

+ The name *bombardier beetle* was coined because early naturalists saw the puff of gases and heard the popping produced by the beetle and likened it to gunfire. Bombardier is a rank in the British army and refers to those soldiers involved with using large artillery guns.

Further Reading: Dean, J., Aneshansley, D. J., Edgerton, H. E., and Eisner, T. Defensive spray of the bombardier beetle: a biological pulse jet. *Science* 248, (1990) 1219–21; Eisner, T. The protective role of the spray mechanism of the bombardier beetle, *Brachynus ballistarius* Lec. *Journal of Insect Physiology* 2, (1958) 215; Eisner, T., and Aneshansley, D. J. Spray aiming in bombardier beetles: jet deflection by the Coanda effect. *Science* 215, (1982) 83; Eisner, T., and Aneshansley, D. J. Spray aiming in the bombardier beetle: photographic evidence. *Proceedings of the National Academy of Sciences* 96, (1999) 9705; Eisner, T., Jones, T. H., Aneshansley, D. J., Tschinkel, W. R., Silberglied, R. E., and Meinwald J. Chemistry of defensive secretions of bombardier beetles (Brachinini, Metriini, Ozaenini, Paussini). *Journal of Insect Physiology* 23, (1977) 1383.

BUSHY-TAILED WOOD RAT

Bushy-Tailed Wood Rat—A bushy-tailed wood rat carrying a camper's watch to add to its midden. (Mike Shanahan)

Bushy-Tailed Wood Rat—An adult specimen captured in a trap. (Tom Haney)

Scientific name: *Neotoma cinerea*
Scientific classification:
    Phylum: Chordata
    Class: Mammalia
    Order: Rodentia
    Family: Cricetidae

What does it look like? As its name suggests, this small mammal looks a lot like a rat; however, it has a more cuddly appearance. The body and to a lesser extent the tail, is covered in a dense pelage, the color of which varies with geographical location but is typically of a buff hue. The fur around the feet is often white. There is a great difference in size between the male and female. Adult males can reach a weight of 600 g, while a big female is only around 350 g. The eyes and ears are large and prominent, and the pointed muzzle bristles with long whiskers.

Where does it live? The bushy-tailed pack rat is native to North America. It ranges from the Canadian Arctic to New Mexico and Arizona. Within this huge geographical range, it is found within a variety of habitats, from the cold, boreal forests of the far north to the hot, dry semidesert scrub habitats of the southern United States. It is often found in mountainous areas and can be seen apparently thriving at altitudes of approximately 4,000 m.

Rodent Rubbish and a Snapshot of the Past

The bushy-tailed wood rat and the numerous other wood rat species also commonly go by the names of pack rat and trade rat. These aliases relate to a very curious feature of the wood rat's behavior. The wood rat, like any other small mammal, is a very nervous creature. It is just the right size to appeal to a huge number of predators, including eagles, owls, bears, foxes, and coyotes to name but a few. Because of this nervous disposition, the wood rat builds itself a shelter into which it can retreat at the first sign of danger. Most other rodents rely on burrows or simple holes to provide a refuge from predators, but the wood rat has opted for an all together more elaborate approach. At the base of a tree in a small cave, underneath a rocky overhang or any other secluded location, the wood rat constructs what is known as a *midden*. To build the midden, the wood rat scours its territory and collects what you and I would simply describe as rubbish. It gathers bits of dead plant material, bones, bits of dead insects, feces, and anything else it can carry back to the midden. During these collecting forays, the animal appears to be easily distracted. If it catches sight of something that looks better than what it already has in its mouth, it will release what it has and grab the new object. This is often the case if it happens on a campsite or if it constructs its midden in human dwellings. Such places may be littered with an abundance of attractive objects, such as wrappers, spoons, bits of cloth, and a selection of other trinkets. As the rodent seems to be quite unselective in the choice of material for its mound, the structure grows rapidly and is soon a sprawling heap. The wood rat's secret, though, is not its hording abilities, but in the way in which it sticks the whole midden together. The food of these animals often contains a high percentage of water and dissolved salts and sugars, which make the rats produce viscous urine very frequently. The rat spreads this fluid liberally over the midden, and the dissolved salts and sugars crystallize out of solution to hold the structure together like some manner of tough varnish. As the midden of the wood rat is coated in this glossy veneer, it can survive for a very long time indeed, especially if it constructed in a cave or underneath a rock overhang where it will be spared from the wind and rain. Middens in these situations have been dated to more than 40,000 years old. The plant and animal fragments that form these ancient relics are essentially fossils, but in an excellent state of preservation. Sifting through the fragments in one of these mounds can tell us a great deal about what the planet was like all those millennia ago. They can tell us what the climate was like and which plants were found where, as well as provide a glimpse of long-dead animals. Who would have thought the humble wood rat with its fondness for collecting rubbish would have provided us with such abundant time capsules showing what life on Earth was like many thousands of years ago?

+ The bushy-tailed wood rat is probably the most well-known wood rat, although there are around 21 other species, all of which are found in North America and Mexico.
+ As you go further north, the average size of the bushy-tailed wood rat increases. This is a phenomenon common to all animals, as a larger body conserves heat more efficiently than a smaller one, giving larger-bodied animals an advantage in cold climates.
+ The actual nest of the wood rat—the place where it sleeps and nurtures its young—is within the fortified confines of the midden. In one of these nests, a female bushy-tailed wood rat may give birth to three litters of young per year, each of which may contain as many as six young. However, the normal number is three young as the female only has four teats.
+ The word *midden* is believed to be Scandinavian in origin and is also used in archaeology to describe a mound of discarded food items at the site of an ancient human settlement.
+ Although the midden protects the wood rat from a number of predators, animals such as Gila monsters and badgers can use their powerful claws to access the midden and the nest within.
+ It is often the case that a midden constructed in a good location will be used by successive generations of wood rats, eventually forming a huge, rambling structure. Some of these huge, ancient heaps can be more than 1.5 m across and contain huge numbers of plant and animal fragments.
+ Because of the sheltered conditions that it offers, the midden may be occupied by reptiles when it is time for them to ovewinter. The rattlesnake, normally a predator of the wood rat is a common lodger.

Further Reading: Finley, R. B. "Woodrat ecology and behavior and the interpretation of paleomiddens." In Betancourt, J. L., van Devender, T. R., and Martin, P. S. (eds.) *Packrat Middens: The Last 40,000 Years of Biotic Change.* University of Arizona Press, Tucson, 1990.

ELECTRIC EEL

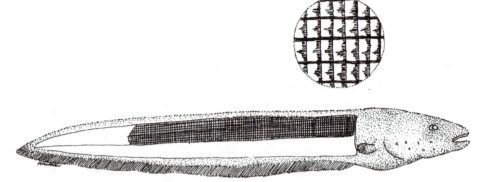

Electric Eel—Cutaway of the electric eel's body, showing the arrangement of the muscle batteries and a close up of the stacked plates in the electric organ. (Mike Shanahan)

Scientific name: *Electrophorus electricus*
Scientific classification:
  Phylum: Chordata
  Class: Actinopterygii
  Order: Gymnotiformes
  Family: Gymnotidae

What does it look like? The adult electric eel is a long, snakelike animal reaching lengths of 2.5 m and weights of 20 kg. In keeping with their serpentine form, they lack tail, pectoral, and dorsal fins; however, the anal fin is well formed and extends almost the whole length of the body. The whole body is cylindrical, with a flattened head and a large mouth. The skin of this fish ranges from grey to brownish/black with a white/yellow patch beneath the chin and throat.

Where does it live? The electric eel is native to the rivers and lakes of northern South America. It is found in the Guyanas, Orinico, and lower portions of the Amazon River. They are bottom dwelling creatures, lurking amongst the mud and detritus of rivers and swamps, typically in heavily shaded areas.

An Animal that Knows How to Shock

The electric eel of South America is a sluggish creature that spends much of its time resting on the bottom or slinking through the muddy waters of its home. It doesn't really need to be a fast mover, for it has a cunning means of defending itself and capturing food. At a glance, there seems to be very little about this fish that is special, and yet the electric eel, as its name suggests, has some amazing adaptations. The vital organs of the eel are all contained within the front section of the animal and only take up about 20 percent of the fish's internal volume, considerably less than other fish.

The rest of the eel isn't just a fatty lump or empty space but is given over to three separate groups of modified muscles that have long since given up contracting to provide movement. These are the fish's batteries, its electrical organs. The cells in the fish's main electrical organ form 5,000–6,000 disks, stacked like plates. The tiny electrical impulses produced by each of these plates are harnessed and channeled to produce a flow of electrical energy that is used by the fish in a number of ways. Normally, the electricity generated by the fish is very weak and is produced by the Sach's organ as a pulsating signal. This is the eel's sixth sense, the so-called electrosense, which is distorted by inanimate objects in the water and can therefore be used as a way to navigate through the cloudy waters of swamps and slow moving rivers, a real advantage when the visibility is close to zero. This sixth sense is also used to locate food. All organisms have an electrical field, and it these auras that the eel also detects with its electrosense, exploiting them to home in on unsuspecting aquatic animals. Aside from navigation and locating food, the electrical field can be cranked up to maximum by the main and Hunter's organs when the eel is in danger or senses the close proximity of a sizable snack. In a burst of short, sharp discharges, the eel can generate 650 volts with a 1 ampere current. This pulsing burst of electricity travels through the water and is potent enough to kill a human or even a horse over a distance of 6 m. It takes the fish a lot of energy to generate these high voltages, but they can do it intermittently for at least an hour, maintaining the strength of each shock.

+ Confusingly, the electric eel is not really an eel; it just happens to look like one. It is in fact a type of knife fish.
+ The ability to generate, store, and use electricity is not limited to the electric eel. Many other fish share this skill. The only species able to generate voltages even partially comparable to the electric eel are the African electric catfish and the torpedo rays, of which there are 69 species. There are many other types of weakly electric fish, such as the stargazers, elephant noses, many types of knife fish and skate. In all of these, except the electric eel, the electric catfish, and the torpedo ray, the electricity is used to locate food or to navigate. Electricity generation is relatively rare in marine fish, but is more common in those freshwater species inhabiting murky water. The only mammal known to have an electrosense is the duck-billed platypus.
+ The electric eel, like many other electrical fish is nocturnal. It will hunt at night for anything small enough to be sucked into its capacious maw.
+ The battery-like abilities of the electric eel is not the only interesting characteristic of this fish. It is also an air breather and obtains around 80 percent of its oxygen from surfacing and gulping air. The oxygen enters the blood through the lining of the mouth. Air breathing enables the fish to live in water where there is very little dissolved oxygen.
+ The electric eel is also unusual for it breeding behavior. In the dry season, a male eel makes a nest from his saliva into which the female lays her eggs. As many as 17,000 young will hatch from the eggs in one nest.
+ The powers of the electric fish are difficult to miss, and they have been known since ancient times. The ancient Romans would stand in a shallow bath with a torpedo ray and be electrocuted to alleviate the symptoms of gout.
+ Electric eels have always been high on the list of brave (or stupid) zoo collectors. Catching an electric eel is not easy, so the only solution was to make the eels tire themselves out with continual discharging. Some unlucky horses or mules were driven into a pool of water harboring a number of the fish where they would have received

enough shocks to at least knock them out. The fish's batteries would eventually drain, allowing the collectors to go in with nets and stout boots.

Further Reading: de Almeida, V.A.M. *Fishes of the Amazon and Their Environment: Physiological and Biochemical Aspects.* Springer, New York 1995; Moller, P. *Electric Fishes: History and Behavior.* Chapman & Hall, New York 1995.

GLAUCUS

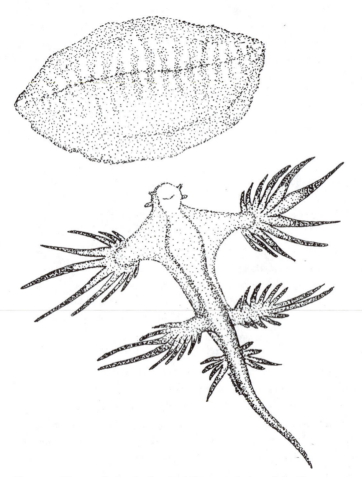

Glaucus—Glaucus closing in for the kill on a relative of the Portuguese Man-of-War. (Mike Shanahan)

Scientific name: *Glaucus atlanticus*
Scientific classification:
   Phylum: Mollusca
   Class: Gastropda
   Order: Nudibranchia
   Family: Aeolidiidae

What does it look like? *Glaucus* is a very odd looking animal. The sides of the tapering body bear six appendages, which branch out into thin, tentacle-like projections known as *cerata*. These contain thin saclike outgrowths of the animal's digestive canal. The mollusk is boldly colored with a blue or blue-and-white back and a silvery underside, making it look as though it is wearing sportswear. Fully grown, they are around 5–8 cm in length.

Where does it live? *Glaucus* is an animal of the open water. It is found around the world in temperate and tropical seas.

Turning the Tables in an Upside Down World

We have already been introduced to the Portuguese man-of-war and have seen the means by which this floating colony of animals catches food and protects itself. As with many defenses in the animal kingdom, they do not offer complete protection. The nemesis of the Portuguese man-of-war is *Glaucus*. Like the colonial cnidarian, this sea slug also floats around, but in a very novel fashion. The animal sucks air into its stomach, which acts like a buoyancy aid to keep it at the surface. Also, using its muscular foot, it can cling to the bottom of the surface film created when air meets water (the meniscus). Here, at the very surface of the water, it is at the mercy of the winds and currents and is pushed and pulled around the seas. Obviously, when the sea is calm and it is no longer buffeted by winds and tugged by currents, the sea slug can concentrate on hunting. Using very sensitive organs, the sea slug can detect the slight taint in the water indicating the presence of a nearby Portuguese man-of-war or one of it close relatives. *Glaucus* slowly makes for the source of this scent, which eventually leads to its favorite food. If the prey in question is a small specimen, a large *Glaucus* may simply swallow it in one mouthful, using its capacious cavern of a mouth to envelope the colonial creature. Should the prey be on the larger side, *Glaucus* will simply nibble its fishing tentacles, the ones that carry the most potent nematocysts, the stinging cells that are capable of delivering very painful and even fatal stings to much larger animals. The sea slug is unperturbed and eats the tentacles with gusto until they are little more than stubs. Eating a very venomous creature is impressive enough, but *Glaucus* goes one step further and turns the tables completely on its prey. It eats the tentacles containing the stinging cells and digests everything but the stinging cells themselves. Exactly how its digestive system can discern between the two is not known, but the complete, undischarged stinging cells are carried away from the stomach and digestive sacs on conveyor belts of minute cilia. After feeding, the sea slug will groom its appendages, distributing its new stinging cells. Eventually, these cells find their way to the fingerlike tentacles of the sea slug (the cerata), which each contain a thin outgrowth of the gut. Slowly, they are edged to the tips of the cerata, into a saclike structure where they will be nourished and primed to defend the sea slug from its own predators.

+ The sea slugs, or nudibranchs as they are known, are a very diverse and successful group of animals. Over 3,000 species have been identified, ranging in size from 0.4 to 60 cm, and as with any other marine, backboneless creatures, there must be many more to be identified. Some of the nudibranch species found in tropical waters are amazingly colorful and have common names, such as pajama slug because of their bold stripes. Many species are grazers, using their rasping mouthparts to scrape algae from corals and rocks, while others are active predators, feeding on a range of marine organisms.

- *Glaucus* is also unusual as it spends all its time upside down. The camouflage patterns that you would therefore expect to find on its back are found on its front and vice versa.
- The ability of *Glaucus* and its relatives to exploit and recycle the defensive apparatus of their cnidarian prey is but one of a whole selection of defensive adaptations exhibited by the sea slugs, which have lost the protective shell of their relatives and live, often exposed, in marine habitats. Some species that feed on sandy sponges have the same color and texture as their food, enabling them to blend in perfectly when they are sitting on a large sponge. Other species are excellent mimics of poisonous, marine flatworms. Secretions are another favored tactic, and some species can exude distasteful or noxious chemicals from their skin glands. Normally, sea slugs are slow movers, but when threatened, they can swim for short distances with wild thrashings of their bodies.
- The beautiful appearance of these animals makes them attractive to people with marine aquariums, but depending on the size of the tank, a single sea slug can kill all the fish inhabitants, such are its defensive abilities.
- The projections (cerata) of sea slugs are not only used as a handy store for nematocysts. They are probably also important in digestion, increasing the surface area through which nutrients can be absorbed, and as they contain blood, they also function as gills. The food eaten by the sea slug will pass through the cerata, showing the color of the food that it has been eating. This may be very important in camouflage.
- *Glaucus* is not the only specialist predator of floating colonial animals like the Portuguese man-of-war. There is also a species of snail, called the violet snail (*Janthina janthina*), that is another odd animal of the open ocean. This animal produces a raft of bubbles, which it coats with mucus. The mucus hardens to produce a gas-filled float. It finds the floating colonies of Portuguese man-of-war and similar colonial animals and devours them. The bubble raft is also used as a nest for the eggs, which are deposited on its upper surface.

Further Reading: Behrens, D. W., and Hermosillo, A. *Eastern Pacific Nudibranchs. A Guide to the Opisthobranchs from Alaska to Central America.* Sea Challengers, Monterey, CA 2005.

GOLIATH TARANTULA

Scientific name: *Theraphosa blondi*
Scientific classification:
  Phylum: Arthropoda
  Class: Arachnida
  Order: Araneae
  Family: Theraphosidae

What does it look like? The goliath tarantula is brown to very dark brown. The long hairs on the abdomen and legs are tinged pink/red. The front part of the body and the base of the limbs have a rather velvety appearance. Fully grown, the spider may have a leg span of more than 30 cm. Females are larger than males, with a heavier abdomen and relatively shorter legs.

Where does it live? The native habitat of this spider is the tropical Atlantic coast of South America, including Brazil, Venezuela, French Guiana, Surinam, and Guyana. The natural habitat is the forest floor, particularly in marshy areas or swampy areas where the spider prefers to locate its burrow.

Goliath Tarantula—A goliath tarantula flicking its defensive hairs at a hungry coati. (Mike Shanahan)

Bristling with Defenses

For obvious reasons, tarantulas strike fear into the hearts of many people. They are big, they are hairy, and they stalk around in a rather menacing manner. The largest of all of them is the goliath tarantula from South America. This massive, long-lived arachnid can have a leg span of more than 30 cm and weigh around 120 g, which is big for a land-dwelling animal that doesn't have a backbone. Although the goliath tarantula and its relatives are scary looking, there is no real reason to fear them. They do have large fangs, but their venom is very mild compared to that of much smaller spiders and is no more painful than a bee sting. Although humans are often afraid of these spiders, in their native environment of the South American coastal rain forests, they fall prey to many forest animals, for whom such a large, protein-packed meal would be a real find. To protect itself in these forests, the goliath tarantula has evolved some unique ways of defending itself. Should it find itself confronted by a predator, the first thing this spider does is to warn the foe by rubbing its bristle clad legs together, making a rasping sound. If the predator has come across one of these spiders before, it knows what follows these warning stridulations and backs off. If the predator is naïve, then it presses on and further tests the spider. The big tarantula must then resort to its next line of defense. It presents its back end to the predator, and using its back legs, kicks off a cloud of small, fine hairs from its abdomen that shoot like miniature harpoons toward the face of whatever animal is bothering it. These hairs lodge in the skin of the predator, causing a painful and irritating sensation much like very potent itching powder. The predator, with these tiny needles stuck in its nose and around its eyes, rapidly loses interest in the spider as it tries to dislodge the sharp bristles. With any luck, this will be enough to deter even the most foolhardy enemy, but should the harassment continue, the spider may have to resort to its last line of defense before it gets gobbled up. The spider rears up, appearing larger, and bares its formidable fangs. Little does the spider's foe know that these impressive looking weapons only inject weak venom. Should all these measures prove ineffective the spider may be in big trouble. Unless it can find a hidey-hole, it is destined to end up as a midnight snack for a lucky predator.

- More than 800 species of tarantula have been identified, and not all of them are large. Some are as little as 2.5 cm long. They are limited to tropical and subtropical regions.
- The name *tarantula* was actually first used in Italy for a type of European wolf spider whose bite was believed to be fatal unless the victim did a wild and manic dance, which became known as the *tarantella*. The name *tarantula* was then used to describe any large spider, and so the name for these big, hairy tropical species has stuck.
- These big spiders are also the most long-lived of all the spiders. Females can live for more than 15 years and possibly longer.
- Tarantulas are also among the most primitive spiders. The accolade of most primitive goes to a type of trap-door spider found in Southeast Asia. Like all primitive spiders, the tarantula bites by rearing up and driving its fangs into its prey.
- Big tarantulas are known as bird-eating spiders, but this type of behavior is very rare. This name stems from old illustrations that depicted tropical tarantulas looming over the bodies of birds they had just captured. Historical illustrators rarely based their drawings on living specimens and so depended on the stories handed down by explorers. These stories were often prone to exaggeration and embellishment. The goliath tarantula has been known to catch and eat fully grown mice as well as frogs and lizards.
- All of the primitive spiders have very poor eyesight. Their eyes are tiny and are only good to tell the difference between light and dark. The tarantula's most developed sense is touch. Its body is covered with hairs, some of which act like tiny wind vanes, picking up minute fluctuations in air pressure that could signal the presence of another animal. Exploring the world with their heightened sense of touch, they can make blindingly fast and accurate strikes at suitable prey.
- Almost all tarantulas are nocturnal. During the day, they rest in burrows, holes in trees or plants, and beneath logs. When hunting, some species rarely move far from their burrow entrance, instead waiting for prey to blunder by.
- The native Indians in South America are fond of eating the goliath spider. They are well aware of its defenses and are therefore very careful when capturing and subduing them. The bright orange eggs of the female are especially favored.
- Like most tarantulas, the goliath is a popular pet. They can be easily maintained in relatively small containers, and as they are not the most active of creatures, their food intake is amazingly small. However, they are large spiders and can be aggressive.

Further Reading: Conniff, R. *Spineless Wonders: Strange Tales from the Invertebrate World.* Owl Books, New York 1997; Conniff, R. Tarantulas. *National Geographic* (1996) 98–115; Cooke, J.A.L., Roth, V. D., and Miller, F. H. The urticating hairs of theraphosid spiders. *American Museum Novitates* (1972) 2498; Preston-Mafham, R. *The Book of Spiders and Scorpions.* Crescent Books, New York 1991.

HONEY BADGER

Honey Badger—A honey guide helps a honey badger find a bee's nest. (Mike Shanahan)

Scientific name: *Mellivora capensis*
Scientific classification:
   Phylum: Chordata
   Class: Mammalia
   Order: Carnivora
   Family: Mustelidae

What does it look like? The honey badger is a stocky creature with powerful limbs tipped with long claws. The claws on the forefeet may be 40 mm long. A white stripe runs the length of the body, separating the silvery, coarse fur of the back from the black pelage of the rest of the body. The fur is longest on the tail and hind limbs. It has small eyes and no external ears. When fully grown, they are over 1 m in length, with fully grown males weighing in at 14 kg. Females are much smaller than males.

Where does it live? The honey badger is an animal of sub-Saharan Africa, but it is also to be found through the Middle East to southern Russia, and even as far east as India and Nepal. They are found in a variety of habitats, from semidesert to rain forest.

Tough and Tenacious

Of the great diversity of mammals on the African continent, the honey badger, also known as the *ratel*, is one of the most notorious. They are often referred to as the "meanest animal in the world." For a smallish animal, the honey badger has quite a reputation to live up to. Its notoriety stems from its tenacity and unyielding nature. The first interesting aspect of the animal's toughness is it apparent ability to withstand the venom of the various snake species that it feeds on in southern Africa. There are reliable reports of this animal shaking off the bites of venomous snakes as if they were no more than disconcerting bee stings. The ratel's ability to survive a dose of venoms stems from the difficulty the snake may have in puncturing its very tough, loose, rubbery skin and some as yet unknown metabolic adaptation enabling the ratel to render the venom harmless.

The ratel is a fervent fan of honey, and its partial immunity to venom comes in very handy as it breaks into the nests of honeybees, who unsurprisingly react with appropriate aggression. The ratel can take many, many stings before it retreats. Not only is the honey badger fond of hunting and eating venomous snakes and stealing honey from angry bees, it is also disregarding of animals many times its own size. Lions and leopards have been known to kill and eat honey badgers, but on the whole, they will give them a wide birth. An angered honey badger is not a pretty sight. Apart from charging the animal it perceives to be a threat, it may also use chemical weapons and discharge a foul-smelling odor from its anal glands. Most potential predators find this particularly repugnant and will keep well out of range of the honey badger's strong jaws and claws. Should the honey badger be forced to sink its teeth into a potential aggressor, it is said that it will often go for one particular part, the scrotum. It is said that it latches on to the pendulous purse of the scrotum with a tenacious grip and eventually castrates the bewildered victim. It is difficult to know if this is exaggeration, but the ratel is definitely not an animal to be messed with.

- The honey badger is not related to either the American or European badgers. It looks quite similar to them, but it differs in many respects. The badgers and the honey badger are all included in the family Mustelidae, a large group of carnivorous mammals that includes the badgers, weasels, tayra, wolverine, otters, and so forth.
- The honey badger, although a powerful and fearless predator, is essentially an omnivore. It will eat a huge variety of foods, including reptiles, rodents, birds, insects, and carrion. It will also take fruits, berries, roots, plants, and eggs. To obtain the latter, the ratel will often scale trees and steal the eggs from the nest.
- Although the honey badger is obviously a very bold and tough animal, some of the stories surrounding it have obviously been embellished. The honey badger is difficult to find in the wild and even harder to study; therefore much of what is known is based on anecdotes.
- The honey badger is predominantly solitary, although small family groups of up to three individuals are occasionally seen. They are nomadic and range over huge areas, which for adult males may be large as 600 km^2.
- Reproduction in the honey badger is a long-winded affair. Only one to two young are produced every 18 months or so. The female has sole responsibility for the young, and they will stay with her for around 14 months.
- The very low birth rate of honey badgers makes them extremely vulnerable to hunting and habitat destruction. They are often scorned and hunted by farmers who own commercial beehives and believe the ratel to be a threat to their livelihoods. Many ratels are killed by farmers or stung to death by bees after becoming snared in a hive trap.
- One of the most interesting relationships in the honey badger's life, which is still poorly understood, is its interaction with a bird called the honeyguide. It is said that these two animals work together to exploit honeybee nests. The bird goes into the bush scouting for nests and calls when it finds one, attracting the honey badger to the scene. The bird, what with its meager beak and claws, would have no hope of breaking open the nest, but its partner, the honey badger, equipped with powerful claws and

tolerance to stings, can break open the nest and extract the honey-filled combs. As a reward for finding the nest, the bird is allowed access to the sweet honey.

+ Other birds have been observed in close proximity to honey badgers, apparently following them around. Several individuals of the pale, chanting goshawk of South Africa can often be seen tailing the honey badger. It is supposed that the ratel, being such an active digger, flushes rodents and other small animals from cover, allowing the birds of prey to catch them.

Further Reading: Killingly, P., and Long, J. *The Badgers of the World*. Charl C. Thomas, Springfield, IL 1983; Neal, J., and Chesseman, R. *Badgers*. Cambridge University Press, Cambridge 1996; Neal, J., and Ernest, T. *The Natural History of Badgers*. Croom Helm, London 1986.

HOODED PITOHUI

Hooded Pitohui—A marsupial is sick after capturing and trying to eat a hooded pitohui. (Mike Shanahan)

Hooded Pitohui—A specimen taken from a mist net in the hand of its captor. (John D. Dumbacher)

Scientific name: *Pitohui dichrous*
Scientific classification:
   Phylum: Chordata
   Class: Aves
   Order: Passeriformes
   Family: Pachycephalidae
What does it look like? The hooded pitohui is about the size of a thrush (about 15–20 cm long), with a black head, wings, and tail in sharp contrast to the orange or brick red hue of the plumage on the body.

Where does it live? The hooded pitohui is a forest bird and can be found over much of New Guinea.

Not a Fowl, but Definitely Foul

In 1989, a researcher studying birds of paradise in the forests of New Guinea was checking the mist nets set to trap the animals he was looking for. As was often the case, a black and red bird, the hooded pitohui, was struggling to free itself from the fine mesh. In freeing the bird from the net, the researcher received a couple of scratches, which he tried to soothe by sucking them. After a couple of minutes, he felt a numbing and tingling sensation in his lips and mouth. Had the bird been responsible for this? Later, to confirm his suspicions, he plucked a feather from another ensnared hooded pitohui and licked it. There was no doubt the feather contained some kind of toxin, and it turned out to be present not only in the plumage but also in the skin. Inadvertently, the young researcher had stumbled across a poisonous bird. Although toxins are produced by many animals, birds were not considered among their numbers, and it took quite a while for the young researcher to convince people that what he found was genuine. Eventually, an extract from the skin and feathers of a dead hooded pitohui was prepared, and the active ingredient was found to be a toxin very similar to that found in poison-arrow frogs, a chemical known as homobatrachotoxin (BTX). What was a small forest bird doing with one of nature's most potent poisons in its feathers and skin, and more intriguingly, where was the toxin coming from? The poison-arrow frogs of the neotropics secrete BTX; yet it is not known how they do it. It is possible it is synthesized in the tiny frog's body, but it is widely believed that the toxin, or at least the building blocks for it, is to be found in the amphibian's diet. The source of the toxin or its building blocks has not yet been identified, and pitohui researchers are having similar problems determining where the bird's toxin originates from. It has been proposed that some forest plants produce precursors for these chemicals and use them as defensive chemicals to deter the chomping jaws of herbivorous insects. In an interesting twist of nature, the insects have turned the tables on the plants and, instead of being deterred, are actually able to incorporate the poisons into their own tissues, modifying them for their own needs and using them as weapons against their own predators. It is possible that the birds have done something similar and are capable of eating otherwise distasteful insects and using the toxins to defend themselves. In 2004, it was found that choresine beetles, small, brightly colored tropical insects, contain BTX, and it is known that hooded pitohuis in the wild feed on a huge range of insects, some of which are known to be choresine beetles and their kin. This fascinating picture is not yet complete, and it may never be fully understood, such is the complexity of tropical ecosystems.

+ Six species of pitohui are known from the forests of New Guinea, and all of them can be found quite easily; some even frequent the forests around the capital of New Guinea, Port Moresby.
+ The rusty, black crested and variable pitohuis are also known to be poisonous, but only the latter species is on a par with the hooded variety.
+ It is possible the pitohuis are only poisonous for certain parts of the year. Perhaps, the dietary source of the BTX is not available all year-round. This and many other questions are yet to be answered.
+ In some parts of their range, the hooded pitohuis are so loaded with toxin that even handling them causes sneezing and watery eyes.

- Whatever the source of the bird's toxin, it is undoubtedly used as defense to deter predators. The bird's bold plumage is a warning sign to potential predators: "eat me and you will regret it!" As the hooded pitohui's breast skin and feathers contain a lot of toxin, the eggs may be coated with the BTX when they are being brooded in the nest. The forests of New Guinea are home to many egg thieves, especially reptiles that specialize in nest raiding. A noxious coating would protect the vulnerable eggs, thereby allowing the parents to forage for food.
- The New Guinea people call the pitohuis rubbish birds as they are not good for eating. In desperate times they can be eaten, but the feathers and skin have to be removed before the flesh is coated with charcoal and then roasted.
- After the pitohui's toxic nature was discovered, another New Guinea bird, the blue-capped ifrita was also found to contain high levels of BTX. The local people of New Guinea call the ifrita the *nanisani*, which is the name they give to the tingling, numbing feeling a person feels if they put their fingers in their mouth after touching one of these birds. The name *nanisani* is also used for the blue and orange choresine beetles, as they cause the same sensation.
- In the swampy lowlands of New Guinea, large flocks of several black and brown bird species are often led by a few members of the toxic pitohui species, suggesting a complex social arrangement between these birds. It is likely that the nonpoisonous birds are relying on the protection afforded by the pitohui's bold plumage and toxins.

Further Reading: Dumbacher, J. P. The evolution of toxicity in Pitohuis: I. Effects of homobatrachotoxin on chewing lice (Order Phthiraptera). *Auk* 116, (1999) 957–63; Dumbacher, J. P., Beehler, B. M., Spande, T. F., Garraffo, H. M., and Daly, J. W. Homobatrachotoxin in the genus *Pitohui*: Chemical defense in birds? *Science* 258, (1992) 799–801; Dumbacher, J. P., Beehler, B. M., Spande, T. F., Garraffo, H. M., and Daly, J. W. Pitohui: How toxic and to whom? *Science* 259, (1993) 582–83.

MIMIC OCTOPUS

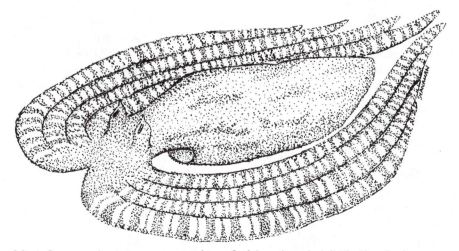

Mimic Octopus—A mimic octopus mimicking a flatfish on the seabed. (Mike Shanahan)

Scientific name: *Thaumoctopus mimicus*
Scientific classification:
    Phylum: Mollusca
    Class: Cephalopoda
    Order: Octopoda
    Family: Octopodidae
What does it look like? Fully grown, this octopus has an arm span of around 60 cm. Its body
    is patterned with brown and white stripes and spots.
Where does it live? The distribution of the mimic octopus seems to be restricted to the
    Indo-Malaysian archipelago. It is found in water 2–12 m deep, often near the mouths of
    rivers where the seabed is silty or sandy.

A Master of Disguise

The camouflage abilities of octopuses and their relatives, the squids and cuttlefish, are second to none as they are able to quickly change their color to blend in with their background. One species, the mimic octopus, is the undisputed king when it comes to disguise. Not only can it change its color as and when it needs to, but it can also use its flexible body to adopt the appearance and movements of a variety of sea creatures.

In its normal foraging guise, the octopus inches along the seabed searching for suitable prey, exploring tunnels and burrows with its long, sensitive arms. Any crustaceans, worms, or small fish quaking in the burrows will try to swim for safety, only to be ensnared by the webbing between the upper parts of the octopus's arms.

Although the mimic octopus is a predator, the bare, silty seabed is a dangerous place for a large, soft-bodied animal. Predators abound, and many could easily make short work of this master of disguise. Should the octopus spy danger, it draws its arms into a leaflike shape, changes color to match the seabed, and swims off with undulations of its body. The posture, color, and particularly the movement are startlingly similar to a number of flatfish found in the same area. These flatfish have venom glands at the base of their dorsal and anal fins, and many predators give them a wide berth.

In situations when the predator isn't fooled by the old flatfish trick, it swims up from the seabed and splays its arms wide. Cruising slowly through the water in this posture, it looks for all intents and purposes like a lionfish brandishing its venomous spines. The approaching predator has dealt with lionfish before, knowing their stings to be particularly painful, so it swims off and searches for easier pickings. The mimic octopus continues on its foraging rounds and then accidentally swims through the breeding ground of a damselfish, which happens to be fiercely territorial. The fish doesn't take kindly to this intruder, and it goes on the offensive. With an angry damselfish bearing down on it, the octopus uses yet another of its impersonations and makes for the nearest hole. It changes color and pattern and sticks six of its arms into the hole, leaving two at the surface heading off in different directions, waving sinuously in the water. Hey presto—a convincing impersonation of a banded sea snake, an animal that will quite happily eat a damselfish. The aggressive fish gets the message and backs off.

As splendid as these impersonations are, they are not the octopus's entire repertoire. Some predators may be invulnerable to the poisonous spines of flatfish and lionfish and unfazed by the venomous bite of a sea snake. In these circumstances, the octopus may swim to the surface, fully extend its many arms, and float slowly back toward the seabed in much the same way as certain jellyfish found in the same waters. Even if the stinging cells of a large jellyfish are not enough to deter some hungry predators, the octopus takes to the seafloor, and on a mound of silt, it raises its arms above its body to give a very convincing impression of a large, stinging anemone.

+ The mimic octopus was only officially discovered in 1998, and its amazing repertoire of impersonations was only discovered in 2001.

+ Although octopuses, squid, and cuttlefish are probably the most adept camouflage artists in the animal kingdom, mimicry appears to be very rare. In most cases, blending into the background may be a far more effective defensive tactic than trying to look like something else, especially if the impersonation is a poor one. Many species of octopus live in areas where there are numerous places to conceal themselves, such as reefs or the rocky seabed. Should they find themselves caught short in the open, they can change color in the blink of an eye to match their backdrop. The mimic octopus does not have this option. It lives in a habitat where cover is in short supply, so the ability to impersonate a number of dangerous neighbors is a very desirable trait. It is likely that the octopus picks its disguise, depending on the threat. Very few animals can do this to the same extent as the mimic octopus.

+ Color changing in octopuses is made possible by specialized skin cells called *chromatophotophores*. These cells can be stretched and squeezed by muscles around their perimeter. When the muscles contract, the cell is drawn out into a large, flat plate. In this state, the pigment in the cell covers a large area. When the muscles relax, the cell contracts, and the pigment spot is much less obvious to the observer. The pigment in these cells may be yellow, orange, red, blue, or black. All of these colors may occur in groups or layers. The visual effect of these color cells is enhanced by underlying reflective layers called *iridocytes*, which vary the shade and the tone of the color directed toward the observer. With this elegant system ,the animal can communicate with a riotous display of shimmering colors or blend into the background.

+ The mimic octopus is also the only known octopus species that traverses tunnels and burrows to conceal itself from predators and to search for food. When surveying its surroundings, it may adopt a sentinel posture in a suitable burrow entrance with only its eyes and head sticking out of the hole.

+ The octopuses and their kin are, arguably, the pinnacle of invertebrate evolution. Their sophisticated anatomy, complex behavior, and learning abilities distinguish them from the vast majority of other invertebrates. Although they are familiar to everyone, there is great deal still to learn about these fascinating animals.

Further Reading: Norman, M. D., Finn, J., and Tregenza, T. Dynamic mimicry in an Indo-Malayan octopus. *Proceedings of the Royal Society (Series B)* 268, (2001) 1755–58.

SEA CUCUMBERS

Scientific name: Holothurian
Scientific classification:
Phylum: Echinoderma
Class: Holothuroidea
Order: various
Family: various

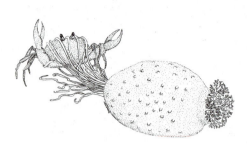

Sea Cucumbers—A sea cucumber rupturing its body and ensnaring a hungry crab in its sticky internal organs. (Mike Shanahan)

Sea Cucumbers—A sea cucumber lying on the seabed in the Caribbean. (Bart Hazes)

What do they look like? Sea cucumbers range from spherical animals to wormlike creatures. The smallest are less than 3 cm in length, while the largest can be more than 1 m long with a diameter of 24 cm. Most of the common species are between 10 and 30 cm in length. At the front end, there is a crown of tentacles surrounding the mouth. The underside of the body bears small, flexible structures known as tube feet. Their colors are normally quite drab, including black, brown, and olive green, although some species can have bold colors and patterns.

Where do they live? The sea cucumbers are exclusively marine and can be found in almost every oceanic environment, including shoreline habitats and the deepest oceanic trenches. They are at their most diverse in the shallow water of tropical coral reefs.

Pop Yourself to Protect Yourself

Most people would be hard pressed to identify a sea cucumber as an animal even if it was held under their nose. Superficially, they look like some sort of exotic root vegetable. They have none of the normal features we associate with animals. Where are the limbs, the eyes, or the ears? They have none of these. Nonetheless, they are animals, and they certainly rank very highly in the bizarre charts. Sea cucumbers are unassuming creatures. They mill around very slowly in their marine habitats using their tentacles to transfer edible matter from the seabed to their mouth. To give an idea of how sluggish these animals are, a large specimen will be rushed off its countless tube feet if it manages to cover 4 m in 24 hours. Many species are far more movement averse than this. They simply find a suitable crevice and wriggle their way in. As they are large soft-bodied animals severely lacking in anything like a burst of speed, evolution has been very imaginative in dreaming up a range of defenses for them.

Most sea cucumbers rely on hiding. A hefty sea cucumber can squeeze in through a very tight gap, and it can do this thanks to the unique connective tissue beneath its skin. This catch collagen as it is known can be loosened and tightened at will. When the animal wants to slip through a small gap, it can essentially liquefy its body and pour into the space. To keep itself safe in these crevices and cracks, the sea cucumber hooks up all its collagen fibers, making its extraction a challenge for even the most determined predator.

This is one way the sea cucumbers can defend themselves, but it is far from the most impressive in their collection. Most species of sea cucumber, if they feel threatened or stressed, can voluntarily explode, spraying a potential aggressor with their internal organs. During these defensive explosions, the body can pop at the front or rear, and the sea cucumber's enemy will get a face full of sticky gunk. What may seem like quite a drastic measure does the sea cucumber no lasting harm. Its tattered body regenerates, and all the missing organs are replaced. If this is not enough, some sea cucumbers have yet another tactic. Within their body they have a number of thin tubules, and when the animal is threatened, it ruptures its own anus, and the tubules stream out of the body. They are very sticky, and on contact with sea water, they expand to many times their original size. For a hungry crab, getting ensnared in these sticky bindings may be the last thing it ever does. The finishing touch to this anal explosion is the release of a toxic chemical called *holothurin*, which has similar properties to soap. Like a miniature chemical explosion, the holothurin can kill any animals in the immediate vicinity that are unable to escape.

+ There are around 900 species of sea cucumber in the world's oceans, and like their relatives, the starfish, brittle stars, urchins, and crinoids, they are very ancient animals, with a fossil history extending back at least 400 million years. Because many sea cucumbers live at great depths, it is very likely many more species are yet to be identified.
+ Sea cucumbers are also unusual in the way they breathe. They don't have the normal gill-like structures that are normally found in marine animals. Oxygen is obtained, and carbon dioxide discarded through the anus. Delicate organs, called respiratory trees, sprout from near the animal's anus taking up a good proportion of the internal body space. The pulsating rear end of the animal slowly fills and empties these channels like a gentle pump.
+ Like any nook in the ocean, the respiratory trees of the sea cucumber make excellent dwellings, and a little fish, the pearl fish, spends its time, when it isn't foraging, inside the sea cucumber. It is not known if the sea cucumber gets anything in return for supplying the fish with a safe place to live, but the fish cannot come and go with carefree ease. It can only sneak into the respiratory trees when the sea cucumber opens its anus—then, tail first, it darts for the hole. It is believed that some species of pearl fish may actually be parasitic on sea cucumbers, feeding on their internal organs after gaining access to the respiratory trees.
+ In many parts of the world, the cooked and dried body wall of certain sea cucumbers is used to make trepang and bêche-de-mer, a flavorsome delicacy. Fishermen use the toxins produced by sea cucumbers to stupefy fish and lure them from their refuges in coral reefs. The sticky, defensive tubules of some species are also used to make improvised bandages. More recently, the numerous chemicals produced by sea cucumbers have attracted the attention of pharmaceutical companies, as some of these compounds show promise as antimicrobial, anticoagulant, and anti-inflammatory medicines.

SLOW LORIS

Scientific name: *Nycticebus coucang*
Scientific classification:
    Phylum: Chordata
    Class: Mammalia
    Order: Primates
    Family: Lorisidae

Slow Loris—A slow loris licking at the elbow skin that produces a defensive secretion. (Mike Shanahan)

What does it look like? The slow loris is a medium-sized primate. The body is around 30 cm long, while the tail is reduced to a mere stump of no more than 5 cm in length. When fully grown, they can weigh up to 2 kg. The short, dense fur is variable in color but is typically of an ash-gray hue. A dark dorsal line divides the fur on the head into two branches, which surround the eyes. Their limbs are long and powerful, and their tiny hands have well-developed thumbs. The front-facing eyes are very large, and the ears are almost concealed amid its fur.

Where does it live? The slow loris and its numerous subspecies are found from Bangladesh to Vietnam. They are arboreal animals, preferring the tops of tress in forested areas.

When a Bark Is Nothing Compared to the Bite

The slender loris is a comical, timid creature. The word *loris* is actually Dutch for "clown," whereas the Indonesian name *malu-malu* can be translated as "shy one." The appearance and slow, exact movements of this primate give the impression of a very cuddly and docile creature. They move with extreme caution along the branches of their arboreal home, surveying their territory and looking for tasty morsels to eat.

Appearances in nature can be deceptive, and this is especially true for the slow loris. This vulnerable-looking little primate is more than capable of looking after itself and has some unique adaptations to deter even the fiercest predator. Should a slow loris be confronted by an arboreal predator, such as a civet, it stops in its already slow tracks and remains motionless. The circulatory system of the hands and feet allows the primate to grip for many hours without losing sensation. If this simple trick doesn't dupe the interested predator, the slow loris will fold its arms over its head in a way that makes its look as though it is holding its ears. It then laps at the skin on the inside of its elbow, the so-called branchial region. The secretions produced by the glands in this area actually become toxic when they are taken into the mouth and mixed with saliva. Should the predator approach within striking distance, the slow loris will lunge at the aggressor with surprising speed in an attempt to bite it. The teeth of the slow loris are a savage looking affair, and its jaws muscles are very strong. The long and pointed canines can deliver a very nasty bite, while the smaller teeth at the very front of the jaw deliver a cocktail of toxin and saliva to the wound. The bites are particularly painful as human victims can vouch for, and the activated toxin can cause allergic reactions in humans, which can cause death.

The slow loris's unique means of defending itself can also be adapted to protect its young. A female loris, feeling the need to forage, will leave her young gripping onto a branch, but before she leaves it to find food, she mixes her elbow secretions with her saliva and coats the youngster with it. Any opportunistic predators that try to snatch the baby loris will be left feeling sick with a bad taste in their mouth. The pungent odor of the elbow secretions, which is likened to sweaty socks, may also act as a deterrent, keeping all but the hungriest predators at bay.

+ The slow loris, the slender loris of Sri Lanka, the pottos, the bush babies, and the lemurs are all known as wet-nosed primates. They are related to the more well-known primates, the monkeys and apes, but the two groups first went their own evolutionary way millions of years ago.
+ The eyes of the slow loris are huge. Each one weighs about 3 g, which is a lot more than the brain of the animal. They are so big that they cannot move within their orbits, and so the only way the loris can shift its gaze is by moving its head.
+ Due to their appearance, the slow loris and the slender loris are often taken from the wild to be kept as pets. They are easily frightened and will bite and scratch when afraid. Unfortunate owners soon learn that these charming-looking creatures are more than a handful. They cannot be tamed, and as they require warm conditions and specific types of food, they don't last very long as house pets.
+ The lorises are carnivores, and in the wild, they catch and eat a wide variety of animal prey including invertebrates, reptiles, amphibians, and possibly small birds and mammals. Their chameleon-like progress through the trees allows them to sneak up on unwary prey before making a fast and accurate lunge.
+ The thumbs of the slow loris are very opposable, much more so than our own. This provides a sure grip and enables them to clamber along even thin branches with remarkable ease.
+ They make a variety of sounds when communicating with one another and when alarmed. These calls range from buzzing growls to chirps and whistles.
+ The lorises and their relatives are rarely seen by humans in their native habitat. They are occasionally hunted and eaten, but the greatest threat to these fascinating primates is the destruction of their forest home.

Further Reading: Krane, S., Itagaki, Y., Nakanishi, K., and Weldon, P. J. "Venom" of the slow loris: Sequence similarity of prosimian skin gland protein and Fel d 1 cat allergen. *Naturwissenschaften* 90, (2006) 60–62.

SPRINGTAILS

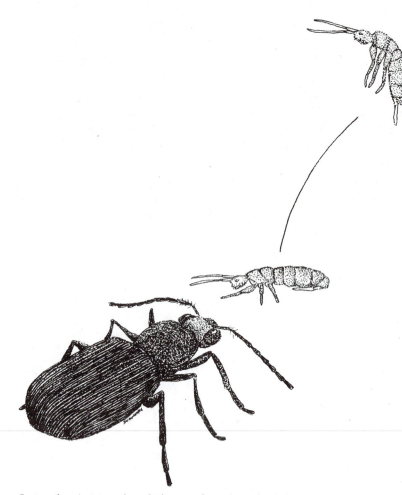

Springtails—A springtail avoids the jaws of a predatory beetle by using its springy furcula. (Mike Shanahan)

Scientific name: Collembola
Scientific classification:
    Phylum: Chordata
    Class: Entognatha
    Order: Collembola
    Family: various
What do they look like? The springtails are very small animals. The largest are only around
    10 mm long, while the smallest are very tiny, measuring around 0.2 mm. They have three
    body parts: the head, thorax, and abdomen. On the head is a pair of simple antenna and a
    pair of very simple eyes. Three pairs of shortish, simple legs sprout from the underside of
    the thorax. The long abdomen is composed of six segments, and at its end, there is structure
    known as the *furcula*. The whole body of these minute animals is clothed in fine scales,
    which can sometimes be very brightly colored.

Where do they live? The springtails are found all around the world in almost every conceivable habitat.

Springing to a Great Escape

Most people have probably seen a springtail. Whenever a large stone is lifted from its resting place or when a log is rolled over, the small specks that jump all over the place are, in fact, springtails. Although commonly encountered living beneath debris, springtails actually dwell in a very wide variety of habitats. Some species can be found in the canopies of trees; others loiter beneath seaweed on the beach. There are springtails that live out their whole lives on the surface film of freshwater and salt water or in the parched, unforgiving deserts of Australia. There are even springtails that eek out a living in the frozen wastes of Antarctica. In the whole diverse world of the terrestrial creepy crawly, it is the springtails whose range reaches the farthest south. In these inhospitable, frozen deserts, the springtails have to tolerate temperatures as low as −60°C. Few other creatures can tolerate such extreme conditions. Not only are the springtails found all over the world, but they are found in huge numbers. It has been estimated that an average square meter of grassland or woodland is home to around 40,000 of these small animals, although in some areas, the number is more likely to be in the region of 200,000. In the tropics, a square meter of canopy habitat accommodates at least 150 species of springtail. The massive population densities of springtails make them a very important component in all terrestrial ecosystems. They feed on organic matter in the soil, including decaying plant and animal matter, fungal filaments, and just about whatever detritus they can get their mouthparts into. Their feeding activities are crucial in the recycling process. The ceaseless, slow, almost imperceptible degradation of once-living tissue frees up nutrients for plant growth, the basis of all land-based food webs. Depending on the location, springtails can be responsible for around one-third of this recycling process.

What contributes to the success of these minute, easily overlooked animals? First, their small size allows them to live in large numbers out of the gaze of potential predators. They can crawl in the small gaps and fissures in the soil, where larger animal can't really get at them. They also have a range of defensive tactics to evade danger. The most impressive of these is their ability to jump high into the air at the slightest sign of danger. At their back end, they have a long thin structure, the *furcula*. This spring can be folded beneath the body, where it is held in place with a catchlike mechanism. Releasing the furcula propels the springtail upward with violent force. In a fraction of second, they can propel themselves many times their own body length. Their reflexes are similarly rapid, and some species can respond to the investigative probings of a predator in a smidgen over 18 milliseconds. The scales of the springtails are also an effective means of defense. A hungry beetle may try to grab the springtail in its powerful jaws, but all it ends up with is a mouthful of scales as the springtail leaps away to safety. Other springtails produce very noxious chemicals, which are more than enough to deter most potential predators. All in all, the combination of small size, good defenses, and unfussy feeding habits make the springtails among the most successful animals on the planet.

- Around 6,000 species of springtail are currently known, but it has been estimated that there could be more than 50,000 living species. They are the most widespread of all the six-legged creatures; their range encompasses all of the continents, even Antarctica.

- It was presumed for a long time that the springtails represented the ancestors of the insects. They have six legs and three body parts, but that it seems is where the similarity ends. It appears that the ancestors of the insects and the modern springtails parted company a very long time ago, even before the forebears of insects and crustaceans took to their separate evolutionary paths.
- The oldest-known six-legged arthropod is a springtail fossil from rocks in Scotland, some 400 million years old. Springtails have even been found in amber more than 40 million years old. These entombed specimens do not differ that much from their descendents.
- Those species of springtail that spend their lives in the complete darkness of subterranean and cave habitats have lost their pigmentation. They also lack eyes, which would be nothing more than an extravagance in perpetual darkness.
- Some species of springtail form dense aggregations containing many millions of individuals. A swarm in Austria of dark-bodied springtails was so huge that local people called the fire brigade to deal with what they believed to be an oil spill.
- Springtail jumps are haphazard to say the least, but there is one species of springtail that has small inflatable sacs on its antenna that exude a sticky secretion. When a series of jumps is careering out of control, the springtail can head-butt the ground and bring itself to a halt.

Further Reading: Coleman, D. C., Crossley, D. A., Jr., and Hendrix, P. F. *Fundamentals of Soil Ecology.* Academic Press, San Diego, CA 1995; Hopkin, S. P. *Biology of the Springtails (Insecta: Collembola).* Oxford University Press, Oxford 1997; Rusek, J. Biodiversity of Collembola and their functional role in the ecosystem. *Biodiversity and Conservation 7,* (1998) 1207–19.

3

THE QUEST FOR FOOD

ANT LIONS

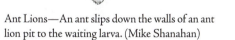

Ant Lions—An ant slips down the walls of an ant lion pit to the waiting larva. (Mike Shanahan)

Ant lions—An adult ant lion, a very different looking animal compared with the ground-dwelling larva. (Ross Piper)

Scientific name: many species
Scientific classification:
    Phylum: Arthropoda
    Class: Insecta
    Order: Neuroptera
    Family: Myrmeleontidae
What do they look like? Adult ant lions look very much like dragonflies. They have very large, ornately patterned wings that fold over the body when the animal is at rest. The abdomen is long and slender and the legs are quite short, but sturdy. The eyes are very large and there

is a pair of short, elegantly curving antennae. The largest ant lions have a wingspan of more than 16 cm, while that of the smallest is a mere 2 cm. The heavy bodied larvae have well developed legs, a tapering neck, a great, fat body, and a broad head with curving mouthparts. They are normally pale with dark markings.

Where do they live? Ant lions are found all over the world, with the exception of the polar regions. They are more diverse in warm, dry areas. In the United States, the largest number of species is found in the southwestern states. The larval habitat is warm, sheltered slopes, the ground beneath rocky overhangs and rocky or woody crevices.

On the Slippery Slope to an Unpleasant End

The handsome appearance of an adult ant lion is a stark contrast to the grotesque larva, but as with many insects, it is the rarely seen and ugly immature stage that is the most fascinating. After all, in insects with a larval stage, the adult is little more than a mating and dispersing machine. The larva's responsibility is to eat and grow and it is the pursuit of food that has spawned such an array of sizes, shapes and behaviors. The larva of many ant lion species are simply sit and wait predators, patiently letting the hours or days role by with their huge, curved mouthparts cocked, ready for an insect to come within striking distance. Some species have taken this sit and wait technique a stage further and construct a conical pit in which to entrap prey. The young larva builds this trap by walking backwards in ever-decreasing circles, forming a small doodle in the sand or soft soil. As the circle gets smaller and smaller, the young ant lion gets deeper and deeper. Soon, a well formed pit has been made and the larva settles down at the bottom, almost completely obscured except, perhaps, for the tips of its impressive, pronged mouthparts. The places where these animals construct their traps are just the sort of places coveted by other insects, such as ants, for their warmth. An ant, scurrying across the sand, rushing back to its nest, may fail to notice the innocuous looking pit. It tumbles into the base of the trap, reorientates itself and attempts to walk up the smooth sides of the depression. No sooner has it started its ascent when grains of sand rain down on it. The ant lion larva at the bottom of its trap has detected the struggling prey and using flicks of its head, hurls sand at the unfortunate prey. This shower of sand creates miniature land slides in the walls of the pit and the prey slips slowly and steadily to the waiting jaws of its tormentor. The eating apparatus of the ant lion is unusual in that it has no distinct mouth, much like a spider. The contents of the prey have to be liquidized by the injection of powerful enzymes and then sucked up through a channel formed by the snug fit of the mouthparts. A liquid diet has an advantage in that there is minimal waste and what left over matter there is accumulates in the larva for it has no anus to get rid of it from anyway.

After as long as three years in its pit, growing fat on insect victims the larva is ready to pupate. It spins itself a silken cocoon using some of the waste matter, accumulated in its gut and goes through the body altering process of metamorphosis. After a month the adult hatches, empties its stomach of the remainder of the accumulated waste and struggles to the surface. It climbs the nearest perch and inflates its body and wings with air, waiting for its outer surface to harden and become air worthy. In around 20 minutes everything is ready, and the adult insect flutters off on its maiden flight, completely unrecognizable as a formidable, miniature sand monster.

> ♦ There are around 2,000 species of ant lion distributed around the world. The largest species are found in Africa and Madagascar. They are related to the lacewings and dobsonflies.

- Ant lions, especially the species forming pits are sometimes known as doodlebugs for their habitat of tracing a path in sand or loose earth when they are constructing a trap, just like a little doodle.
- Adult ant lions live for little over a month. This is just enough time to mate and for the females to deposit her eggs in ground suitable for her offspring.
- In suitable areas, the ground will be pitted with the traps of ant lion larva, undoubtedly a gauntlet for any small, ground dwelling invertebrates.
- Adult ant lions are weak flyers and many are nocturnal. Despite their poor aeronautical skills they are predatory and will catch smaller insects on the wing. They will also eat pollen.

Further Reading: Griffiths, D. The feeding biology of ant-lion larvae: prey capture, handling and utilization. *Journal of Animal Ecology* 49, (1980) 99–125; Napolitano, J. F. Predatory behavior of a pit-making antlion, *Myrmeleon mobilis* (Neuroptera: Myrmeleontidae). *Florida Entomologist* 81, (1998) 562–66; New, T. R. A review of the biology of Neuroptera Planipennia. *Neuroptera International* Suppl 1, (1986) 1–57.

Go Look!

In the United States the only species of ant lion to make the little, conical pits is *Myrmeleon immaculatus*. This is the archetypal doodlebug and it can be found in large populations throughout the eastern and southern states. They can be found in sandy, sheltered habitats, such as blowouts in grassland or dune areas, vacant lots and scrub. The pits of the larvae will be immediately obvious if you are in the right place. You can either watch the larvae in their natural setting by tossing them a small insect or two, or you can carefully dig out the animal and take it home where it can easily be kept in a suitably sized contained part filled with sand. Take a look at their unusual, almost grotesque appearance, particularly their impressive jaws. The young ant lion will require feeding, but they will eat almost any small ground dwelling creepy crawly. Watch the sand flicking behavior to bring the prey within range of the fearsome jaws. When they no longer pay an interest in food, the larvae may be pupating. Now is the time to insert a stick into the sand and surround the container with a sleeve of light mesh or net, closed at the top to prevent the escape of the adult. After three weeks to a month, the adult will emerge and haul out of the sand before climbing the stick to harden and inflate its body. The adult can then be returned to the wild where it will feed and mate.

AYE-AYE

Scientific name: *Daubentonia madagascariensis*
Scientific classification:
   Phylum: Chordata
   Class: Mammalia
   Order: Primates
   Family: Daubentoniidae
What does it look like? This is the largest of the nocturnal primates. They can grow to 30–37 cm from head to body, with a long tail measuring between 44 and 53 cm. They weigh approximately 2.5 kg when fully grown. It is a rather shaggy looking beast with long black fur grizzled with white and grey. It has an odd head with large luminous eyes and a large pair of cuplike ears. The head is carried on a monkey-like body.
Where does it live? The aye-aye is found only in Madagascar. It is an arboreal animal and is found in the forests on the eastern fringe of the country. It is limited to the northern parts of the island.

Aye-Aye—An aye-aye uses its long finger to probe a hole in a tree for juicy grubs. (Mike Shanahan)

Long Digits Make Light Work

The disappearing forests of Madagascar are home to a myriad of strange mammals, one of which is the aye-aye. This peculiar creature mystified zoologists for years as they could not decide what it was. Its head and teeth are rodent-like, and for a long time it was thought to be some manner of giant squirrel. Eventually people realized it was a primate, but it took even more time for people to realize how much of an oddity it was. Perhaps its strangest feature is its hands. The middle finger on each hand is thin and bony and up to three times longer than the other fingers. These modified fingers are the aye-aye's feeding tools. At night, anywhere between 30 minutes and 3 hours after sunset, the aye-aye leaves its daytime hidey-hole of a nest built in the large fork of a tree. It leaps and clambers from branch to branch with acrobatic ease, its highly sensitive ears moving to and fro, listening for the slightest rustle or stirring that may indicate danger, or the movements of its dinner. It leaps to a branch and homes in on a vanishingly faint scraping coming from within the tree. To confirm its initial suspicions, the primate taps the wood and listens in much the same way as a builder finds cavities in a wall. A distinct hollow sound gives a positive answer to its hunch. Using its large incisors the aye-aye scrapes away the surface bark eventually revealing what seems to be a tunnel. Somewhere in the depths of the branch there is a juicy beetle grub happily gnawing its way through the wood oblivious to the imminent danger. The aye-aye brings its so-called witch's finger into play and probes the tunnel. The finger is thin and mobile, and like a delicate tool, it can tease the plump grub from its tunnel into the waiting maw of the expectant aye-aye.

- The aye-aye is a lemur. The lemurs are a specific type of primate found only on the island of Madagascar. There are 52 living species, although several of the larger species have become extinct due to loss of habitat and hunting. The larger species of lemur are diurnal, while the smaller ones are mostly active during the night. The aye-aye is by far the most bizarre lemur.

- There are no woodpeckers in Madagascar, leaving a space in the ecological web for a creature capable of penetrating wood and capturing the larvae of wood boring insects. Behold the aye-aye! Instead of large teeth for gnawing and a bony digit for probing the woodpecker has a sharp beak for hammering and a long tongue for grabbing larvae.

- The teeth of the aye-aye are what most confused early naturalists. They grow throughout the life of the animal, like those of a rodent. They must grow continuously as the animal is a feverish gnawer of bark and wood.

- Apart from their genitals and their weight, male and female aye-ayes show no sexual dimorphism.

- Even a newborn aye-aye has excellent climbing and clinging abilities, which is a must when you emerge into the world 10 m above the ground.

- An aye-aye nest is not used permanently. Every few days an individual will build a new one in another large tree fork.

- Unlike some other lemurs the aye-aye is predominantly solitary. They only come together purposefully to breed and for the rest of the time they treat each other with disdain. Males protect territories and interlopers in these territories will be met with a frosty reception.

- All lemurs are regarded with derision by the natives of Madagascar who share their forest home. They are thought to be evil omens and harbingers of doom. There are a few rare exceptions to these widespread superstitions and some tribes actually believe the lemurs to be bringers of good. Unsurprisingly, the appearance and behavior of the aye-aye afford it the lion's share of these ill feelings. It is said by some tribes that if an aye-aye points at you with it elongate middle finger you are certain to die very soon. People of the Saklava tribe claim the furtive and evil aye-aye will enter a hut and kill the occupants by puncturing the arteries in the chest using its spindly finger. Due to this very unfortunate bad press, the aye-aye is often killed on sight, even though it is a completely harmless, yet fascinating creature of the night.

- The aye-aye is one of the many animals and plants that are threatened by the habitat destruction that continues in Madagascar at an alarming rate. In 1985, only 34 percent of the islands original forest cover remained and only 6 percent of this was protected. Even today, in these more conservation aware times, Madagascar's wildlife is still under extreme threat. This pressing situation is made even more desperate by the fact that much of Madagascar's flora and fauna is unique. The chunk of land that today forms the island split from the continent of Africa many millions of year ago. The life it carried with it evolved in isolation forming a selection of unique species.

BOLAS SPIDERS

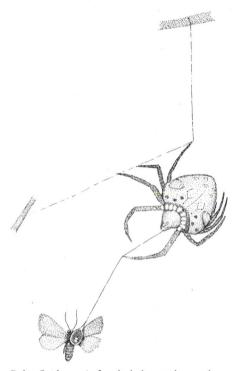

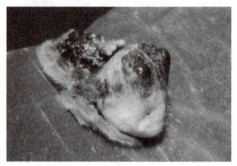

Bolas Spiders—A female bolas spider catches a moth with its minimalist, but ingenious web. (Mike Shanahan)

Bolas Spiders—An adult female bolas spider photographed in daytime pose where she mimics a bird dropping. (W. Mike Howell)

Scientific name: *Mastophora* species
Scientific classification:
    Phylum: Arthropoda
    Class: Arachnida
    Order: Araneae
    Family: Araneidae

What do they look like? The bolas spiders are typical orb web weavers. The females have a large, globular abdomen that is normally white and patterned with darker colors. The carapace is small and of a dark color. The male is much smaller than the female, with an abdomen that is in proportion to the rest of its body.

Where do they live? Bolas spiders are found around the world, but no species are found in Europe. They are found in a range of different habitats, although they all require tall vegetation to catch their prey from.

A Deceiving Arachnid with an Accurate Aim

By day, the female bolas spider sits motionless on a leaf or branch. When night falls, the spider prepares to hunt, but unlike other related spiders, the web of the bolas spider is very minimal indeed. First she trails a horizontal, nonsticky thread of silk on the underside of a twig or leaf.

Suspended from this support line by two of her legs she then extrudes another silk thread about 2.5 cm in length. This second thread is sticky and the free end of it is finished off with a very sticky blob of silk. This second thread is the bolas. This is the full extent of her web and with it complete she dangles by a couple of her free legs from the support line, while one of the front legs supports the bolas. She will hang there, motionless, for 15 minutes and if no potential prey presents itself she will reel in the bolas and eat it. It is possible the female does this because the bolas has lost its stickiness. However, unperturbed, she makes another bolas and adopts the same posture. This time she hears the familiar sound of fluttering moth wings. During the first few hours of the night, cutworm moths are active, the caterpillars of which are crop pests. The spider has very poor eyesight, but it is the flapping of moth wings she has been waiting for. As the moth approaches, the spider uses her front leg to twirl the bolas. Interestingly, the moth appears to be drawn to the spider and as it flies past it gets slapped with the bolas and is stuck fast to the sticky globule at the end of the spider's snare. Trapped, the moth is reeled in and the spider begins to feed. It has recently been discovered the spider can emit a pheromone mimicking the scent of a female cutworm moth. This odor lures male cutworm moths within range of the bolas.

When she has finished her meal the spider may spin another bolas and try her luck for more cutworm moths. This busy early evening schedule is topped off with a rest before the late evening session begins. At around midnight, other moth species, particularly the smoky tetanolita are on the wing and the sound of their flight triggers the spider into action once more, so she sets about spinning another bolas. The process is the same as before with the moth being drawn to the spider and then becoming ensnared in the whirling bolas. It is thought the two moth species have very similar sex pheromones, but they happen to be active at different times of the night. The bolas spider with its hunting strategy can target both moth species over the period of one night, simply by whirling its bolas when both moths are active.

+ There are several species of spider around the world that use a bolas to trap prey. There are species in both North and South America and in other parts of the world, such as Australia. This suggests the bizarre bolas strategy evolved on a number of occasions.

+ When at rest, the adult female bolas spider looks remarkably like a bird dropping, thanks to its large, globular abdomen and brownish carapace. This is a form of defensive mimicry as the animals that prey on succulent spiders pay very little attention to bird droppings.

+ Web building amongst spiders has it roots in the burrows constructed by the primitive spiders, which often have silken threads radiating out from the burrow entrance to alert the spider inside to potential prey. Over time, this basic system would have been adapted into the amazing structures constructed by orb-web spiders we see today. In several groups of spiders the standard web design has evolved and is often reduced. The most extreme example of web reduction amongst the orb-web weavers is seen in the bolas spiders.

+ The term *bolas* refers to the ranching and hunting tool of South American gauchos and the eskimos, which consisted of a length of rope or leather with a weight at each end. The user whirled the bolas above his head before releasing it at the legs of cattle or game.

+ The feeding strategy used by the bolas spider is a form of aggressive mimicry. The spider mimics the smell of female moths to attract male moths so it can catch and eat them. There are many examples of aggressive mimicry in the animal kingdom. A well known one is the anglerfish, where the first spine of the dorsal fin is greatly modified with its tip resembling a small worm.

+ Juvenile bolas spiders do not construct the snare of the adults, but they are able to produce pheromones to attract small flies, which they can pounce on. They normally do this by lurking on the lower side of a leaf near its edge. They wait until the fly approaches along the edge of the leaf before pouncing.

Further Reading: Gemeno, C., Yeargan, K. V., and Haynes, K. F. Aggressive chemical mimicry by the bolas spider *Mastophora hutchinsoni*: Identification and quantification of a major prey's sex pheromone components in the spider's volatile emissions. *Journal of Chemical Ecology* 26, (2000) 1235–43; Stowe, M. K. Chemical mimicry: Bolas spiders emit components of moth prey species sex pheromones. *Science* 236, (1987) 964; Yeargan, K. V. Biology of bolas spiders. *Annual Review of Entomology* 39, (1994) 81–99.

BULLDOG BAT

Scientific name: *Noctilio leporinus*
Scientific classification:
   Phylum: Chordata
   Class: Mammalia
   Order: Chiroptera
   Family: Noctilionidae

What does it look like? In terms of looks, the bulldog bat leaves a lot to be desired. Its wrinkled, pink, puglike nose makes it look as though it has been involved in an accident. Its small, beady eyes are situated beneath a pair of large, pointy ears and the membrane forming its wings is thin enough to show the immensely lengthened digits that form the scaffold of its flying structures. Its back bears a reddish-grey fur, while its underside is silvery grey. Its feet are large with sharp, curved claws. The toes and the claws of the bulldog bat are flattened from side-to-side, reducing the drag as they are trawled through the water. It is a small mammal with a body length of around 10 cm and a weight of only 70 g. Its wingspan is around 30 cm.

Where does it live? The bulldog bat is found in Central and South America from Mexico all the way down to Argentina. There are also populations in Trinidad and the Antilles. Its preferred habitats are forests and mangrove swamps where it frequents sheltered pools, slow moving rivers and sheltered coastal lagoons. Occasionally, it is seen over open water.

A Lesson in Fishing on the Wing

During the day the bulldog bat sleeps, grooms and washes itself in a suitable hideaway, such as a tree hole, or a rocky cleft. A good roosting site is hard to find and is often used by many bats. With the setting of the sun, the bats become restless and prepare for the night's activities. They leave the safety of the roosts and make for their hunting grounds. Pools, lakes and slow moving rivers are the bats favored places to find food. They crisscross the water, only illuminated by the moon and stars, using their power of echolocation to detect the faintest ripples on the water's surface that may betray the presence of fish below. The bat produces pulses of ultrasound that

Bulldog Bat—A bulldog bat keeps a good grip on the fish it has just caught with its trawling claws. (Mike Shanahan)

Bulldog Bat—A bulldog bat dangling from a branch, clearly showing its wide, long-clawed trawling feet. (Phil Myers)

are emitted from the nose and directed and focused by the complex, convoluted skin of the nose, which gives the animal such a grotesque appearance. These pulses of sound bounce off whatever they are aimed at and it these echoes that the bat uses to build a picture of its surroundings. It is 'seeing' with sound. As soon as it detects some ripples it descends to the waters surface and dips its feet in so that its long, curved claws are trawling approximately 2 or 3 cm below the surface. With its fishing hooks in the water, the bat makes a sweep of 30 cm to 3 m before ascending and turning to make a return sweep. The second time, the bat's claws make contact with a slippery fish and before the prey can react the bat scoops it from the water and straight to its jaws, which bristle with needle-like teeth. It makes for a suitable perch and dangles from its claws to feed on the fish. It first chews the fish to break it into large pieces, which are then transferred to cheek pouches to be chewed again before finally being swallowed. The bat is a small mammal with a fast metabolic rate that requires a lot of food, so as soon as the first meal of the night has been ingested the bat takes to the air again and continues trawling the water. In a single night the bat may catch and eat 20–30 small fish.

- Bats are a very diverse group of mammals. At least one in every five mammal species is a bat. There are two, distinct types of bat, the microbats (microchiroptera) and the megabats (megachiroptera—flying foxes, fruit bats). Although the two groups look quite similar, their relationship is complex. The fossil record of bats is very poor, making it very difficult to understand their ancestry. It was once thought the microchiroptera evolved from an insectivorous mammal, while the megchiroptera were once believed to have descended from a primitive, primate-like mammal, but this has largely been discredited.

+ There are other mammals that can glide using membranes of skin, but the bats are the only mammals with functional wings that can be flapped. The thin membrane of the bat wing allows the bones to be clearly seen. The wing is just a modified limb with an elongated forearm and ridiculously long digits.

+ Bats, particularly the microchiroptera are much maligned by humans due to their nocturnal activities, unsavory appearance and predilection for living in caves and other dark corners. The study of bats has helped to dispel many of the myths associated with them. In 1793, the Italian, Lazzaro Spallanzani, showed that a blinded bat could still hunt effectively, but one that could not hear was disorientated. This was the first real indication of echolocation. It wasn't until 145 years later that sensitive microphones were developed that allowed the American, Donald Griffin, to listen in on the ultrasonic sound pulses produced by bats. Humans can hear sounds with frequencies of between 20 and 20,000 Hz. Any sound above this range is known as ultrasonic and bats are sensitive to a range of less than 100 Hz up to 200,000 Hz. These rapid, high-pitched clicks are used to navigate and find food in the dark. Even in the pitch dark of a cave, the bat has little trouble finding its way. This fantastic adaptation is one reason for their success.

+ The majority of bats are predatory, although the larger flying foxes and some of the smaller bats are herbivorous and feed mainly on fruit and nectar. The predatory microchiroptera hunt mainly nocturnal insects. Moths are a favorite snack, and bats have been hunting them for so long that they grown wise to the tactics used by these nighttime hunters. Some moths are able to hear the distinctive hunting sound pulses produced by an approaching bat and go into an evasive dive just before the bat strikes.

+ Where they occur, bats are incredibly important components of the ecosystem. They not only consume huge numbers of insects every night, but in the tropics they are also responsible for the pollination and seed dispersal of many hundreds of plant species.

+ In many western societies the bat is associated with evil, yet in many countries, such as China bats are associated with longevity and happiness.

+ With a body length of 2.9–3.3 cm and tipping the scales at 1.5–2 g, Kitti's hog-nosed bat is probably the smallest mammal in the world. It is about the size of a big bumblebee and was only discovered in 1974. Like many bat species it is critically endangered.

Further Reading: Fenton, M. B. *Bats*, revised edition. Facts On File Inc, New York 2002; Hutson, T. *Bats*. Colin Baxter Photography Ltd, UK. 2000; Neuweiler, G. *Biology of Bats*. Oxford University Press, Oxford 2000.

CANDIRÚ

Scientific name: *Vandellia cirrhosa*
Scientific classification:
  Phylum: Chordata
  Class: Actinopterygii
  Order: Siluriformes
  Family: Trichomycteridae

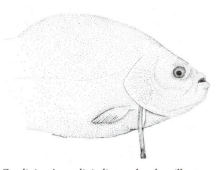

Candirú—A candirú slips under the gill cover of a larger, Amazonian fish. (Mike Shanahan)

Candirú—An adult candirú after a blood meal, showing its distended belly. (Ivan Sazima)

What does it look like? Candirús are small fish. Adults grow to only 15 cm with a rather small head and a belly that can appear distended, especially after a heavy meal. The body is also rather translucent making it quite difficult to see in the water. There are short, sensory barbels around the head, together with short backward pointing spines on the gill covers.

Where does it live? The candirú is native to South America where it can be found in the slow flowing waters of the Amazon and Orinoco rivers and their tributaries.

Reaching the Parts Other Fish Cannot Reach

Few animals are shrouded in as much myth and fable as the candirú fish. For many years, travelers to the Amazon have told stories of a fish with some very grisly habits. Stories aside, the candirú is a very interesting fish. It is one of the only known, wholly parasitic vertebrates. It depends on other living animals, normally bigger fish, for its sustenance. In the cloudy, sluggish waters of the lower Amazon, the candirú spends much of its time, on or near the bottom. The water is permeated by a whole range of different odors. Some emanate from decaying matter littering the river bottom, while others drift from the bodies of living animals. Tiny molecules of ammonia in bodily fluids find their way into the candirú's nostrils. To this small fish these odors are the nasal equivalent of a dinner bell. It follows the trail of molecules to its source, which turns out to be a hefty Amazonian river fish, many times the size of the candirú. Not wanting to alarm its quarry the candirú swims tentatively up to the area behind the head, where the large gill covers open and close circulating water to the fish's breathing structures. As the gill covers open the candirú seizes its moment and darts under, into the big fish's gill cavity. The candirú sinks its jaws into the blood rich tissues of the gills and feeds on the flesh and blood. The big fish is powerless. It must simply bear what surely must be a very painful experience as the little parasitic candirú gnaws its gills. The host's torment will only cease when the candirú has a full belly and slips from beneath the gill covers to rest on the river bed where it will digest its meal.

This is the candirú's natural behavior, but the local people of the Amazon and those who have heard stories tell of a different behavior guaranteed to send a shiver through the spine of even the most hardened river bather. In the same way the candirú finds its normal quarry, it may follow scent trails to find a person at the end of it instead of a big fish. Urea in the person's urine is thought to be what gets the candirú's nostrils twitching, but instead of a gill cover to swim beneath, it finds the urethra, the tube from which urine is passed, or perhaps the anus, and darts right up, in to the darkness. It feeds on the soft tissues until replete or until the victim's protestations force it to question its choice of quarry. Unfortunately, the passages it swam up are a tight fit and if it tries to back out, the backward pointing spines on its body keep it fixed in place. This is how the stories go at least and up until rather recently this is all they were, stories. However, an unfortunate man in Brazil was bathing in the river and decided he needed to urinate. No sooner had he started to relieve himself than a candirú shot straight up his urethra. A surgeon in the nearby city of Manaus had to operate on the man to dislodge the fish from within his penis. So, it seems there is some truth in the legends surrounding the candirú.

+ The candirú is a type of pencil catfish of which there are more than 180 species. In actual fact, the huge expanse of the Amazon basin with its huge variety of freshwater habitats is home to at least 700 species of freshwater fish. Undoubtedly, the cloudy waters and huge expanses of impenetrable waterways of the Amazon make this a haven for many more fish that are, as yet, unknown to science.
+ Several, very similar species of pencil catfish may have the same parasitic behavior as the candirú.
+ Apart from the name candirú, the fish is also known as the carnero, canero, or vampire catfish.
+ The candirú, like its relatives shuns the light and burrows in to the mud and silt of the river bed, only emerging to feed and find a mate. The other species in this family feed on small invertebrates they find in the river detritus.
+ Locals know of the candirú's tendencies so well that there is apparently a herbal remedy for a well lodged candirú. The leaves of the jagua plant (*Genipa americana*) and the fruit of another Amazonian plant are combined to make a preparation inserted into where the fish has lodged itself. The active ingredients in the plants, which are very acidic, allegedly dissolve the body of the fish. This miracle cure has not been investigated, so it is difficult to know whether it is true. It is also said that this preparation can be used to dissolve kidney stones. The chances of a candirú entering the penis, vagina or rectum of a person are very slim indeed and it should be stressed only one documented case is known.
+ Apart from the grisly feeding behavior of the candirú very little is known about the rest of its life. This is also the case for the vast majority of Amazonian fish. Many are caught for eating, but there is scant information on what they do in the wild.

Further Reading: Gudger, E. W. Bookshelf browsing on the alleged penetration of the human urethra by an Amazonian catfish called candiru. *American Journal of Surgery* 8, (1930) 170–88; Herman, J. B. Candiru: Urinophilic catfish—Its gift to urology. *Urology* 1 (1973) 265–67; Vinton, K. W., and Stickler, W. H. The Carnero, a fish parasite of man and possibly animals. *American Journal of Surgery* 54, (1941) 511.

COMMON CHAMELEON

Scientific name: *Chamaeleo chamaeleon*
Scientific classification:
  Phylum: Chordata
  Class: Reptilia
  Order: Squamata
  Family: Chamaeleonidae

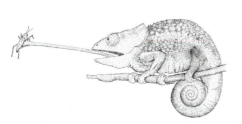

Common Chameleon—A poised chameleon strikes at a resting insect with its amazing tongue. (Mike Shanahan)

Common Chameleon—The grasping digits and highly mobile eyes of this lizard can be clearly seen in this photograph. (Martin Rejzek)

What does it look like? The common chameleon is a medium sized lizard reaching lengths of around 60 cm, half of which is the prehensile tail that is held in a coil when not in use. The body of the chameleon is flattened from side to side, while on the feet there are five digits forming excellent claspers. The head of the chameleon is large with bony keels making it look as though it is wearing a helmet. The eyes are large, independently mobile and mounted in small turrets. The mouth is large and gives the animal a sort of cartoon look.

Where does it live? The common chameleon is found around the Mediterranean as far north as the south of Spain and as far east as Israel. It prefers forested and wooded areas where it dwells in trees and bushes.

When Only a Tongue Will Do

The chameleon must rate as one of the most unusual and entertaining reptiles. It is a comical looking creature, yet its amusing appearance belies it effectiveness as a predator. The chameleon is a diurnal stalker of any insect or spider that shares it arboreal home. The digits are arranged on its feet so that it grips a branch with two claws on one side and three on the other, giving it secure purchase on even flimsy twigs. The tail is also prehensile and the tip is wrapped around the branch to act like a safety line. Haste means nothing to the chameleon. It is an animal of slow, measured strides. The eyes of the chameleon are perhaps the most peculiar among the vertebrates. They are mounted in what look like tiny turrets and it can move each one separately. When it is scanning the nearby vegetation for a potential meal they swivel in all directions, enabling the lizard to see all around without moving its head and alerting an unwary snack to its presence. The eyes, appearing to have a life of their own, do nothing to give the chameleon a serious, predatory look. However, when one of the constantly rolling eyes has spotted something of interest it becomes apparent what an effective little hunter this animal is. It begins edging up the branch it is on towards the prey with the stealth of an assassin. Not only are its movements very slow and measured, but it also sways very gently to fool the prey that it is part of the tree moving with the breeze. Its body is also flattened from side to side and it attempts to approach the prey so that it faces it head on, presenting a small target for a flighty insect to fix its gaze on. Not only is the chameleon an expert stalker, it is also a master of disguise. Special skin cells can change size, altering the color and tone of the animal. It can't change from green to some gaudy shade of pink, but it can make subtle modifications that help it blend in with the vegetation, or even give it markings to blend in with the patterns of the light and shade in the sun-dappled canopy. All of these fantastic adaptations bamboozle the prey and before long the chameleon is within range of its chosen victim. The most elegant part of the chameleon's hunting technique is that is doesn't even have to be on top of the prey to strike. The final trick in the reptile's repertoire comes into play when it is about a body length from the insect or spider it means to eat. Its eyes stop rolling and both zero in on the target, giving it a stereoscopic view. Its mouth opens very slowly to reveal a glistening, pinkish blob of a tongue. In less time than it takes to blink, the chameleon leans forward and the tongue springs from the mouth, the sticky club at its end adheres to the unlucky victim and before it knows what is happening it is hurtling towards the waiting maw of the reptile. The whole act is over in a fraction of a second and it is often difficult to tell if the chameleon has been successful until you see a faintly quivering insect leg poking out from the lizard's mouth.

- There are around 80 species of chameleon. They are reptiles of the Old World and are found in Africa, Madagascar, Europe, and Asia. They all have the same basic body plan with prehensile feet and tail, the swiveling eyes and projecting tongue. The smallest species have a body length of around 5 cm, while the largest are more than 60 cm long. Some species bear horns on their head giving them the look of a miniature dinosaur.
- When both of the chameleon's eyes are focused on a target they give it excellent perception of depth. This is the key element in their hunting success as it must be able to accurately gauge the precise distance to its prey.
- The projectile tongue of the chameleon is a complex system of bone, muscle and sinew. At the base of the tongue is a bone and this is shot forward giving the tongue the initial momentum it needs to reach the prey quickly. At the tip of the elastic tongue there is a muscular, club-like structure, covered in thick mucus that forms a suction cup. This attaches the tongue to the prey enabling it to be reeled in.
- The word chameleon is synonymous with deception and 'blending in'. Its color changing abilities are thanks to the unique properties of its skin. There are two types of color cells in the skin: red and yellow. Beneath these cells is a reflecting layer of blue and white, underneath this there is a layer of brown pigment. Depending on what color the chameleon needs to be, the skin cells get bigger or smaller, modifying the color of light that is reflected from the underlying layers. The camouflage skills used to hunt prey are the same used to avoid detection by predators, but if they are spotted their last line of defense is to straighten their limbs, inflate themselves, hiss and sway about in an attempt to look intimidating.
- The chameleons are supremely adapted to an arboreal existence and they are not at home on the ground; however, females have to descend to the ground to lay their eggs in the soil at the base of a tree.

CONE SHELLS

Scientific name: *Conus* species
Scientific classification:
    Phylum: Mollusca
    Class: Gastropoda
    Order: Sorbeoconcha
    Family: Conidae

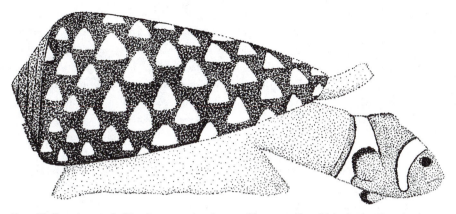

Cone Shells—A cone shell has harpooned and is engulfing a small reef fish. (Mike Shanahan)

What do they look like? Cone shells are snails with exquisitely patterned shells, normally consisting of brown decorations against a light background. The shell can be as much as 23 cm long. The spire of the shell is not very obvious, giving it the appearance of a short, stubby cigar. The only parts of the animal visible outside are the large, muscular foot, the long proboscis, which is essentially its mouth and its eye stalks.

Where do they live? The cone shells are animals of tropical and subtropical waters, typically in the western Atlantic and the Indo-Pacific oceans. They are often found on coral reefs.

A Snail with a Nasty Surprise in Its Shell

When we think of snails we normally think of the humble garden snail—slow moving, benign creatures. However, in the seas, many snails are fierce predators, able to tackle prey larger than themselves. The cone shell is one such example. Active cone shells move around their habitat using their muscular foot and using their siphon to "taste" the water for potential prey. When prey is within range, the snail loads its proboscis with a small harpoon, which is actually a single tooth from its radula (the snail equivalent of jaws). The harpoon is long, grooved and has a barb at one end and is retained at the entrance of the proboscis by the radula muscle. Once in place the proboscis is shot forward and impales the prey on the harpoon, which remains connected to the snail by way of a slender cord of tissue. A large muscular bulb and very long poison duct inside the snail are used to pump venom into the prey through the hollow harpoon. This venom is very potent indeed, containing various neurotoxic and pain-killing chemicals. In no more than 1–2 seconds the prey is immobilized and as the harpoon is still attached it can be drawn back towards the animal where it is engulfed by the capacious proboscis, before slipping into the snail's stomach. The prey can be sizeable, including small fish and large polychaete worms. In those cone shells that feed on other snails, there is no chance of the prey escaping; therefore the harpoon is freed from the proboscis. In the species feeding on very active prey, the snail is often concealed in the sand, where it waits until the prey is directly above it before thrusting its harpoon into the soft underbelly of the fish or worm.

- There are approximately 500 species of cone shell throughout the world's tropical oceans and all exhibit this unique feeding strategy. As the shells of these animals are very attractive, they have been coveted by collectors for centuries. One species, the glory of the seas, was very highly prized, until large populations of it were discovered.

- In recent years, the venom of the cone shell has attracted much interest from scientists who were intrigued by the speed of its effect. It was also discovered that the venom contains certain pain-killing components. These pain killers prevent the animal from struggling before the paralysis takes hold. The neurotoxic compounds found in cone shell venom offer potential new therapies for a range of human conditions, including Alzheimer's disease, Parkinson's disease and epilepsy. A pain killer developed from the cone shell has already been tested and is believed to be many times more effective than morphine for the relief of pain. It is likely that with the passage of time, more medical applications for these compounds will be discovered.

- Due to the interest in cone shells, both from a medical and aesthetic point of view, there have been over 30 recorded deaths, following a person carelessly picking up one of these shells and receiving a sting, even through gloves and wetsuits. The venom of one species, the cigarette snail, is said to be so fast acting the victim will have just enough time to have a smoke, before they expire. Unlike snake venom there is no known antivenom and the only treatment involves a life support machine until the body has had time to breakdown the venom.

- Cone shells are not the only marine snails with an interesting way of catching prey. The helmet shells (*Cassis* species) use sulfuric acid secretions to cut a hole in sea urchins, a process taking no more than 10 minutes. Once the urchin's defenses are breached the snail inserts its proboscis to feed on the flesh inside. The drills (*Urosalpinx* species and *Rapana* species) are snails that use a combination of rasping radula action and acidic secretions to wreak havoc in oyster beds. It takes these snails about 8 hours to penetrate a shell 2 mm thick. The proboscis can then be inserted through the drill hole allowing the radula to break up the soft tissues of the prey.

COOKIE-CUTTER SHARK

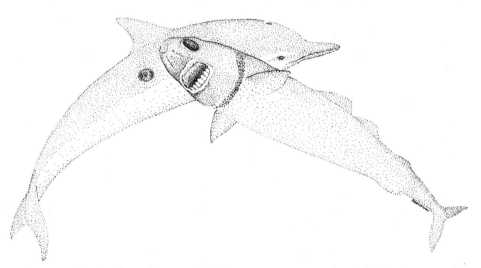

Cookie-Cutter Shark—The cookie-cutter shark shows its impressive teeth and the damage it can inflict on a dolphin (not to scale). (Mike Shanahan)

Scientific name: *Isistius brasiliensis*
Scientific classification:
    Phylum: Chordata
    Class: Chondrichthyes
    Order: Squaliformes
    Family: Dalatiidae
What does it look like? The cookie-cutter is a small shark. An adult is only around 50 cm long and as is the way with sharks the female is larger than the male. The slender, tapering body lacks an anal fin and its color ranges from grey to grey-brown. The head is compact with eyes near the front. The lips can form an effective suction cup and part to reveal an impressive complement of teeth.
Where does it live? Specimens of this shark have been found in the Atlantic and Pacific, normally in tropical waters. The furthest north it has been found is Japan, while to the south it reaches at least to Southern Australia. It seems to be more commonly encountered off islands, but has also been fished up from open water. It lives at least 100 m below the waves and is probably able to plumb depths of at least 3 km.

A Sneaky, Sharp-Toothed Menace from the Deep

The cookie-cutter shark, with its flabby little body, is far from the most elegant of the sharks, but as with the majority of these marine animals it is a consummate predator and what it lacks in bulk it more than makes up for in technique. As it is an animal of the deep sea it has evolved some bizarre ways of snatching a meal. In one of the most alien of earthly habitats, many animals can generate light, either as a means of illuminating prey, communication, or deception. Special light producing cells are found on the bodies of these animals. The cookie-cutter shark can also produce its own light; its underside, apart from a collar behind its head, gives off a pale green/blue luminescence. From below the shark, apart from its collar, would simply blend in to the diffuse, pale light scattering into the depths from above. To a large predatory animal swimming below the shark the collar would stand out like a beacon, resembling to all intents and purposes a small fish. This optical illusion would be even more effective if the big animal was swimming below a shoal of cookie-cutters. Fooled into thinking these shapes are some tasty morsels the large predatory animal, be it a large fish or marine mammal would head straight for it. Like the marine equivalent of a matador, the tiny shark deftly swims out of the way before the predator strikes and using the aggressor's forward momentum for added purchase the shark latches onto the animal with its suction-cup lips before plunging its teeth into its victim's flesh. The dentition of the shark is by no means run of the mill. Its capacious mouth bristles with 25–32 rows of pointed, triangular teeth, the front set of which look as though they belong in a miniature man-trap. The lower set are larger than the uppers and as they sink into the prey, its forward momentum and a quick pirouette gouge out a mouth-sized plug of skin, fat and muscle. This lump of tissue is cradled by the upper teeth and then hooked by the smaller upper teeth before being wolfed down. The scar left by one of these attacks resembles the neat depression left in a melon after it has been assaulted with a melon scoop.

    ♦   The cookie-cutter and its close relative, the large-toothed cookie-cutter, are types of dogfish with unique predatory tactics bordering on the parasitic. As they live at such great depths little is known about them. The only known specimens have been those found in trawl nets.

- Although the cookie-cutter is a small shark it is far from the smallest species. The spined pygmy shark is reputed to be the smallest, measuring a mere 22 cm.
- Female cookie-cutter sharks produce eggs that hatch inside their body. The offspring are nourished by their large yolk sac. Six to 12 young are produced at a time and upon emerging from their mother, the young are fully independent and can hunt almost straight away.
- The skeleton of the cookie-cutter shark is quite heavily calcified, which probably helps it regulate its buoyancy in the ocean depths. Its diet contains little in the way of calcium, therefore when its worn teeth are ready to be replaced they are swallowed and digested for the calcium they contain. In many other sharks, the teeth are simply discarded and fall to the seabed.
- The liver of this shark is large and oily and is an effective buoyancy aid so the animal can float in the water column without much effort.
- Many types of sea creature bear the scars of a meeting with a cookie-cutter shark. Even submarines are not safe from their gnashing teeth. The rubber-covered sonar domes of these vessels are occasionally pocked with the feeding gouges of cookie-cutters who would have been undoubtedly disappointed to find themselves with a mouth-full of rubber instead of delicious blubber.
- Like most deep-sea animals, the cookie-cutter shark makes daily, vertical migrations over distances in excess of 2 km. During the day, they remain in deep water, in what is known as the deep scattering layer, but during the night they swim up into shallower water, perhaps to within a few meters of the surface.
- The deep scattering layer is inhabited by many animals and is an important foraging area for tuna, porpoises, and so forth. It is when these animals visit this twilight zone that they fall prey to the fearsome jaws of the cookie-cutter.
- Although the cookie-cutter is specialized to feed on larger marine animals it will also eat squids, crustaceans, and other marine invertebrates.
- It has been reported that a cookie-cutter shark gives off its eerie blue-green glow up to 3 hours after it has died.

Further Reading: Widder, E. A. A predatory use of counterillumination by the squaloid shark, *Isistius brasiliensis. Environmental Biology of Fishes* 53, (1998) 267–73.

EGG-EATING SNAKE

Scientific name: *Dasypeltis scabra*
Scientific classification:
    Phylum: Chordata
    Class: Reptilia
    Order: Squamata
    Family: Colubridae

What does it look like? The egg-eating snake is a thin-bodied species that can attain a size of around 75 cm. They range in color from gray to brown. Along there back they have dark chevrons and or squares/blotches.

Where does it live? This snake is a native to Southern Africa. It is found in a variety of habitats, but does not frequent deserts or forest areas where the canopy is completely closed. It is often encountered in areas of thorny scrub with rocky outcrops offering numerous crevices for shelter.

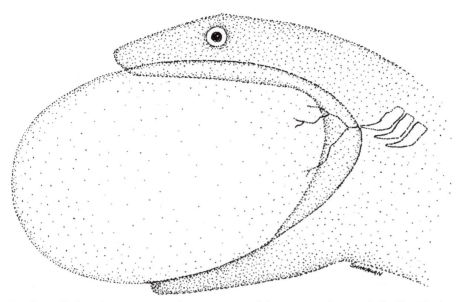

Egg-Eating Snake—An egg-eating snake gets its very mobile jaws around an egg. (Mike Shanahan)

Eyes Bigger than Its Belly…or Perhaps Not

The word *snake* conjures up images of animals that subdue active prey with crushing coils or deadly venom, but nature never misses an opportunity and there is even a snake that feeds exclusively on eggs. The common egg-eating snake, with its slight, flexible body can easily negotiate the thin limbs and smaller branches of trees in the hunt for bird's eggs. Its flickering tongue draws air into its mouth and into contact with the Jacobson's organ. This small organ functions like a nose, picking up the subtle hint of egg odors that may be carried in the air. As soon as it picks up the scent it will make straight for the source. The egg-eater is not a large snake, so if it chances upon a nest containing some big eggs it will have to continue its search. As soon as it finds a clutch of suitably sized eggs the snake will sniff each one carefully. Even from this preliminary investigation, the small serpent can assess whether the egg is healthy or rotten and how developed it is. Obviously, a rotten egg is out of the question and an egg with a well-developed chick inside would be too difficult to eat. What it really wants is a healthy egg with a small embryo and plenty of runny yolk and albumin. When it has identified such as an egg, it coils its long, thin body around it to hold it in place and proceeds to work its elastic head and jaws around the encapsulated meal. The bones forming the skull and the lower jaw of the snake are held together very loosely enabling the animal to get its gums around objects that are far wider than its head. With meticulous care the snake edges the egg into its mouth. Exerting too much force and breaking the egg at this point would be most unfortunate as the nutritious fluids would ebb away. As soon as the egg has been edged down into the throat the mouth can be re-aligned and closed. With its mouth closed and the egg halfway down its throat, the snake looks like it has swallowed a ball and to the human bystander the whole process looks quite painful. The muscles of the snake and peristaltic waves in its throat force the egg further along until it comes into contact with some specially adapted vertebrae. On their inner surface these bones bear a rasping, saw like edge and after being repeatedly brushed against them the egg cracks and its contents slide into the snake's belly. Using its muscular throat the snake squeezes the shattered egg to completely empty it of its contents. The remnants are compacted into a small, elongate pellet, which is duly regurgitated.

+ The common egg-eating snake is one of six species, divided into the African and Indian species. There are five African egg-eating snakes and one Indian species, which is quite rare. The African species are all found in sub-Saharan Africa.
+ Egg-eaters don't have teeth as they would rupture the egg shell before it passed safely into the throat.
+ A snake's head and jaws are unique in the animal kingdom. The bones are held together so loosely that the whole structure can flex tremendously allowing large objects to be swallowed. Unlike all other terrestrial vertebrates the bone of the lower jaw is actually composed of two separate pieces linked together by connective tissue. These modifications allow snakes to work their head and mandible over the prey.
+ The modified feeding apparatus of the snakes allows them to escape the constraints on meal size imposed by their thin bodies. Some of the vipers have eyes that are far larger than their belly. They can swallow and digest prey items that are larger than themselves, the only problem being that such a large meal takes a lot of digesting. The prey can rot before it is digested and the snake can die as a result.
+ As snakes can take in a large amount of food at once and are very efficient at converting it into body mass they can survive for long periods between meals. In captivity egg-eating snakes only need to eat about once every month, sometimes even less.
+ When threatened, egg-eating snakes coil and uncoil, rubbing their scales against one another. This produces a hissing or rasping sound and is accompanied by rapid lunges with the open mouth.
+ Because of its unusual feeding habits the egg-eating snake is a favorite among snake fanciers everywhere. They are quite difficult to keep in a vivarium as a constant supply of small eggs is required. They have not been reared in captivity, so all specimens in zoos and private collections have been taken from the wild, which has important implications for the populations of these snakes in their natural habitats. The Indian species is the most endangered egg eater, due to the rapidly expanding human population in this part of the world.

FAT INNKEEPER

Scientific name: *Urechis caupo*
Scientific classification:
    Phylum: Echiura
    Order: Xenopneusta
    Family: Urechidae
What does it look like? This spoon worm resembles a great, fat, uncooked sausage. Fully grown it can be 50 cm long. Its plump, pink body has no distinctive features to speak of, apart from a short, tonguelike projection at its head end a circle of spines on its rear end.
Where does it live? This animal can be found along the Pacific Coast from Alaska to California. It is a specialist burrow dwelling creature that makes its home in the soft mud

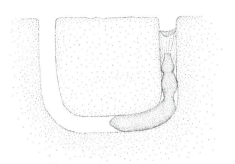

Fat Innkeeper—The mucus funnel produced by the fat innkeeper allows it to trap food from the water that it pumps through its burrow. (Mike Shanahan)

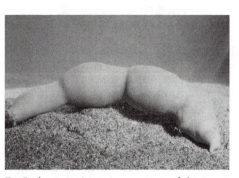

Fat Innkeeper—A captive specimen of this spoon worm removed from its burrow. (Lynn M. Hansen)

of tidal flats. Closely related species, similar in appearance and behavior, are found on the coasts of Chile, Australia, China, Korea, Japan, and Russia.

Finding Food the Spoon Worm Way

The fat innkeeper is an odd beast for a number of reasons. It feeds in a peculiar way, it doesn't breathe conventionally and it has some liberal attitudes to cohabitation. Young innkeepers take up residence in the burrow of an adult and will excavate their own lair as they grow. The burrow of a fully-grown specimen is U shaped. It descends vertically into the mud for 50 cm or so, and continues for 15–100 cm horizontally before striking for the surface. The animal spends its time in one corner of this burrow, but can move backward and forward as it pleases. The burrow is not only a safe place to hide from the many predators that would dearly love to sink their teeth or beaks into a deliciously plump snack, but it is also an integral part of the innkeeper's feeding technique. When the innkeeper is hungry it exudes a cone shaped net of mucus from its head end. The opening of this mucus cone is near the burrow entrance so any water coming into the animal's lair will pass through the fine mesh of this net. With its net cast, the fat innkeeper begins to pump. A rhythmic undulation along the animal's body pulls water into the burrow and pumps it out the exit at a rate of 18 L per hour. As this pumped water is being strained by the mucus net, its fine mesh will trap particles of food. After a lot of pumping the net will be clogged with edible tidbits and instead of picking it clean the innkeeper simply detaches the heavily laden net and uses its short proboscis to grab hold of it. Slowly and steadily it works the whole net into its mouth and eats it before secreting a new one. The innkeeper's burrow enables it to feed without expending a lot of energy on movement, however as the water in the burrow may be quite stagnant, especially at low tide, it must have an efficient means of extracting oxygen from the water. It must have some form of gill. It does and they are up its bottom. Water from the burrow is sucked into its rear end where oxygen and carbon dioxide diffuse across the thin membrane of the hindgut. Gas exchange complete, but in an odd place!

- As with any habitat, refuges on the mudflats are at a premium. The U-shaped burrow of the fat innkeeper is one such safe haven and it attracts a small assemblage of animals

Go Look!

Mudflats are very interesting habitats wherever you are in the world, but most of what goes on in these places happens beneath the surface. They support a huge diversity of animals, from microscopic invertebrates to wading birds. Of course, a mudflat comes alive when the tide is in and when the tide is out you can walk on the mud and see evidence of frantic activity. The mud is pocked marked with huge numbers of holes, many of which will house some invertebrate far down in the mud. There are also tracks left when the owners were foraging for food during high tide. Small mounds of what look like brown spaghetti are worm casts, left by worms after their gut has processed it for edible particles. Dig into the mud with a spade and the rotten egg smell of the black layer beneath the surface will hit your nose. Digging will expose the burrows of many animals and you may be lucky to see a large worm retreating down a burrow. Place a small spade-full of mud into a large white tray with some seawater and you will see a range of different organisms: small clams, other mollusks and many worms. A small amount of this mud in a petri dish with seawater examined under a microscope will reveal a miniature universe of tiny animals and single-celled organisms that spend their entire life between the mud and sand particles.

looking for a place to live. The name *fat innkeeper* was given to this animal because of its corpulent appearance and the fact is shares its home with a number of lodgers. The first of these is a small, filter-feeding crab, which feeds in the tunnel alongside the second lodger: a scale worm, which probably feeds on any scraps left by the landlord. Thirdly, there is a clam that burrows into the mud of the burrow and filters the water for nourishment. These three animals live exclusively in the burrow, although the exact nature of the cohabitation is unknown. Perhaps the lodgers are simply taking advantage of the sanctuary, or maybe they give the landlord something in return. There is an additional part-time tenant, a goby, which flits in and out of the tunnel as it pleases.

+ In the places where they occur, fat innkeepers are common animals. They are an important component of the mudflat ecosystem as their burrows provide a home for other animals and irrigate the sediment, improving its oxygen content.

Needless to say an animal like this also has its fair share of predators, such as the leopard shark (*Triakis semifasciata*), which purses its lips around the burrow entrance and gives an almighty suck to dislodge the fat, pink worm.

+ The fat innkeeper and other burrow dwelling animals of the mudflats are favorites of fishermen who use them as bait. They search the mudflats for the burrows of these animals and suck them out with a large syringe type device.

+ Mudflats are very productive habitats. The abundance of organic matter in the mud is processed by a multitude of bacteria, breaking it down in the absence of oxygen, forming toxic hydrogen sulfide gas and giving the mud a distinctive stench of rotten eggs. When the tide is in the gas percolating into the water is diluted, but when the tide recedes any standing water, such as that in burrows will quickly become noxious. Animals, such as the innkeeper, have cells that detoxify this gas enabling them to survive in their burrows when the tide is out.

Further Reading: Julian, D., Chang, M. L., Judd, J. R., and Arp, A. J. Influence of environmental factors on burrow irrigation and oxygen consumption in the mudflat invertebrate *Urechis caupo*. *Marine Biology* 139, (2001) 163–73; Osovitz, C. J., and Julian, D. Burrow irrigation behavior of *Urechis caupo*, a filter-feeding marine invertebrate, in its natural habitat. *Marine Ecology-Progress Series* 245, (2002) 149–55.

GHARIAL

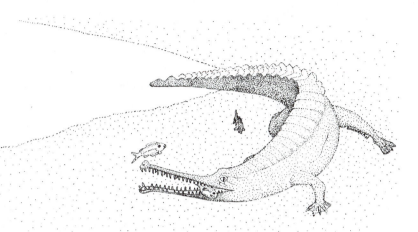

Gharial—A gharial corrals fish between its body and the river bank. (Mike Shanahan)

Scientific name: *Gavialis gangeticus*
Scientific classification:
    Phylum: Chordata
    Class: Reptilia
    Order: Crocodilia
    Family: Gavialidae
What does it look like? An adult male gharial can be much as 6.5 m long and weigh several
    hundred kilograms. Females are a great deal smaller than males, but they are still impressive
    looking crocodilians. The snout is very long and thin and is armed with numerous needle-
    sharp teeth. The body is very heavy, yet the limbs are weak. The tail is large and powerful
    and is flattened from side to side.
Where does it live? The gharial is found in various Asian river systems, including the Indus
    (Pakistan), the Brahmaputra (Bangladesh, Bhutan and India), the Ganges (Bangladesh, India
    and Nepal), the Mahanadi (India), and the Ayeyarwady Kaladan (Myanmar). It prefers the
    deep, fast flowing portions of these rivers.

Fishing—Crocodilian Style

The Indian subcontinent and the surrounding areas are dissected by several very large river
systems. It is the icy peaks of the Himalayas where much of this water originates. Over the eons
these rivers have carved their way through the rock of the land forming gorges, valleys and flood-
plains. It is in these narrow gorges, where the water flows deep and fast, that the strangest of all
crocodilians makes its home. The gharial, sometimes known as the gavial, is the most specialized of all these ancient reptiles. It is for example the most aquatic of the crocodilians, spending
almost all of its time in the water. They only haul out to bask in the warming rays of the sun or
to lay eggs. Their bodies are not really built for a life out of the water—the legs are not robust

enough or in the right position to carry the weight of the animal on land. To move about out of the water the gharial must resort to belly sliding and although it can slip along at quite a pace when it needs to, this is hardly an efficient means of locomotion. Thus, the gharial is never found far from water. Although awkward and lumbering on the land the gharial is the most agile of the crocodilians in the water. Its large tail propels the animal swiftly and easily through the water. The gharial's love of the water is closely bound to its diet, which is more specialized than any other crocodilian. As an adult a gharial preys exclusively on fish. Crocodiles and alligators would find it difficult subsisting exclusively on fish as they are too quick and can quite easily evade the snapping jaws of these reptiles. The gharial is equipped with a host of adaptations that make it an almost unparalleled dispatcher of river fish. The snout is long and thin, which means that it can be swung through the water with the minimum of resistance. Not only do the jaws slice through the water, but they are bristling with many needlelike teeth giving excellent purchase on the slippery prey. A typical hunting strategy employed by the gharial is to make a fish corral from the riverbank and its body. Fish inadvertently swim into this trap and are rapidly snared in the toothy jaws. To swallow the fish impaled on its jaws the gharial lifts its head from the water and with a sideways jerk the fish falls off the teeth and deep into the mouth.

For much of the year, gharials concern themselves with fishing and basking, but from November to January the cycle of eating and basking is interspersed with courtship and mating. A big bull gharial defends a harem of several females and he mates with all of them, aggressively driving off hopeful suitors and fighting with other mature bulls who like the look of his females. Several months after mating, usually around March to April, the females leave the water to lay their eggs. They excavate holes in the sandy riverbanks, depositing somewhere between 35 and 60 eggs before carefully backfilling the nests. During the incubation period the female diligently guards the nest and after around 60 days her offspring hatch. She assists them in leaving the nest and stays with them until the arrival of the monsoon rains. The heavy and torrential rains flood the nest sites and the young are carried downstream to start their own lives.

+ The gharials are quite distinct from the other crocodilians (crocodiles, alligators and caimans) and they are represented by two species, the second being the false gharial, which is found in six river systems in Malaysia and Sumatra.
+ Male gharial have a lump on the end of their snout that grows with age. This boss as it is known partially restricts the nostrils to the extent where the breath of the animal sounds like a buzz or hiss. The males use this sound during courtship and also during aggressive exchanges with other males.
+ The crocodilians are a very ancient group of animals. The earliest known representative of this group is a fossil from the late Triassic, somewhere between 200 and 230 million years ago. It is now widely accepted they are the closest living relatives of birds, which evolved from the other ancient lineage of reptiles, the dinosaurs.
+ Just how the crocodilians survived the cataclysmic events that spelled the end of the age of the dinosaurs is a mystery. Perhaps they survived because they are normally generalists and were able to survive on the carcasses of other animals, seeking refuge in the relatively sheltered environment of their aquatic habitat. Following the demise of the dinosaurs the crocodilians diversified and for a while they competed with mammals and birds for the dominance of the terrestrial ecosystems. They were outcompeted eventually and receded to the semiaquatic habitat they are so well adapted to.

+ The gharials, with their thin jaws have lost the tremendous bite pressure of their broad-snouted relatives. A moderately sized crocodile (3–4 m long) can bite with a force of around 960 kg. Compare this to the spotted hyena, which is also featured in this book.
+ The gharial is such a specialized animal that its numbers and range have been heavily affected by human activities. Dams have been built, waters polluted and riverside habitats developed for construction or agriculture. In the 1970s the species came within a hair's breadth of following the dinosaurs to extinction, but state protection and tireless conservation have allowed it to regain some of it former glory, especially in India. Unfortunately, in other countries the situation is not quite as rosy and the gharial remains on the cusp of extinction in these areas.

Further Reading: Thorbjarnarson, J. B. Notes on the feeding behavior of the gharial (*Gavialis gangeticus*) under semi-natural conditions. *Journal of Herpetology* 24, (1990) 99–100.

GIANT ANTEATER

Giant Anteater—A giant anteater has broken into an ant's nest and is using its long, sticky tongue to catch the inhabitants. (Mike Shanahan)

Scientific name: *Myrmecophaga tridactyla*
Scientific classification:
　Phylum: Chordata
　Class: Mammalia
　Order: Xenarthra
　Family: Myrmecophagidae

What does it look like? The giant anteater is an elongate mammal covered in shaggy gray, white and black fur. A stripe of black fur runs diagonally from the shoulder to the back. The very long, bushy tail can be used to cover the animal's body. Its small head tapers down into a long conical snout. The powerful front legs end in long, curved claws, but to keep these off the ground the anteater walks on its knuckles and the sides of its hands giving it a limping, shuffling gait.

Where does it live? The giant anteater is found in Central America and east of the Andes in South America. It can be found in both forest and grassland habitats.

You Need More than One Ant to Make a Meal

Ants, due to their large nests chock-full of workers, succulent grubs and pupae, are relished by a number of predators. Perhaps the most specialized of these is the giant anteater. This ambling ant vacuum is the scourge of these insects in Central and South America.

Even in adulthood the long, conical snout of the giant anteater contains no teeth whatsoever, but these are not important for an ant specialist. What is more important is being able to extract these insects from their heavily fortified, subterranean nests. To get access to these underground galleries and chambers the anteater makes use of its formidable claws, which sprout like miniature sickles from its hands. Using its powerful shoulder, back and arm muscles the anteater breaks into the nest, but is careful not to cause too much damage. With the nest breached it brings its snout into play. As the snout is long and thin it can easily be inserted into the hole left by the claws. The ants rush from all corners of the nest to protect their home from this hungry interloper. Their resistance is, of course, futile. They are caught in the piston like motion of one the most tremendous tongues in the animal kingdom. This straplike muscle can be pushed out a remarkable 60 cm. A big giant anteater's body may only be 120 cm long; therefore, this massively long tongue, its sheath, and the muscles that retract it are all anchored to the animal's breastbone. In the one minute in which the anteater feeds at a nest the full length of the tongue may be inserted 150 times into the nest. The salivary glands of the animal also produce large quantities of very thick, sticky saliva. As the tongue flicks in and out of the nest, workers and their immature brothers and sisters stick to it and are drawn back to the anteater's mouth. The roof of the mouth bears a number of tough bumps, which squash the prey and begin to the grind them up. As they are toothless, further grinding takes place in the muscular stomach. Stones and other debris, ingested as the tongue probes the nest, also find their way into the stomach and assist with the blending process.

The anteater is very careful not to destroy a nest completely. In the minute or so in which it probes a nest with its massively long tongue it may only snare 140 ants. The anteater needs at least 100 times this amount to survive, but in order to avoid exploiting its prey too heavily it visits a number of nests in its territory each day. As forest habitats are speckled with ant nests, 5–10 adult anteaters may inhabit one square mile. Grassland, on the other hand, is nowhere near as productive and a single anteater may require a territory of almost 25 km². To visit all the ant nests in such a large area it obviously has to spend a lots of its time walking.

- There are four species of anteater, the giant anteater, northern tamandua, southern tamandua and silky anteater. All the species are found in Central and South America. The nocturnal tamanduas spend a good deal of their time in the trees and the rat-sized silky anteater is a specialist tree-dweller, of which very little is known. The favored prey of the tamanduas is termites, whereas the silky anteater likes to feast on arboreal ants. They are related to those other South American mammalian oddities, the sloths and armadillos.

♦ At 32.7°C, the body temperature of the giant anteater, is one of the lowest of all terrestrial mammals. This and its slow rate of metabolism means it is far from the most active mammal. They spend around 15 hours a day asleep in a shallow depression scraped in the ground. Their huge, shaggy tail is used as a blanket and they are warned of approaching danger by their sensitive ears and nose.

♦ The fully grown giant anteater is one of the largest South American mammals and the only animals that prey on it are pumas and jaguars. Although the anteater is a sluggish, docile-looking creature it is more than capable of defending itself. More often than not it will flee, but it is hardly fleet of foot and may have to resort to combat to protect itself. In this case it rears up onto its back legs and slashes at the aggressor with its large claws. Its last, ditch defensive tactic is to embrace the threatening animal in its very powerful arms. There are documented cases of dogs, large cats and even humans being killed in a so-called anteater hug.

♦ The female giant anteater normally gives birth to a single young. To give birth, the female stands upright, using her large tail as a prop. The newborn infant is able to cling to the fur on its mother's back straightaway and she tends her new offspring by licking it. The young anteater clings onto its mother for up to a year and at the age of around two years it will be ready to start breaking and entering ant's nests by itself.

Further Reading: Naples, V. Morphology, evolution, and function of feeding in the giant anteater. *Journal of Zoology* 249, (1999) 19–41; Shaw, J. H., Machado-Neto, J., and Carter, T. S. Behavior of free-living giant anteaters. *Biotropica* 19, (1987) 255–59.

HARPY EAGLE

Scientific name: *Harpia harpyja*
Scientific classification:
　Phylum: Chordata
　Class: Aves
　Order: Falconiformes
　Family: Accipitridae

What does it look like? The average weight of a fully grown female harpy eagle is just over 8 kg, although captive specimens have been known to exceed 12 kg. Males are considerably smaller, averaging just under 5 kg. The wing span of a fully grown bird is around 2 m. The plumage on their back is black in contrast to the white undersides. On the eagle's head is an erectable crest of feathers.

Where does it live? The harpy eagle is native to the neotropics. Its range extends from southern Central America to Northern Argentina. Its habitat is pristine, lowland rain forest.

A Phantom of the Forest

High above the shaded floor of a South American rain forest a dark shape glides silently over the canopy. Such snatched glimpses of what is, perversely, one of the world's largest and most powerful birds of prey are not unusual. This formidable animal is one of the world's least known eagles. The only chance for scientists to study this bird is when they chance upon a nest and can observe

Harpy Eagle—The harpy eagle's mastery of the air allows it to pluck sloths and other animals from the branches of the rainforest canopy. (Mike Shanahan)

Harpy Eagle—An adult of this species with its crest of head feathers folded flat. (U.S. Fish and Wildlife Service)

the comings and goings from this lofty platform. The nest of a harpy eagle is a large, but rough affair in the upper reaches of one of the South American Forest's tallest trees, the kapok or ceiba tree. On a thick mattress of sticks the female eagle usually deposits a single egg. Rarely, she may lay a second egg, but in these cases it is neglected when the first egg hatches. Like many eagles, the harpy mates for life and the male and female work together to rear the single, essentially helpless, down covered young.

The harpy eagle is an expert carnivore and the forests in which it lives are rich in all sorts of arboreal animal life. Sloths make slow progress around the canopy, monkeys move through the branches in chattering troops, and countless bird species search the trees for fruit. All of these and more are on the Harpy's menu. Its favored prey is the sloth. Catching one of these sluggish creatures sounds easy, but they are very well camouflaged and spend most of their time in amongst the thick canopy where not even the aerially accomplished harpy eagle can make hunting sorties. To catch such elusive prey the eagle sits on a perch commanding excellent views of the surrounding forest. It may sit and wait for a long time, patiently watching and waiting. Its eyes give it an amazingly sharp view of the forest, incomprehensible to us. The pupils are very large, permitting enormous amounts of light to hit the sensitive retina without distortion. It is also believed, that in the same manner as owls, the eagle can pinpoint the source of sounds thanks to the crest of feathers on its head. It raises these feathers every time it hears an unusual sound.

With its finely tuned senses the eagle sits tight, waiting for an opportunity. Sloths will often venture out on a limb in order to bask in the tropical sun. It is then the eagle makes its move. It pitches forward from its perch and extends it wings, which are shorter, but broader than other similarly sized eagles. It slices through the air towards the painfully unaware quarry and as it gets within striking distance brings its ferocious talons to bear. The claws of this eagle are about 13 cm long, as long as a grizzly bear's and are so big they look slightly out of place. The talons grab the sloth in a vicelike grip, puncturing its major organs and killing it swiftly. All of this happens midflight in an almost fluid action. With its stout wings the harpy eagle is the heavy lifter of the bird world and is able to carry three-quarters of its own body weight back to the nest. Once back at the nest, the female takes time to remove some fur from the prey before tearing off bite size pieces of flesh and feeding them to her waiting young. After six months of this tender care the young is ready to leave the nest, but its parents will continue to feed it for another 6–10 months. Although these eagles only breed once every 2–3 years, the parental care they invest ensures the next generation of these majestic birds will be well prepared for the rigors of life in the neotropical rain forests.

+ There are around 64 species of eagle. The harpy eagle is one of the largest along with the Philippine eagle and Steller's sea eagle. They are all carnivores with hooked bills and powerful talons. The species of open habitat have huge wing spans for soaring, while forest species have stouter wings for improved maneuverability.

+ The name harpy comes from the creatures of Greek mythology, which had the body of an eagle and the face of a human, normally a woman. The mythological harpies would grab people and carry them off to the underworld, Hades. Although the harpy eagle could not fly off with a person, their fearsome claws and ability to pick prey from the canopy inspired their naming after these mythological creatures.

+ Although the harpy eagle would never go out of its way to attack a human they can be very aggressive especially if their nests are disturbed.

+ By a good stroke of fortune, natives in the forests where these birds live consider it bad luck to cut down the stately kapok tree, safeguarding the eagle's nesting sites.

+ Although the native Indians are not a threat to these eagles, development in these areas is responsible for the destruction of huge swathes of forest each year, fragmenting pristine habitats and making it more difficult for these birds to find prey for themselves and their young. Hunting is also reducing their numbers. They are even killed by bird collectors who see them as a threat to the macaws, which are coveted by the pet trade.

Further Reading: Rettig, N. Harpy eagle. *National Geographic* 187, (1995) 40–49.

KIWIS

Scientific name: *Apteryx* species
Scientific classification:
  Phylum: Chordata
  Class: Aves
  Order: Struthioniformes
  Family: Apterygidae
What do they look like? Kiwis have a large, stocky body, a small head and thick legs with powerful claws. The legs are positioned far back on the body, giving the bird a rather ungainly

Kiwis—A female kiwi probing the earth for a grub with a cutaway showing just how big her egg grows. (Mike Shanahan)

appearance. The beak is very long and straight and the feathers are so fine they look like fur. Long, sensory whiskers grow from the base of the beak. Fully grown, adults range from 25 cm in height and 1.3 kg in weight (little spotted kiwi) to more than 44 cm and more than 3 kg (great spotted kiwi).

Where do they live? Kiwis are found only in New Zealand and some of the islands off its coast. They are animals of the forest floor.

A Bird with Its Feet on the Ground

New Zealand is unique as apart from three species of bat (one of which is extinct) there are no native mammals whatsoever. The birds of this antipodean land evolved to the fill the various ecological spaces that are filled in other parts of the world by mammals. The role of a nocturnal, insectivorous animal, much like a hedgehog, has been filled in New Zealand by the Kiwi.

The kiwi, with its brownish, yet glossy pelage is a creature of the night. It emerges from its daytime lair to skulk around in the undergrowth looking for food. They are completely bound to the forest floor. They have lost the ability to fly and even if they wanted to their wings are no more than tiny stubs beneath the hairlike feathers. Not only are their wings small and useless, but millions of years of adaptation to a ground dwelling existence has left the kiwis without the large keel that runs the length of the breastbone in flying birds. The large muscles required for powered flight have to be attached to this bone. Also, as they no longer take to the air, the kiwis have lost the delicate, hollow bones of other birds. The bones are large and heavy and are filled with marrow.

Although the kiwis gave up their flying abilities they are very adept, forest floor foragers. They are expert hunters of juicy grubs and other soil dwelling invertebrates. They use their beak like a very sensitive probe, inserting it into the soil in an attempt to detect the faint odors that betray the presence of prey. Unlike all other birds the nostrils of the kiwi are at the very tip of its elongate beak and its sense of smell is very sensitive indeed. When it has sniffed out a suitable smelling item it will quickly wheedle the tasty morsel from the soil with the skill of a chopstick expert. On this rich diet of invertebrates and other foodstuffs, like fruit, amphibians and crustaceans the kiwis can invest the energy needed to produce its big egg, which is probably a quirk of evolution. The modern kiwis more than likely evolved from much larger ancestors, which produced big eggs. As the kiwis diversified to fill different niches on the forest floor, they shrank but were saddled with the internal arrangement of their forebears—a setup geared to producing large eggs.

The eggs produced by these birds are huge. Although a female kiwi is about the same size as a domestic chicken the egg she produces is not much smaller than an ostrich's. In fact, relative to her size, the female kiwi lays the largest egg of any living bird species. When the egg is fully developed it takes up a good proportion of the female's body cavity and can weigh as much as 450 g. A chicken egg, in comparison, weighs about 45 g. It is so large, the female can only waddle with her massively swollen belly almost touching the floor. As the female has the eye-watering responsibility of delivering this huge egg, the male has to do the brooding. The egg is laid in a burrow and during the day the father stays with it, brooding it with a bare patch of skin on his belly. This brood patch transfers warmth from the male to the egg, but when night falls the male takes his leave and conceals the entrance to the burrow before he goes off to forage. He keeps up this demanding schedule for as many as 80 days until the young hatches.

+ There are three species of kiwi, the great spotted, the brown and the little spotted, although some experts argue there may be six species. All of these species are found in or around New Zealand. There are also three species of very similar birds known as *tokoeka* (Southern tokoeka, Stewart Island tokoeka and the Haast tokoeka). It was presumed for a long time the closest relatives of the kiwis were the giant extinct birds of New Zealand known as *moas*, but evidence from DNA shows the closest relatives of kiwis are the emus and cassowaries. This suggests the ancestors of the kiwi arrived on New Zealand from somewhere in Australia a long time after the ancestors of the moas.

+ The kiwi is also unusual for its sense of smell, as most birds, with the exception of some sea birds, have quite poor olfactory abilities.

+ Kiwis are often monogamous animals, mating with the same partner for life, which can be as long as 20 years. Only in areas where there are high densities of kiwi does this situation of happy families break down as males and females may mate with other birds apart from their long-term partner.

+ The huge egg of the kiwi places a lot of pressure on the female. For the 30 days it takes the egg to grow the female has to eat three times her normal amount of food. Two to 3 days before the egg is laid there is little space left inside the female for the stomach and she is forced to fast.

+ It is not uncommon for a female kiwi to produce more than one egg, but the second, or even third, is laid around 25 days after the preceding egg. Females can also lay more than one clutch of eggs, with brown kiwis able to lay 2–3 clutches per year.

+ Not only is the egg of the kiwi particularly large, but it is also very rich in yolk. An average bird egg contains about 35 percent yolk, but a kiwi egg contains around 65 percent. For the first week of its life the newly hatched chick survives entirely on this nutritious substance.

+ Kiwis, very successful animals in the absence of mammals, were brought to the edge of extinction by the introduction of dogs, cats, pigs and stoats to New Zealand. Their inability to fly makes them easy pickings for carnivorous mammals used to pursuing wary, flying birds. Some of the species are now restricted to heavily protected habitat or islands off the coast of New Zealand.

LUMINOUS GNAT

Luminous Gnat—In its hammock of mucus and silk, a luminous gnat larva has set a number of trap threads to ensnare its prey. (Mike Shanahan)

Luminous Gnat—A larvae of this fly species in its hammock of mucus. The mucus trap threads are clearly visible. (David Merritt)

Scientific name: *Arachnocampa luminosa*
Scientific classification:
    Phylum: Arthropoda
    Class: Insecta
    Order: Diptera
    Family: Mycetophilidae
What does it look like? The grub of the luminous gnat is the creature we are interested in here. They are maggotlike creatures, pale in color and almost transparent. The cuticle on the head is harder than elsewhere on the body giving the animal what looks like a brown helmet. The fully grown grub is about 3 cm long.
Where does it live? The larvae and the adults of this fly species are found only in a few places in New Zealand.

Keep Away from the Lights

The Waitomo caves in New Zealand are a labyrinth of tunnels, caves and grottos fashioned over millions of years by the erosive power of water as it percolated steadily through fissures and cracks. Caves are natural refuges. They offer their inhabitants shelter from the extremes of the climate and lots of hidey-holes in which to rear young. In some of the larger, vaulted grottos of the Waitomo cave system the roof is dotted with a shimmering field of bluish white lights. The first time observer could be forgiven for thinking he was looking at a place where the cave roof had disappeared giving them a clear view of a fantastically clear starscape. These lights are not distant suns out in the cosmos but the glow of myriad fly larvae, each with its own little beacon. Although the cave roof with its multitude of lights is a rather unearthly and serene sight it belies the true purpose of these beacons. The lights are actually lures and by shining in the perpetual darkness of the cave the fly larvae can attract their prey. The light is just one ingredient in the whole prey catching cake.

The grubs hatch from their eggs and construct themselves a nest, a hammock of silk, using specialized glands in their head. They squirm along this hammock and proceed to produce more silken threads. These threads can be long as 40 cm and they hang straight down from the nest. At intervals along the thread there are little droplets of mucus—like tiny pearls on a necklace—each of which was applied by the grub as it produced the strand. These little globules are very sticky and it will soon become clear what these are for.

Below the lights and the sticky, silken strands, the stream that formed the caves meanders its way through the blackness. In the water are numerous, immature insects that have inadvertently drifted into the cave with the current. Some of them, such as caddis flies and mayflies will be ready to the leave the water to begin their brief life as adults in the air. Emerging from the water they are fooled by the darkness in the cave into thinking it is night time and the lights above are stars—the direction to head for is the sky and away from their larval home. They fly feebly towards the lights, but they soon find to their peril that these lights are not what they seem. They soon blunder into one of the silken threads with its beads of sticky mucus and rapidly become entangled while fluttering futilely. The gnat larva feels the struggling of the insect through its hammock and heads to where the unfortunate victim is entangled. It finds the lucky thread and reels it in at up to 2 mm a second using its jaws. The mayfly is powerless to resist as it is drawn inexorably closer to the gnashing jaws of the carnivorous grub. After it reeled the insect in and consumed it the grub will secrete a replacement thread and return to the center of its hammock where it will sit and wait for more unlucky insects to be delivered by the gently flowing stream and enticed by its intricate trap.

- The luminous gnat is a type of fungus gnat. These are generally small flies and worldwide there are at least 3,000 species, but the adults are so short lived and the larvae so hard to find that it is highly likely there are many more species yet to be identified. As the name implies they are fond of fungi, especially the larvae and the adults are often found around fungi, mating and laying eggs.
- Some of the fungus gnats (i.e., the luminous gnat and its relatives) are the only flies that can produce light. Like other light producing animals the luminous fungus gnats rely on the chemical luciferin and the enzyme luciferase. The enzyme breaks down the luciferin, and light is produced.
- In some other relatives of the species described here, the droplets that cling to the silken trapping strands of the nest are actually poisonous, helping to subdue the prey as soon as it becomes entangled.
- The luminous gnat is fond of caves to build its trap, as even the slightest breath of wind would catch the fine, sticky threads and they would become entangled and useless. However, they can be quite common in the forest outside of caves. In this situation they make much shorter fishing lines although it is probable they spend more time removing tangles and making new lines.
- The adults of the luminous gnat are very short lived animals. They do not feed and there only purpose is to find a mate and reproduce. The female can lay around 120 eggs in total, which are deposited in batches. The males die soon after mating and the females die as soon as her eggs have been deposited.

✦ When the grub of the luminous gnat is fully grown it pupates and even then its light still shines. However, the male's glow fades and disappears before he emerges as an adult, whereas the female keeps shining bright. It is thought this acts a beacon for the males to find the females. The males can wait as she emerges from her pupae and tussle for the right to mate with her.

Further Reading: Fulton, B. B. A luminous fly larva with spider traits. *Annals of the Entomological Society of America* 34, (1941) 289–302; Harvey, E. N. *Bioluminescence.* Academic Press, New York 1952; Sivinski, J. Phototropism, bioluminescence and the Diptera. *Florida Entomologist* 81, (1998) 282–92.

MANTIS SHRIMPS

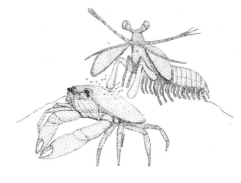

Mantis Shrimps—A mantis shrimp uses its club-like fore-limbs to smash the carapace of an unsuspecting crab. (Mike Shanahan)

Mantis Shrimps—A fantastically colored mantis shrimp photographed outside of its burrow on the coral reef. (Jeff Jeffords)

Scientific name: Stomatopods
Scientific classification:
    Phylum: Arthropoda
    Class: Malacostrata
    Order: Stomatopoda
    Family: various

What do they look like? Mantis shrimps are large invertebrates. The largest species can be as much as 36 cm long, but most are between 10 and 20 cm in length. They are often beautifully colored. They have a flattened body and very large abdomen ending in a shieldlike plate known as the *telson.* Much of the thorax of the animal is protected by a tough carapace and in front of this is the head with its very obvious eyes. In all species the second pair of limbs on the thorax is large, hugely modified and resembles a folded penknife. The appendages beneath the abdomen are adorned with comblike gills.

Where do they live? Most species of mantis shrimp are found in tropical and subtropical regions of the Indian and Pacific oceans. They range from the shores of east Africa to the volcanic Hawaiian Islands where they dwell within rock formations, coral or burrows excavated in the seabed.

A Crustacean that Packs a Mighty Punch

In the clear, aquamarine waters of the tropics every crevice and every hole is the home of some sort of animal. In some of these holes, fleeting glimpses of unblinking eyes mounted on short stalks can be had. Just what animal owns these restless eyes is difficult to know. Slowly, the mystery animal edges cautiously out of its refuge revealing the front of its body, more than enough to identify it as a mantis shrimp—pound for pound one of the fiercest predators in the sea. They are known as mantis shrimps because of the very well developed limbs slung beneath the front part of their body, which resemble the killer weapons of the preying mantis. These insects are adept predators, but the mantis shrimp makes them look like well-mannered clergymen. Basically, there are two types of mantis shrimp. There are the spearers and the smashers. The second pair of legs in both types has evolved into a seemingly innocuous, neatly folded hunting device. In the spearers, the ends of these limbs bear ferocious spines, while the smashers limbs are tipped with what are basically clubs. When some suitable looking prey comes within striking distance the limbs can be flicked out in around 4 milliseconds. In the spearers this is fast enough to impale a fish before it is even aware of the shrimp's presence. In the smashers, the bludgeon is swung at a rate of about 23 m/s (about 80 km/h). This violent force is equivalent to a .22 caliber bullet and is enough to generate a wall of tiny air bubbles. These bubbles collapse as they hit the target, releasing heat, light, and sound. The impact can be seen as a flash of light and heard as a sharp bang. These forces crash over the victim and are enough to shatter the tough shell of a mollusk or crack the resilient carapace of a crab. The victim, killed or at least disabled by this tremendous blow is pulled into the shrimp's lair where it can be eaten at leisure. The punch of a smasher mantis shrimp is enough to fracture the glass of an aquarium and there are reliable reports of glass, 2.5 cm, thick being broken by these heavyweight crustaceans.

To allow the mantis shrimps to use their potent claws effectively, they have what are widely considered to be the most complex eyes in the animal kingdom. Each eye is separated into three different sections, all of which can used to take a good look at the world. Mounted on their flexible stalks, the eyes can be swiveled in all manner of directions. Not only can they form clear images and perceive depth, but the eyes are also equipped with at least sixteen different types of light sensitive cells. Human eyes, in comparison, have only four types. It is difficult to know exactly how these impressive animals view the world. They can see at least 10 times more colors than us, around 100,000, and survey their habitat in four types of ultraviolet light, infrared and polarized light. They must therefore see things invisible to our humble eyes. With such finely tuned senses the mantis shrimp can sit like a brooding sentinel at the entrance of its hideaway, patiently watching and waiting for an unfortunate victim to swim within striking distance of its terrible claws.

- More than 400 species of mantis shrimp have so far been identified.
- The impressive punching power of a mantis shrimp is possible thanks to an arrangement of catches that allow the striking limb to be primed and then cocked before being released.
- The spearers and the smashers often reside in quite different habitats. The former are often found in shallow water, where they excavate burrows in the soft seabed. The latter are more likely to be found in areas where the substrate is harder such as rocky outcrops or coral reefs. In these rocky places, unoccupied nooks and crannies are hard to come by; therefore a smasher will defend its lair vehemently and as result they

tend to be more aggressive than the spearers, who instead of fighting over a hole, will simply scuttle off and build a new one.

- With their complex eyes, the mantis shrimps have evolved some elaborate behaviors. Ritualized fighting for holes and mates is common and individuals use complex signals to communicate with one another, sometimes with fluorescent patterns. During fights, a mantis shrimp can parry the blows of its opponent by lying on its back and using its tough telson as a shield.
- Female mantis shrimps are dedicated parents. They guard and clean their brood of eggs until the tiny planktonic larva hatch and disperse into the open water.
- Mantis shrimps are sometimes known as thumb splitters, as unknowing divers and fishermen have had the skin on their fingers and thumbs broken to the bone by the animal's punch.
- Although they are common animals and among the most important predators in many shallow marine environments, mantis shrimps are poorly understood animals. Most species spend most of their life tucked away in their burrows and holes. This secretive nature means there is a great deal still to learn about these amazing crustaceans.

Further Reading: Cronin, T. W., Marshall, N. J., and Caldwell, R. L. Tunable colour vision in a mantis shrimp. *Nature* 411, (2001) 547–48; Mazel, C. H., Cronin, T. W., Caldwell, R. L., and Marshall, N. J. Fluorescent enhancement of signaling in a mantis shrimp. *Science* 303, (2004) 51; Patek, S. N., and Caldwell, R. L. Extreme impact and cavitation forces of a biological hammer: strike forces of the peacock mantis shrimp *Odontodactylus scyllarus*. *Journal of Experimental Biology* 208, (2005) 3655–64; Patek, S. N., Korff, W. L., and Caldwell, R. L. Mantis shrimp strike at high speeds with a saddle-shaped spring. *Nature* 428, (2004) 819–20.

MEGAMOUTH SHARK

Megamouth Shark—A megamouth shark cruises through the water with its mouth open to engulf plankton. (Mike Shanahan)

Megamouth Shark—A dead megamouth shark washed up on a beach in Australia. Note the flabby body. (B. Hutchins, Western Australian Museum)

Scientific name: *Megachasma pelagios*
Scientific classification:
  Phylum: Chordata
  Class: Chondrichthyes
  Order: Lamniformes
  Family: Megachasmidae

What does it look like? The megamouth is a large shark. Fully grown adults are at least 5.5 m in length and 800 kg in weight. They have a very large, broad head, small eyes and five pairs of gill slits. The back is grey to grey/black and the underside is white. The upper surface of the pectoral and pelvic fins has a distinctive, light margin.

Where does it live? The distribution and range of this animal is poorly known. The few specimens found thus far suggest it is found around the world in tropical and temperate waters. Most specimens have been encountered around Japan. The understanding of their habitat requirements is sketchy, but they are thought to spend most of their time in mid water at a depth of at least 100 m.

A Shark that Sieves Planktonic Soup

Until 1976, the megamouth shark was unknown to science. The inadvertent discovery of such a large, new species remains one of the biggest zoological stories of the late-twentieth century. It is not uncommon for trawlers to turn up several new species of animal, especially during sweeps of deep water and poorly mapped areas, but these are small fry compared to the megamouth shark. The first specimen got itself entangled in the anchor lines of a United States Navy ship, 40 km from Kahuku point in Hawaii. It was a big one, measuring 4.5 m long and weighing in at 750 kg. Its gaping mouth alone was almost a meter across. In the 30 years since the accidental discovery of this mysterious animal, only 36 other megamouth sharks have been caught or sighted. Ten of these encounters have been around the islands of Japan. All megamouth sharks have very flabby bodies and the observations of live animals show this species to be a very slow swimmer. The stomachs of examined sharks have been densely packed with small, marine shrimps and like the other giant sharks, the basking shark and whale shark it is thought to be a filter feeder. In deep water, the shark swims along with its mouth open, sometimes closing its rubbery lips to swallow a mouthful of food that gets snared on sieve like elements known as gill rakers. To increase the success of this rather passive feeding technique the lining of the shark's mouth is silvery and it is possible that in the ocean depths this silvery layer could reflect light. Some specimens have also been said to have light producing tissue in their mouth. Regardless of the shark actually producing its own light or reflecting the dim, scattered light from above, its cavernous maw may glow sufficiently to entice small shrimps, copepods and jellyfish to their death. In the dark oceans, light is often used as a means of attraction and the megamouth shark may be exploiting this to fill its belly.

In October of 1990, a large (4.9 m) male megamouth shark was caught near the surface off Dana point in California. It was towed, alive, back to the local harbor and after deciding what to do with it, was taken back out to sea where a small radio tag was attached to its soft body. Upon release the shark headed downward, apparently none the worse for its stressful experience. The tag enabled it to be tracked over a two day period. This tag gave a fascinating glimpse of the shark's behavior. During the day the shark would cruise at a depth of around 120–160 m, but as the sun set it would ascend and spend the night at depths of between 12 and 25 m. During day and night, its progress was very slow and it chugged along at only 1.5–2.1 km/h. Many animals that live in midwater have this same pattern of vertical migration. Filter feeding animals like the megamouth shark are probably following their invertebrate prey as they move from the ocean depths to the surface and back again. Although the tracking showed the megamouth to be quite a sluggish species it can swim at these low speeds for long periods of time.

Apart from this one fleeting glimpse of the daily life of the megamouth shark, very little else is known about this intriguing animal and much of that has been gleaned from the capture and dissection of dead specimens. What is known is that the megamouth is far from common and if the threats to marine life continue it will be enigmatic beasts such as this that disappear first.

- The megamouth shark is the only representative in its family, although some people think it may actually be closely related to the basking shark.
- The megamouth shark, along with the basking shark and whale shark is the only shark species that has forsaken the normal, predatory tendencies of this group for a life of sedate filter feeding. Sprouting from the gills, these sharks have finger like projections, commonly known as gill rakers. These sieve the water for small, marine organisms in a similar way to the baleen plates of the filter feeding whales. In the basking shark, these rakers drop off in the winter, which means they cannot feed. Just where the sharks go at this time of year is another of zoology's mysteries.
- Only educated guesses have been made of how the megamouth shark reproduces. The ovaries are very similar to other shark species that ovulate large numbers of eggs as food for young developing in the uterus. The females also bear scars, indicative of the nuzzling and biting behavior of copulating males.
- Several megamouth sharks have shown the scars of meetings with the cookie-cutter shark . The megamouth shark may be a favored target of this small, semiparasitic animal. It has a flabby body, which is probably easy to excise lumps from. It swims slowly and it lives in the same places as the cookie-cutter.

Further Reading: Bera, T. M. Some 20th century fish discoveries. *Environmental Biology of Fishes* 50, (1997) 1–12; Taylor, L. R., Compagno, L. J. V., and Strusaker, P. J. Megamouth—a new species, genus, and family of lamnoid shark (*Megachasma pelagios*, family Megachasmidae) from the Hawaiian Islands. *Proceedings of the Californian Academy of Sciences* 43, (1983) 87–110.

PORTIA SPIDER

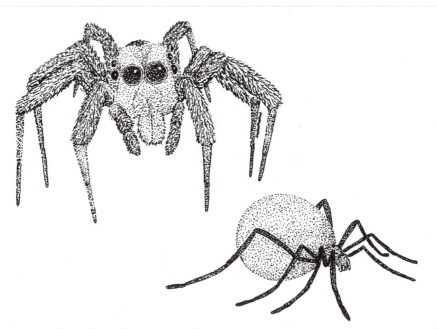

Portia Spider—After stalking its quarry, the Portia spider moves in for the kill. (Mike Shanahan)

Scientific name: *Portia fimbriata*
Scientific classification:
  Phylum: Arthropoda
  Class: Arachnida
  Order: Araneae
  Family: Salticidae
What does it look like? This is a long-legged jumping spider. The front part of its body
  (cephalothorax) is large and angular. The broad, flat face is studded with a huge pair of glossy,
  unblinking eyes each accompanied by a much smaller eye. Four more eyes dot the head. On
  the legs and abdomen there are ornate tufts of brown, white, and black hair. The males are
  only around 5–7 mm long whereas the heavier bodied females are 6–9 mm in length.
Where does it live? The species is native to the tropical forests of Australia, Southeast Asia,
  and the Indian subcontinent, where it spends it time on foliage, on tree trunks and the
  vertical surfaces of boulders and rock ledges.

A Spider that Thinks?

Amid the tangle of the rain forest undergrowth countless minibeasts play out a continual struggle for survival. Something with fangs, sharp mandibles, claws, pincers, or a beak lurks around
every corner. To cope with these varied dangers the hunted have a myriad of defensive adaptations to protect themselves. To hunt effectively in these forests you must be one step ahead and
this is exactly what the *Portia* spiders is. This small spider has taken stalking and hunting and
turned them into art forms. Its favored prey is other spiders, which is unusual in itself as many
small predators give them a wide berth. They have poisonous bites and can be difficult to attack
in their lairs and webs.

To a predator the size of *Portia*, the prey it attacks are lethal, but this does not deter this
amazing little spider. Equipped with camouflaging hairs, making it look like a bit of dirt, the
Portia spider slowly approaches the web of a prey spider, taking advantage of wind and other
disturbances to disguise its intent. It arrives at the web and slowly and tentatively it strums the
silken threads to arouse the occupant's curiosity. It must be very careful not to mimic the futile
struggles of an ensnared fly too effectively as the owner of the web will rush out, brandishing
its fangs. If this was to happen *Portia* would be in trouble. It must be patient and draw its prey
out slowly. Only when the owner of the web is within a few millimeters of *Portia* will it strike,
delivering a fatal bite. In other situations the *Portia* spider high on a perch may spot the orb web
of another potential meal. For anything up to an hour the jumping spider will survey its position, the position of its host and the path it must take to reach it. Its large eyes can form very
clear images, but only small fields of view can be taken in at one time. With some form of mental
image of the path it must take the spider sets off. To get to a position where the prey can be
surprised from the spider may take a very convoluted route, but unlike most other invertebrates
is doesn't lose interest as soon as the prey is out of sight. It sees the path and the objective at its
end. Eventually this smart little spider will get to a position directly above the web and with the
use of a silken drag-line it drops to within lunging distance of the prey. These complex behaviors can be tuned to the species of quarry in question, suggesting this spider with its miniscule
brain is somehow capable of learning and problem solving. The seemingly intelligent behavior
of these spiders baffles scientists. How can an animal with such a small brain be capable of anything more than simple, reflexive behavior? The key may lie in the way in which the spider sees

Go Look!

Jumping spiders can be found in almost any habitat. They are instantly recognizable thanks to their huge, front eyes. They are warmth loving animals and can be seen foraging on the ground, walls and vegetation. Many species are brightly colored or have very bold markings. The males often have huge mouthparts that project a long way in front of the head. It is very easy to keep a jumping spider in captivity to watch its very interesting behavior. Any small transparent container with a lid is suitable and to make it look like a home from home, stones, bark or vegetation can be included in this terrarium. Small insects, like flies are good prey and a hungry spider will stalk a meal as soon as it catches sight of it. Initially they approach the resting prey very slowly in the same way as a cat creeps up on a bird, always keeping their eyes fixed on the target. As soon as they are within range they poise themselves by waggling their body very slightly before making a killer leap. The powerful front legs and the fangs are used to grip the prey until it has been completely sucked dry. The courtship behavior of these animals is also interesting to watch. The males perform a dance by walking in jerky movements and waving their pedipalps at the female. The male must be careful when approaching the female because if his little dance is not up to scratch the female may see him as her next meal.

the world. As its field of view is very limited it must be able to compare a fragment of an image with what it has seen before. The unique way in which the spider sees the world acts a filter, concentrating only on the important points. By comparison, the eyes of an intelligent mammal are flood gates for visual information and it is left to the brain to sort the useful information from the rubbish.

- There are around 20 species of *Portia* spider, and they are found in Southeast Asia, Africa, the Indian subcontinent, China, and Australasia. Like any rain forest animal there are probably many more species to be identified.

- In terms of species the jumping spiders are the most successful of all the spiders. So far, more than 5,000 species have been identified. Almost all of them have forsaken the ability that distinguishes spiders from other animals: web building. Instead of constructing a web to snare their prey they actively hunt, using their excellent eyes and catlike stalking abilities. Like their name suggests they are rather adept at jumping and can leap several body lengths to pin their prey down in a death grip. Unlike most jumping animals that rely on powerful muscles or rubbery proteins to propel themselves the jumping spiders force fluids into their legs making them extend violently. Their accuracy is unerring and they rarely miss their target.

Further Reading: Jackson, R. Eight-legged tricksters: Spiders that specialize in catching other spiders. *BioScience* 42, (1992) 590–98; Jackson, R. A web-building jumping spider. *Scientific American* 253, (1985) 102–110; Jackson, R., and Hallas, S. Comparative biology of jumping spiders. *New Zealand Journal of Zoology* 13, (1986) 423–89; Jackson, R., and Wilcox, R. Spider-eating spiders. *American Scientist* 86, (1998) 350–57.

PURSE-WEB SPIDER

Scientific name: *Atypus affinis*
Scientific classification:
    Phylum: Arthropoda
    Class: Arachnida
    Order: Araneae
    Family: Atypidae

Purse-Web Spider—A female purse-web spider poises to impale her prey through the side of her sock-like web. (Mike Shanahan)

Purse-Web Spider—An adult female rearing up in a defensive posture to show her massive fangs. (Roger Key)

What does it look like? The purse-web spider is one of the primitive mygalomorph spiders, characterized by the way in which the fangs move, amongst other features. The carapace of the animal is dark, while the abdomen is light brown. The legs are relatively short and thick. The fangs (chelicerae) of the spider are huge, approximately one half of the body length. Essentially, the spider resembles a small tarantula. As with almost all spiders the female is bigger than the male, with a fatter abdomen.

Where does it live? This spider prefers warm habitats with low vegetation and loose, sandy soil that it can burrow in to. It is found throughout Western Europe, as far north as Denmark and Sweden. It is often found in coastal locations. It is probably more common than people think.

Living in a Sock

The females of this long-lived spider, with their formidable fangs are a match for many creepy crawlies. The females (and males before they reach maturity) dig into soft soil to make a burrow, which can extend straight down for 15 cm or more. This burrow is lined with fine silk from the spider's spinnerets. The silk lining projects from the mouth of the burrow, forming what looks like a small sock, known as the purse web. Small pieces of vegetation and soil stick to this purse-web, helping it blend in with the surroundings. This camouflage is so effective that the purse web is very difficult to find, but it is often situated in a position where it is not directly exposed to the wind and rain, such as the overhang of a large rock or log. It is normal to find lots of these little 'socks' in the same place, forming a colony. Once the purse web is complete the

Go Look!

This spider is very difficult to find, but should you find a purse web, it is possible to trace the silken tube through the soil. The female, sensing the disturbance will have retreated to the very bottom of her lair. It is not advisable to dig the spider out of her burrow as she will be homeless and may perish trying to find and make a new home. It is possible to lure the spider into a position where it can be seen more easily. To do this, the movement of a prey animal can be imitated on the web with a small twig or a blade of grass. The spider should attack and then her path of retreat can be blocked. A female in the open will, not surprisingly, feel threatened and will rear up in a defensive posture, giving an impressive view of her huge fangs.

female sits tight and waits for her first meal to amble along. Like any ambush predator, the wait can be a long one. Any small insect or spider is potential prey for the purse-web spider and they go about their everyday business unaware of the danger beneath their tiny feet. Should a hapless victim alight on the purse web, the female, poised underneath, will strike, stabbing the victim through the web with her enormous fangs. The victim is skewered on the female's huge fangs and she waits for it to stop struggling before cutting a slit in the web and dragging it into her lair to be consumed. After feeding, the spider leaves the remains of the prey outside the web before repairing the slit and digesting her meal.

The way in which this spider ambushes its prey has important implications when it comes to breeding. An amorous male could easily be mistaken for food if he's clumsy, so in the late summer/autumn the male leaves his own burrow and goes in search of females. His is probably led to the burrow of a receptive female by her scent and eventually he will arrive at her purse web. A blunder by the male now could be disastrous. He tentatively taps on the walls of the purse web, and if the female is receptive, he ventures into the confines of the burrow. The two spiders will mate and even cohabit for a few months until the male dies. The female does not mourn her mate's passing but eats him instead. She then makes her egg sac and hangs it up in her burrow. It is not until the next summer that the eggs hatch and it is not until the subsequent spring that the spiderlings leave the safety of their mother's burrow and wander off to build a burrow of their own.

+ The purse-web spider is one of the primitive mygalomorph spiders. These spiders have fangs that operate in an up and down fashion, while the fangs of the more evolved spiders move in a side-to-side, pincer motion.
+ The ambush tactic of the purse-web spider is the lazy approach to finding food, waiting for the prey to come to it, rather than actively seeking out its victims. Due to this low energy approach to life, the spider's energy requirements are very small indeed and it may be able to survive for long periods without eating anything. Most of the primitive spiders are ambush predators and this slow pace of life often means they are very long lived, compared with many other invertebrates.
+ Young purse-web spiders do not disperse far. They are small and will make a tasty snack for a myriad of predators if they remain on the surface for too long. Those that survive will therefore tend to construct a burrow quite near that of their mother's.
+ In the United States, a similar species constructs its sock using a tree trunk or root as a support. They feed in the same way as their European relatives.

SHREWS

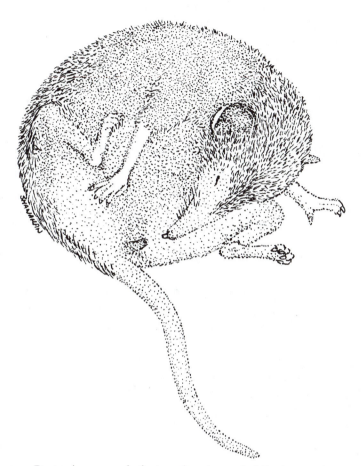

Shrews—During the process of refection a shrew laps and nibbles its everted rectum to digest valuable nutrients. (Mike Shanahan)

Scientific name: many species
Scientific classification:
    Phylum: Chordata
    Class: Mammalia
    Order: Insectivora
    Family: Soricidae
What do they look like? Shrews are small mammals. The smallest, the pygmy white-toothed shrew is one of the smallest mammals with a head to tail length of 3.5–4.8 cm and a weight of around 2 g. The largest species, the African forest shrew is up to 29 cm long and weighs a relatively hefty 35 g. They are mouselike creatures with a long snout, dense fur and small, beady eyes. The fur is normally brownish or grey and the ears are often concealed in this dense pelage. Their feet have five digits.

Where do they live? Shrews are found almost worldwide. Of all the landmasses outside the polar regions, only New Guinea and Australasia have no native shrew species. The majority of shrews are ground dwelling animals, although there are some species that climb and other species that spend a lot of time in water. Predominantly, they are animals of the undergrowth and leaf litter.

Furious and Fast Paced

Although cute and furry the shrews, in reality, are the tigers of the undergrowth. They are secretive little animals and a fleeting glimpse of something small and furry scurrying through the leaf litter is the best view that most people get of one of these insectivores. Shrews are short-lived animals and in this brief existence food is unquestionably their number one priority. The shrews are so small that the ratio of their surface area to volume is the highest of all the mammals. An elephant, by comparison has a small surface area compared to its considerable bulk. The relatively large surface area of the shrew's body means there is a lot of space heat can escape from and its metabolism must work overtime to keep the body running at a standard mammalian temperature. Even when compared to rodents of a similar size, the shrew's metabolic rate is fast. Many shrew species must eat their own body weight in food every day to fuel their internal fires and since they carry only enough food to provide energy for an hour or two it is imperative they feed throughout the day and night, resting in short bursts. The range of their diet is broad, but insects and other invertebrates form the bulk of their diet. Seeds, nuts and other plant matter may also be consumed with gusto, as will small vertebrates like reptiles and amphibians when they can be subdued.

As food is very important to the shrew they must have a reliable way of seeking it out quickly and efficiently. Their sense of sight is dull, but this is more than made up for by their acute senses of smell and hearing. Some species can also use their larynx to generate ultrasonic pulses of sound, which is perhaps a primitive form of the echolocation employed by the bats to find their way. To help dispatch live quarry found with the battery of senses, some shrews have salivary glands that produce venom, giving them a poisonous bite. There are no fangs, but grooves in the teeth channel the venom to the bite wound. The venom paralyses the prey, enabling the shrew to satiate its hunger immediately or cache the prey for later.

To further maximize the nourishment gained from their diet, shrews perform an act known as refection. To do this the animal must curl up to reach its anus with its mouth, sometimes holding the contortion by grabbing its hind limbs with its forefeet. In this uncomfortable pose the shrew laps at its anus until after a few seconds, muscular abdominal contractions cause the rectum to pop out. The shrew nibbles and licks its rectum for a few minutes before it disappears back into the body. The shrew only does this when the intestine has been emptied of feces and dissected refecting specimens contain a milky substance in their intestine containing fat globules and partially digested food. This could be a way of thoroughly digesting its food and obtaining as many trace elements and vitamins as possible.

+   There are 368 shrew species grouped into three subfamilies: the white-toothed shrews, the African white-toothed shrews, and the red-toothed shrews. In terms of species, the shrews are one of the most successful groups of mammals. As they are small and secretive it is more than likely that many more species are yet to be discovered.

+ Some of the identified species are very poorly known. The Sri Lankan shrew and the African forest shrew are only known from a handful of specimens.
+ The salivary glands of the American short-tailed shrew produce enough venom to kill 200 mice by intravenous injection. The components of shrew venom may have applications in medicine and cosmetics. One chemical extracted from their venom may be potentially useful in the treatment of blood pressure and other conditions of the circulatory system, while another compound with paralyzing qualities may be of use in neuromuscular conditions, migraines and even the fight against facial wrinkles—shrew-spit Botox.
+ Some female shrews and their young are also renowned for what is known as caravanning. When the litter leaves the nest with their mother they form a line, each youngster gripping its sibling's rump in its mouth with the one at the front gripping the rear end of its mother. Their grip is very strong and the whole caravan can be lifted off the ground by picking up the mother.
+ The life of the shrew is short and very frantic. The heart pounds inside the tiny chest as many as 1,000 times per minute. Individuals of the temperate species are born in the summer and somehow must survive the winter. Their unusual metabolism and diminutive size makes its impossible for them to enter a state of hibernation like other small mammals. In northern climes they forage in the space below the snow or in tunnel complexes (their own or those of rodents) looking for over-wintering insects and whatever else they can find, including the dead bodies of their own species. If they make it through the winter their teeth will be worn out by the following autumn and they will perish.
+ Glands along the side of many shrews produce a repugnant secretion making them distasteful to many potential predators and scavengers in life and in death. In the autumn, their tiny intact corpses can often be found.

Further Reading: Crowcroft, P. Refection in the common shrew. *Nature* 170, (1952) 627; Dufton, M. J. Venomous mammals. *Pharmacology and Therapeutics* 53, (1992) 199–215; Tomasi, T. E. Function of venom in the short-tailed shrew, *Blarina brevicauda*. *Journal of Mammology* 59, (1978) 852–54.

SPITTING SPIDER

Scientific name: *Scytodes thoracica*
Scientific classification:
   Phylum: Arthropoda
   Class: Arachnida
   Order: Araneae
   Family: Scytodidae
What does it look like? This is a small spider with a body length of around 6 mm. It is yellowish brown with dark markings. The body is divided into two parts carried on eight spindly legs. The front part of the animal is the fused head and thorax, which is large and globular compared to the majority of spiders species. On the front of it there are six small eyes, the jaws and a pair of small feelers called *pedipalps*. The abdomen is roundish and contains most of the animal's organs.

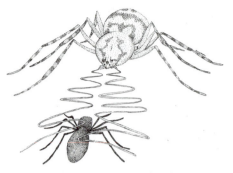

Spitting Spider—A spitting spider glues its prey to the floor with its sticky, poisonous saliva. (Mike Shanahan)

Spitting Spider—An adult female of this species holding her egg sac in her fangs. (W. Mike Howell)

Where does it live? The spitting spider has taken to living with humans in their homes and any building offering some degree of protection from the elements. Originally, it was probably a denizen of natural refuges, such as caves, tree holes, rocky overhangs, and so forth. It is found throughout Eurasia and North America.

No Web, but Still Very Deadly

Spiders are renowned for the complex webs that they weave, but some species have no need of these silken structures as they catch their prey in other ways. The spitting spider is one such species. When darkness falls the spitting spider emerges from its daytime hideaway to look for food, unlike other spiders that can move with considerable speed, this species seems to be more relaxed and wanders about with slow, measured strides. Its eyesight is very poor, limited to sensing the difference between light and dark. Its sense of touch, however, is very well developed. Tiny hairs on the surface of its body detect the faintest change in air pressure that occurs as a consequence of a nearby, resting fly shrugging its wings or grooming itself. The spider homes in on the miniscule pressure waves and is eventually upon the quarry. Any sudden movements now may give the game away and the prey may flee, so the spider must move slowly and with stealth. Initially, it gauges the distance to the prey by stretching one of its legs towards to the hapless insect. About 10–20 mm is the range of choice. If the spider is

Go Look!

Spitting spiders can be found in many homes and outbuildings, especially ones where the occupants are animal friendly. During the day and cold periods, the spider will hide away in any suitable nook or cranny, such as the space behind a cupboard, some old shoes in a utility room, or a rarely used chest of drawers. During the spring and summer months and even on warm winter days it will emerge at night to go hunting. Should you see one on a night-time hunting mission, watch the slow, precise way in which it moves. If you are very lucky you may see it find an insect or small spider and see its relaxed response before unleashing its sticky secret weapon with alarming speed. You could also catch a spitting spider and move it to a transparent box along with a fly or two. Be careful moving this spider as its fragile legs can easily be damaged. Under a dim light the spider will go about its normal business, eventually finding the flies and dispatching them. The see-through container would allow you to see the pattern of the spider's squirted trap. If you keep the spider for any length of time make sure there is something in the box it can hide in.

satisfied it slowly rocks from side to side before dousing the prey with two jets of fluid squirted from small holes in its jaws. The fluid not only contains paralyzing venom, but also a very sticky adhesive, which glues the meal solidly to the surface it was sitting on. The prey has little time to react as the whole spitting sequence is over in a little under 1/700ths of a second. Struggling is futile. The jaws of the arachnid move from side to side very rapidly as the gunk is squirted, covering the prey in a zigzag pattern of sticky poison. With the prey incapacitated, the spider ambles over to it and delivers the killer bite, before settling down to feast.

+ There are approximately 150 species of spitting spider found all over the world. One key to the success of these spiders is their ability to make use of the abundance of habitats provided by humans, supplementing the available natural habitats.
+ The odd appearance of this spider is due to the presence of large glands in its carapace, which produce the venom and glue for squirting. When ready to strike, large muscles either side of the glands force the carapace to contract, squeezing the paralyzing adhesive out through the jaw nozzles.
+ Almost all spiders build a nest of sorts in which to lay their eggs, but once again, the spitting spider has turned its back on elaborate silken constructions and instead weaves a simple net of silk to carry its eggs. This net is attached to the female's fangs and goes everywhere with her.

TRICLADS

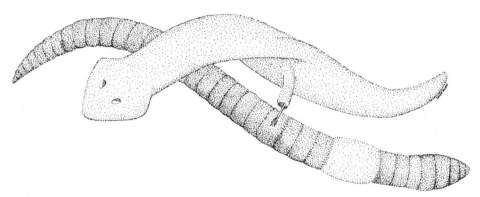

Triclads—A triclad everting its penis from its mouth to stab its prey. (Mike Shanahan)

Scientific name: Turbellarians
Scientific classification:
 Phylum: Platyhelminthes
 Class: Turbellaria
 Order: many
 Family: many
What do they look like? Flatworms are soft bodied, wormlike creatures, which can be oval or elongate with all manner of variation in between. All of them are exceedingly thin and fragile, for example, marine specimens a few centimeters long are no more than 1 mm thick.

There are microscopic species and giant forms reaching 60 cm in length, but most are small animals, a few millimeters long. Some species have arrow-shaped heads, while others bear stubby tentacles. Most species are drably colored in shades of black, brown and gray, but some marine forms are spectacularly patterned in bold colors.

Where do they live? Triclads are aquatic animals, although there are few species that have managed to set up home on land, but only in very moist habitats. There are many freshwater forms, but the vast majority are denizens of the world's seas. In terms of habitat there are a few free-swimming species, but most are bottom-dwellers, living amongst the sand and mud of their watery home. As many species are small they can even go about their daily business in between the grains of sand and mud.

Delicate, but Deadly

The digestive system of most animals has an entrance, the mouth and pharynx, and an exit, the anus. Triclads, because they are either very primitive or very simplified (a bone of contention among zoologists), lack an anus. The gut of these animals is not a tube, but a blind sac. The pharynx is the digestive system's entrance and exit, with the mouth positioned halfway along the body of the animal. This unusual setup has given rise to a range of interesting adaptations for the capture and ingestion of prey. Although paper thin and very fragile, the triclads are accomplished aquatic predators. They glide along the seabed or lake bottom on a carpet of cilia, seemingly floating, tasting the water for the scent of food. Some species, when they chance upon a likely looking meal, can turn their pharynx inside out through their mouth. This technique is particularly useful when the prey in question is protected by a tough shell. The long feeding tube can simply be extended into the prey through a suitable gap. Digestive secretions flow from the projected pharynx and break down the prey's tissues. The resulting soup can be sucked up into the gut of the triclad. In some species, this has gone one stage further and the pharynx has evolved into a capacious bag that can be used to engulf prey far larger than the triclad.

A capacious pharynx is one thing, but the triclad still needs a way of overpowering its prey as it is soft and thin and has no obvious appendages. The penis (cirrus) of

Go Look!

Due to their small size and predilection for aquatic habitats, these very interesting animals are often overlooked. A good way to see these animals for yourself is to take a small amount of water from a pond in the spring or summer. Use a small jar and look for a place where you can easily reach the bottom. Fill the jar with water from near the bottom and take it somewhere you can have a close look. If you hold the jar up to the light you will see a myriad of swimming things in the water, some whizzing about at high speeds, others creeping along the glass. Transfer some of this water into a shallow white or light-colored tray. Many animals will show up on this background and there, gliding serenely on the bottom should be a triclad or one of its relatives. The species in small pools and lakes are usually grey and are around 3 mm or so in length. With the naked eye you will be able to see the arrow-shaped head and the tapering body. With a magnifying glass you should be able to see the dark eyespots. If you have a microscope, transfer one of these specimens to a small, transparent dish and take a look when it is magnified. The large gut should be clearly visible in the semitransparent body and you may be able to make out the stirring cilia on its underside, which look like short, transparent hairs. The water from the pond or lake will contain many other tiny, but fascinating animals, all of which look amazing when you take a look at them through a microscope.

one species is armed with hardened stylets and this has become an organ that is not only used for reproduction. Cruising over the muddy lake-bottom the worm descends on its unwary prey and brandishes its penis out through its mouth. The stiletto like weapon is repeatedly stabbed into the prey to subdue and kill it before feeding commences. If daggerlike genitals are not enough, some other species embrace their prey and use their numerous slime glands to produce an entangling slime, which traps it, allowing the triclad to pin in to the surface using its suckerlike adhesive organs. The slime produced by some of these animals is neurotoxic and quickly paralyses the ensnared prey, allowing the triclad to make short work of it.

- The turbellarians are very successful organisms. They are found in all marine, freshwater habitats and terrestrial habitats where there is sufficient moisture to allow their survival (i.e., tropical rain forests). There are approximately 3,000 known species.
- They are all very thin because oxygen must diffuse across their body wall straight into their tissues. They have no specialized gill-like organs. This system becomes increasingly inefficient as the animal becomes thicker. In general, the larger the triclad the more pronounced the flattening. The large species are less than 1mm thick and correspondingly fragile.
- The largest species of triclad (*Rimacephalus arecepta*) lives in Lake Baikal in Russia where it feeds on dead and dying fish in the depths of this great body of water. It can be as much as 60 cm long.
- On the outside, triclads are very simple, but they show surprising internal complexity with a diverse array of tissues, organs and organ systems. They have a distinct brain and many species have large eye spots on their head.
- Almost all triclads have male and female sexual organs making them hemaphrodites. Individuals come together and will exchange eggs and sperm. In some species this can be quite a rough process as the stiletto-like penis is pushed through the body wall of the partner to inject the sperm.
- Depending on the time of year some species lay summer or autumn eggs. Summer eggs have thin shells and hatch quickly, while autumn eggs have thick shells and are resistant to the ravages of winter, only hatching when fairer conditions return in the spring.

VELVET WORMS

Scientific name: *Onychophorans*
Scientific classification:
  Phylum: Onychophora
  Class: Onychophora
  Order: Onychophorida
  Family: Peripatidae and Peripatopsidae
What do they look like? The velvet worms look like slugs with legs. They have a long body with 14 to 43 pairs of stubby legs (depending on the species and sex) and at the head end there is a pair of antennae. The skin of the animal looks rather velvety, hence the common name. They range in size from 1.4 to 20 cm. Most of the species are relatively small.
Where do they live? All of these animals are found in the Southern Hemisphere. They are known from the tropical regions, such as the East Indies, the Himalayas, the Congo, the

Velvet Worms—A velvet worm rearing up and dousing its prey with two streams of sticky, poisonous saliva. (Mike Shanahan)

Velvet Worms—A small group of velvet worms on the underside of a log. (Robyn Stutchbury, Peripatus Productions Pty Limited)

West Indies, northern South America and temperate regions, including Australia, New Zealand, South Africa, and the Andes. Although their distribution encompasses a range of climatic zones they are always found in association with cool, moist microhabitats.

Soft-Bodied Night Stalkers

The velvet worms were first described in 1826 and at the time they were thought to be related to the slugs, snails, and other mollusks. Today, much more is known about these animals, but they still arouse interest because of their fascinating biology. All velvet worms are predators and they catch their prey in a very interesting way. Most species emerge from their daytime hideaways to hunt for prey in the humid cool of the night. They amble along on their short, fat legs using the sensory organs in their antennae to detect the faint signs of their prey, which includes just about anything smaller than themselves. When they pick up the scent of their prey they will follow the trail until they are almost on top of their quarry. Their eyesight is not the best, but their sense of smell and touch are very well developed. Ready to strike, the velvet worm can employ its secret weapon. Either side of the velvet worm's mouth there is a small fleshy turret. These lumps are connected to large glands inside the animal that produce a sticky slime. With the prey in its sights, the velvet worm rears up and squirts two streams of slime at the unfortunate prey. This slime can be squirted up to 30cm and almost immediately it begins to harden, entangling the prey in a net of sticky threads. With the prey incapacitated, the velvet worm ambles over to it and begins to feed by passing poisonous saliva into the body of the animal, which starts the digestion process. The velvet worm then sucks the partially digested tissues into its mouth.

Apart from their amazing feeding behavior, some species of velvet worm have also evolved a unique way of fertilizing their eggs. The males of these species crawl over the female's body and deposit a small packet of sperm (spermatophore) randomly on her side or back. Over time a female may accumulate several such packets from a variety of males. Somehow, the spermatophores trigger blood cells within the female to dissolve the underlying skin, allowing the sperm to escape from their parcel into the body cavity of the female. Free in the female's blood the sperm must swim for the female's sperm storage organs, where fertilization will take place. This is not the only complexity in the reproductive behavior of this animal. Some species lay eggs. Other species produce eggs that hatch inside the female's body. A few species of velvet worm have dispensed with shelled eggs altogether and give birth to live young. These young are nourished inside the female's body by a secretion produced by the uterus, which reaches the young either through a special membrane or a direct placental-like connection to the uterine wall.

+ There are 110 described species of velvet worm. Their appearance suggests they are living fossils that have remained essentially unchanged for hundreds of millions of years. Fossils, at least 500 million years old, have been found showing the impressions of marine animals that look strikingly similar to the living velvet worms. The earliest fossil showing what appears to be a terrestrial velvet worm is approximately 300 million years old. The locations where these animals are found also indicate great antiquity. The continents in the southern hemisphere were once joined in one gigantic landmass, known as Gondwanaland, but the continents, sitting on their huge continental plates drifted away from one another, carrying their living cargo with them to give us the arrangement of the continents we see today and unusual distributions of the surviving ancient animals.

+ There are two families of velvet worm and they have different distributions. The peripatids are predominantly equatorial and tropical while peripatopsids are all found in what used to be Gondwanaland.

+ Because of their great antiquity, the velvet worms are regarded with great interest by those scientists who investigate how different animals are related. The velvet worms possess characteristics that make some people think they are closely related to the annelid worms, but other characteristics suggest a closer affinity with the arthropods (insects, spiders, crustaceans etc). Some even think they represent a so-called missing link between the annelids and arthropods. Recent findings point to the velvet worms being more closely related to the arthropods than they are to the annelids. Only time and further investigations will reveal the true relationships of the velvet worms.

+ Recently, it has been found that some velvet worms are capable of complex social behavior. Groups of around 15 individuals dominated by an adult female, live, hunt and feed together. An animal's place within the group is determined by bouts of fighting.

+ The skin of the velvet worms contains the substance known as chitin, which is found in the skin of arthropods. As this material does not stretch, an animal with a chitin containing covering must periodically shed its skin. In velvet worms this can occur as often as every 10 days.

+ Velvet worms, unlike many invertebrates have quite a long life span, at least six years in some species.

+ The velvet worms are restricted to humid environments or only emerge during the night when it is cool and humid. One of the reasons for this restriction is that the breathing apparatus of the velvet worm, the trachea, are open to the air, without any form of valve that conserves moisture. Arthropods, like insects have closeable spiracles limiting their water loss.

Further Reading: Read, V. M. St. J., and Hughes, R. N. Feeding behaviour and prey choice in *Macroperipatus torquatus* (Onychophora). *Proceedings of the Royal Society of London (Series B)* 230, (1987) 483–506; Reinhard, J., and Rowell, D. M. Social behaviour in an Australian velvet worm, *Euperipatoides rowelli* (Onychophora: Peripatopsidae). *Journal of the Zoological Society of London* 267, (2005) 1–7; Sunnucks, P., Curach, N., Young, A., French, J., Cameron, R., Briscoe, D. A., and Tait, N. N. Reproductive biology of the onychophoran *Euperipatoides rowelli*. *Journal of Zoology* 250, (2000) 447–60; Sunnucks, P., and Tait, N. N. Velvet worms: Expect the unexpected. *Nature Australia* (2001) 60–69.

4

GETTING FROM A TO B: SOLUTIONS TO THE PROBLEM OF MOVEMENT

BEE HUMMINGBIRD

Bee Hummingbird—The tiny bee humming-bird probes a flower for energy rich nectar. (Mike Shanahan)

Bee Hummingbird—An adult of this tiny bird perching on the end of a very small twig. (Pete Morris)

Scientific name: *Mellisuga helenae*
Scientific classification:
    Phylum: Chordata
    Class: Aves
    Order: Trochiliformes
    Family: Trochilidae
What does it look like? A fully grown male bee hummingbird is around 5.5 cm long and about 1.9 g in weight. Females are larger, with a body length of just over 6 cm and a weight of 2.6 g. They are attractive birds with iridescent plumage. The male is bluish with a whitish grey underside, although during the breeding season his head, chin and throat take on a pink/red hue. The females are more greenish with a white belly.

Where does it live? The range of the bee hummingbird is restricted to the island of Cuba in the Caribbean and the nearby Isla de la Juventud. They are forest animals, preferring the edge habitats around the perimeter of these large tracts of vegetation.

Small, but Perfectly Formed

The bee hummingbird is a zoological wonder. Its beautifully feathered body, little bigger than a large bee, still has all the features, albeit in miniature, that are unmistakably those of a bird. It is one of the smallest warm-blooded animals, yet it has a tiny heart, a brain, little feet and perfectly formed wings. Small size in warm-blooded animals is associated with a furious pace of life and the bee hummingbird is no exception. It does everything at breakneck speed. To fuel such a tiny body the fires of metabolism burn fiercely. The favored food of this bird is the energy rich nectar produced by plants. This sugary solution is just the sort of stuff the bee hummingbird needs to fuel its metabolism. Flowers evolved as a means of attracting insects for the purposes of pollination; therefore they are hardly built to take the weight of a perching bird. The bee hummingbird gets around this problem with some very impressive aerobatics. Its small wings lack the hinged joints of other birds, which means they can be beaten by the powerful wing muscles in a figure-eight pattern. During normal flight the wings beat about 80 times a second, although during its courtship displays they flap 200 times a second. This gives the tiny bird precise control over speed and direction. The flying jewel can fly forward, backward, sideways, and upside down. It can stop dead in the air and hover with mechanical precision. Such delicate skills enable it to probe its favorite nectar flowers with its beak. The thin tongue darts into the flower to lap at the sweet liquid. In one day the bee hummingbird may visit more than 1,500 flowers and in doing so may take on board more than eight times its own weight in liquid. Such a volume for a small animal amounts to a huge number of calories, which in human terms would be more than 150,000 (the normal amount for a human is about 2,500). The bee hummingbird's internal workings operate at a pace akin to the blurring beating of its wings. For example, its heart, which also happens to be the largest, relatively, of any warm-blooded animal, beats at around 1,200 times a minute (our heart rate is about 70 beats per minute). Digestion is similarly rapid. On the odd occasion it consumes an insect, the digestive system can process it in a little over 10 minutes. In larger animals this process can take many hours. Such speedy metabolism gives the bee hummingbird a very high body temperature of around 40°C, the highest of any bird.

Only when the food is at its most abundant can the bee hummingbird interrupt its foraging to look for a mate. During the courtship display the male hovers in front of the female beating his wings at terrific speed before shooting straight into the air to a height of around 15 m, his metallic plumage catching the rays of the sun. The daring little suitor then free falls back to earth stopping his descent right in front of the female using the deftness of his hovering. After mating the female builds a tiny nest constructed from moss, spider's webs and down. This is attached to a small branch and because of the materials used in its construction it blends in perfectly with its surroundings. Into the tiny cuplike nest are usually deposited a pair of tiny, white eggs, little bigger than peas. The female incubates them and in around two to four weeks the young hatch. The young are helpless and are completely dependent on their mother who must up her food forays to feed her offspring.

- The hummingbirds take their name from the sound they make when they fly. The wings beat so fast they make a humming sound. They are restricted to the Americas. They are at their most diverse and numerous in the tropics and the subtropics where the high temperatures and long or never ending growing season enables them to thrive. They are a specialized group and have diversified into more than 300 species.

+ As the metabolism of a hummingbird is so rapid and in need of almost constant fuelling, night time presents something of a problem. They must rest, but as they are not eating they would quickly perish. They overcome this conundrum by going into what is known as torpor. This is where the metabolism slows right down to the point where it is just ticking over so as to conserve energy and stop the tiny warm-blooded creature from dying during the night.

+ The beautiful, iridescent plumage of the hummingbird has long made them a favorite amongst collectors. Aboriginal people used hummingbird feathers to decorate head dresses and so forth and in Victorian high society, the tiny stuffed bodies of these animals would adorn expensive hats. Fortunately, such accessories are no longer fashionable helping to ensure these magnificent birds are protected for future generations to enjoy.

Further Reading: Peters, S. *Bumblebee Hummingbirds of Cuba.* Welschner Books Inc., New York 2000; Terres, J. *Hummingbird Family.* Alfred A. Knopf, New York 1982; Tyrrell, Q. *Hummingbirds of the Caribbean.* Crown Publishers Inc., New York 1990.

COMMON SWIFT

Common Swift—A common swift in flight, showing its very efficient wings. (Mike Shanahan)

Scientific name: *Apus apus*
Scientific classification:
  Phylum: Chordata
  Class: Aves
  Order: Apodiformes
  Family: Apodidae
What does it look like? This small, brown bird is usually around 16 to 17 cm long with a forked tail and a large pair of wings that form a crescent shaped span of around 40 cm when the animal is in flight. They can weigh up to 56 g. On the chin there is a small patch of white feathers. The beak is short, but wide. The feet are small and positioned far back on the body.
Where does it live? The common swift is an animal of the air. It is a migratory species, so it spends the temperate winter in Sub Saharan Africa and flies north in the spring, ranging as far north as Scandinavia and as far east as the Himalayas and Pacific coast of Russia.

Masters of the Air

No other birds can compete with the swifts for sheer mastery of the air. In many ways they are an aerial reflection of certain fish and sharks that ceaselessly cruise the world's oceans. These aquatic animals have a hydrodynamic form that can cut through the water with utmost ease. Although the common swift glides through the air, it too has certain adaptations that allow it to cut through the air with an amazing elegance. The body is compact and streamlined and the feathers pad out and smooth all the angles of the body. The wings are long and taper to a fine tip. This shape is ideal for reducing drag as the bird flies through the air. The bones in the wing of the swift near the body are short and stocky and allow the surface of the wing to be moved with considerable force by the flight muscles, enabling high speeds to be reached. During displays and disputes when they go into steep dives, common swifts can reach speeds of 60 m per second, which can only be matched by certain falcons and other, larger swifts. The bones in the wing that extend from the elbow towards the tip are long and mobile giving the swift amazing agility in the air. Typically, the common swift flies at around 5 to 14 m per second for hunting purposes. Like a filter-feeding fish, with its mouth wide open, the common swift flies through the air collecting tiny insects as it goes. If it is collecting these insects for nestlings it will hold them in a saliva bound ball in its throat. Nestlings are very hungry and each bird in a breeding pair may have to deliver 40 helpings of insects to the young every day. Building the nest, brooding the eggs and feeding the young are the only times that a common swift stops flying. It is phenomenal because it does everything else on the wing. It feeds by catching aerial insects, it drinks by skimming water from the surface of a lake or puddle and it sleeps in the air and even mates in the air. More amazing still is the fact that a fledgling common swift does not breed until its third or fourth year, which means that when it does come to set up a nest it will be the first time it has folded its wings for two or three years. The distances these birds cover in a single year are astonishing and are probably 200,000 km at the very least. Most confusing among the swift's aerial abilities is its sleep flying. How does it fly and sleep at the same time? Again, it is difficult to know for sure as it is difficult to record a swift when it is sleeping. Sleep, in most animals, is a natural relaxed state when they have reduced awareness of their surroundings. Sleeping and flying don't mix—just ask a pilot. Somehow the swift has found a solution to this problem and it is thought that it can rest one half of its brain at a time. While one half of the brain rests, the other half takes over all the functions and vice versa. During these bouts of sleep the bird may only be capable of simple flying and may be forced to ascend where it can fly in steady circles for a while. The common swift probably divides its sleeping into small naps instead of one long, continuous slumber.

+ There are around 96 species of swifts, swiflets, and needletails, all of which have the same general body plan. They are all expert flyers.
+ Swifts resemble those other masters of the air, the swallows and the martins, but in actual fact they are not closely related, they just happen to look the same because they do the same thing. This is another example of convergent evolution.
+ The nest of the common swift is composed of material that it finds on the wing, such as feathers, dry grass, straw, dead leaves, winged seeds, flower petals, and paper scraps. Nest building can take sometime especially if conditions are calm and there has been insufficient wind to lift suitable material into the air.
+ The appearance and the flying abilities of the swifts are a consequence of hunting aerial insects. Many types of insect and spider spend at least some of their life in the air together with bacteria, viruses, fungi, and protists. This so-called soup of small

creatures and microorganisms is known as aeroplankton and many different animals depend on it.

+   Bad weather makes it very difficult for common swifts to hunt, which would be disastrous for the ever-gluttonous nestlings. Fortunately, the chicks have an adaptation that allows them to drop their body temperature and to enter a form of torpor. In this state of slowed metabolism the chicks lose their wild hunger, for a while, at least.

+   Naturally, the swift is an animal of rock faces and cliffs, but human habitations have given them a whole new range of nesting sites to exploit.

+   Certain species of cave swiftlet from Southeast Asia build their nest entirely from saliva. These nests are the key ingredient of bird's nest soup, which is a delicacy in oriental cuisine. Nests collected from caves using towering bamboo scaffolds command a higher price than those obtained from purpose-built nest houses. The nest in the soup has a gelatinous texture and is it said to be good for general health, however, in some people the soup can cause the excessive secretion of stomach acid.

+   Swifts are utterly reliant on insects for food and are therefore at risk from the indiscriminate use of insecticides in modern, intensive agriculture.

EMPEROR PENGUIN

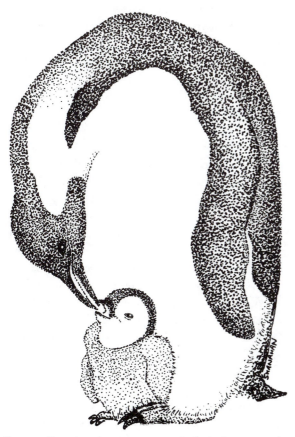

Emperor Penguin—An emperor penguin feeds the young it has nurtured in its remote breeding grounds. (Mike Shanahan)

Scientific name: *Aptenodytes forsteri*
Scientific classification:
> Phylum: Chordata
> Class: Aves
> Order: Sphenisciformes
> Family: Spheniscidae

What does it look like? This is the tallest and the heaviest of the penguins. Adults can be up to 115 cm tall and weigh anywhere between 22 and 37 kg. Its stature is complemented by its glorious deep black and shimmering white plumage accented by a golden patch of feathers on either side of its neck. The wings are reduced to flippers. The legs are very short and positioned at the very back of the animal so it can stand and walk upright.

Where does it live? The emperor penguin is restricted to the cold waters of the Antarctic. They are rarely found north of 65 degrees south and only leave the water for significant lengths of time to breed when they take up residence on the pack ice of the continental shelf or islands in waters of the far south.

An Antarctic Trek without a Sled

The Emperor penguin is a master of survival. Annually, it spends large tracts of time in one of the most inhospitable places on earth and at the most unpleasant time of year. The place is Antarctica and the time is winter. The story begins in March or April when the males leave their true home, the sea, to begin a long, slow walk inland. Their progress is impeded by the fact they don't really have the legs for it. With a combination of awkward waddling or sledging, where they lie on their belly and kick themselves along, they eventually reach their rookeries. They may have shuffled and slid up to 120 km. The males are soon joined by the females who started their migration a little later. When the females reach the rookeries the males begin posturing and calling to attract a mate and if successful they form a monogamous relationship for the breeding season. Soon after they have forged their bond they mate and in May or early June (at the height of the southern winter) a single large egg tipping the scales at around 450 g is laid by the female. For the time being the female's work is done and she carefully passes the egg to the male who balances it on his feet and drapes it with a large roll of skin. The female makes the arduous journey back to the sea to stock up on food and for the next 64 days the male broods the single, large egg. During this time he does not eat, but sustains himself throughout the bitterly cold winter on deposits of fat laid down when he was feeding out at sea. For much of the incubatory period the male sleeps as a way of conserving energy and all the males in the rookery, abandoned by their mates, huddle together for warmth forming a tight, milling throng where every male has a go in the relatively warm middle. They really need to huddle. The weather is numbingly cold and the wind can reach speeds of 200 km/h. After approximately two months, the female returns from the sea and relieves the male of his egg-sitting duties. Unerringly, the female can locate her mate in the throng by his call, in a cacophony of hundreds or thousands of calling males. With the utmost care he passes the egg to the female and heads to sea. Not a morsel has passed his lips in over 110 days. Several more weeks elapse and the male returns, his belly full, to his mate and baby. The female has been feeding the chick, now hatched from its egg, on regurgitated food and the male can now do the same. The chicks have to grow rapidly and before long they have moved off their parent's feet and form crèches on the ice with other downy youngsters. The parents make regular trips to the sea to collect food for their chick, locating their offspring on their returns by its calls. By the time they are 150 days old the chicks will have almost lost their downy covering and many will have been abandoned, left alone to find their way to the sea where they will take to the waves, only returning to land five years hence, to breed themselves.

+ The penguins are a group of 17 species of bird, which are superbly adapted to an aquatic way of life. On land they are very ungainly creatures but they glide through the water with graceful ease, slicing through this dense medium, propelling themselves with their stubby, paddle-like wings. Their plumage also retains a layer of air, acting as buoyancy aid. Fossils show that the modifications to the standard bird form and way of life, very evident in the penguins, are very old—at least 40 million years and possibly as much as 65 million years.

+ All penguins are found in the southern hemisphere. Not many species of penguin are actually found as far south as the emperor and the Galapagos penguin lives on the equator year-round—but in the relatively cold, rich waters of the Antarctic Humboldt Current. The penguins of cold environments have dense plumage and thick blubber to help them survive the harsh conditions of these very southerly conditions.

+ The bird cousins of the penguins, the auks, exploit similar niches in the northern hemisphere. The largest species of auk, the great auk, was around 75 cm tall. The last pair was killed in 1844 in Iceland. Auks resemble penguins physically and in color patterns. Their resemblance is an excellent example of convergent evolution.

+ Not only is the emperor penguin a hardy bird, it is also an animal capable of prodigious diving feats, reaching depths far out of the reach of other birds. The diet of these penguins consists of crustaceans, fish and cephalopods and during their hunting missions they can dive to depths of 560 m and hold their breath for 20 minutes.

+ Penguins of cold environments have a complex circulatory system in their appendages to avoid the ravages of frostbite and hypothermia. Blood returning from a foot in contact with the cold ice is warmed up by the descending blood in a heat-exchange-type system. This ensures the animal's core temperature is not lowered by the frosty conditions.

+ There are thought to be approximately 200,000 breeding pairs of emperor penguin alive today. These may seem like a lot of animals, but they are severely at risk from the effects of climate change.

Further Reading: Deguine, J. *Emperor Penguin: Bird of the Antarctic.* The Stephen Greene Press, Burlington, VT 1974; Rivolier, J. *Emperor Penguins.* Elek Books, London 1956; Williams, T. *The Penguins.* Oxford University Press, Oxford 1995.

EUROPEAN EEL

Scientific name: *Anguilla anguilla*
Scientific classification:
    Phylum: Chordata
    Class: Actinopterygii
    Order: Anguilliformes
    Family: Anguillidae
What does it look like? An adult European eel is a long, snakelike fish; however, during its development it goes through several stages that bear little or no resemblance to the adult.

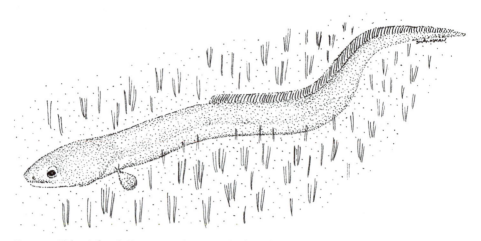

European Eel—A female European eel migrates back to the ocean crossing land where she has to. (Mike Shanahan)

Adult females can be 1.5 m in length and 2 kg in weight. Males, on the other hand, are much smaller.

Where does it live? This eel spends most of its time in bodies of freshwater throughout Europe, however when fully grown it takes to the sea in order to spawn.

Where There Is an Eel There Is a Way

The life history of the eel is one of the most remarkable and bizarre in the whole animal kingdom. For centuries, it baffled scientists and today there are still unanswered questions. The eel has always been coveted by Europeans and it has always been known that during the fall, large numbers of adult eels would descend the streams and estuaries and head out to sea never to return. Then, each spring vast numbers of young eels, known as *elvers*, would mysteriously appear in their millions in estuaries, heading upstream. It was assumed that the eels must reproduce somewhere out at sea, but the location of this spawning ground was completely unknown. It took many decades and funds from the Carlsberg foundation to provide some of the answers. Small animals, frequently encountered in nets out at sea were thought to be small fish completely unrelated to the eels. These tiny, leaf-like creatures are in fact the larvae of eels and a Danish scientist spent much of his career trying to find the smallest of these larvae, which would indicate their birthplace. His search led him across the Atlantic to a warm, calm patch of ocean called the Sargasso Sea. It seems the adult eels swim the 6,000 km from their home streams to the Sargasso Sea over a period of 1–2 months and apparently at great depth. At the start of this migration the body of the eel undergoes massive changes. Its gut dissolves, so the prodigious swim must be fuelled with the deposits it laid down when it was feeding in freshwater. Its eyes become bigger and pigments within them change so the fish can see in the dim light of the ocean and the sides of it body take on a silvery luster as a camouflage to help it evade the many predators that will be waiting for it in the open ocean. The mammoth migration to the breeding grounds exhausts the eels, so much so that once they have spawned, the adults die. The eggs, nurtured in the calm, warm waters of the Sargasso Sea hatch into miniscule larvae that must begin the arduous task of swimming back to the rivers, lakes and streams where their parents came from, following the taste of their home waterways. Such a small creature is relished by predators and huge numbers of them are gobbled up. After two years of perilous and slow progress, the larvae will be in the middle of the Atlantic, and a further year will take them to the coastal waters of Europe. Here, they change from leaflike animals into creatures more reminiscent of eels. These elvers begin to swim upstream, but the males go no further than the estuaries or coastal rivers, whereas the females continue inland, sometimes for hundreds of miles, overcoming barriers with fishy tenacity. The females feed and grow for 8–15 years before heading downstream to join the males where they will be unerringly drawn to complete their life cycle in the distant Sargasso Sea.

+ There are approximately 400 species of true eel. They range in size from 10 cm to 3 m monsters weighing more than 100 kg.
+ The European eel is a nocturnal predator, feeding on small animals such as fish, arthropods, crustaceans and mollusks.
+ The European eel is very similar to the American eel, but the American species migrates from freshwater bodies all along the eastern coast of North America to the same spawning ground as the European eel: the Sargasso Sea. The distance that this species travels to and from its breeding grounds is much less than that traveled by its European relative.
+ A substance in the blood of eels is poisonous to some fish and mammals, but cooking destroys the toxin.
+ When eels are migrating they can traverse considerable barriers to reach their destination. Adult females returning to the sea will commonly leave the water and slither across patches of dry land. Their skin produces copious quantities of mucus that aids their passage. When elvers are migrating upstream they will overcome small barriers by piling their bodies up against the obstacle until they start pouring over its top.
+ The Sargasso Sea, an area of ocean in the Atlantic, roughly 3,200 km long and 1,100 km wide is a mysterious place. It exists because it falls in the confluence of several ocean currents that form a large area of calm, slowly rotating water. The area moves, tracking the surrounding currents and at its surface there are huge expanses of seaweed known as *Sargassum*. The water is warm and very salty and it supports an assemblage of animals that live amongst the floating weed. Large fish are rare as there isn't the food to support them making the Sargasso Sea a refuge for young eels. With a changing climate it is uncertain what will become of the ocean currents in the Atlantic, especially the Gulf Stream. Should any of these currents lose strength or stop, the Sargasso Sea may cease to exist.
+ There have been eel fisheries for thousands of years as the animal is a popular food item. They are caught as adults or as elvers. A common name for elvers when they are around 45 mm long is *glasseels*. In recent years, the numbers of these juvenile eels caught in places like Epney on the River Severn in the United Kingdom have declined dramatically, so much so that in 1997, the demand for them in Asia could not be met and huge sums were changing hands for available stocks. Some dealers were paying more than $1,100 per kg of glasseels. The reasons for this decline are not understood. It could be part of a natural long term trend, or human activities could be severely reducing their numbers and preventing them from reaching their spawning grounds.

Further Reading: Lecomte-Finiger, R. The early life of the European eel. *Nature* 370, (1994) 424–25; Sinha, V., and Jones, J. *The European Freshwater Eel.* Liverpool University Press, Liverpool 1975;

Tsukamoto, K., Nakai, I., and Tesch, W. Do all freshwater eels migrate? *Nature* 396, (1998) 635–36; Van Ginneken, V., and Van Den Thilart, G. Physiology: Eel fat stores are enough to reach the Sargasso. *Nature* 403, (2000) 156–57.

FLYING DRAGONS

Flying Dragons—One of the smaller flying dragon species in the hands of a biologist who is teasing the ribs apart to show the skin gliding surface. (Jim McGuire)

Flying Dragons—A flying dragon extends its ribs for a controlled glide. (Mike Shanahan)

Scientific name: *Draco* species
Scientific classification:
   Phylum: Chordata
   Class: Reptilia
   Order: Squamata
   Family: Agamidae

What do they look like? Flying dragons are small, thin bodied lizards with long, tapering tails and slender legs ending in five sharp-clawed digits. Most species are approximately 20 cm long, but some can reach nearly 40 cm. Much of this length is tail. In most species females are larger than males. They are colorful creatures and both sexes of a few species have a bright fold of skin beneath their head called a gular flap, which can be extended to function like a small flag. The flap is always present in males, but absent in females of many species.

Where do they live? Flying dragons are native to the forests of southern India and Indomalaysia where they scurry up and down trees, rarely venturing to the ground.

Glide to Get Where You Are Going

In the lush Malaysian forest a flash of color darts from the trunk of a tree and alights on another tree several meters away. The bright colors disappear and whatever made it is now camouflaged against the bark making it difficult to see. The animal scampers up the tree and its movements reveal it to be a lizard, but no ordinary lizard. It is one of the only reptiles that can take to the air. It cannot fly like a bird, bat or insect, but its gliding abilities are second to none. What makes this reptile unique are the structures it uses to glide. Flying animals and most gliding ones have modified forelimbs that function as wings or flaps of skin (patagium) stretched between the limbs forming a kind of parachute. Flying lizards have elongate ribs protruding from their body covered with skin, forming fan like wings. The ribs are mobile and for much of time the wings are held tight against the body and the lizard looks relatively normal. However, moving around its habitat, hunting and chasing mates the animal may take to the air with a daredevil leap. It spreads its rib wings and glides to another tree. The glides of the flying dragon are very elegant aeronautical maneuvers. Initially, the animal goes into steep dive at an angle of around 45° before leveling out and using the momentum from the dive to carry it some distance horizontally. When the landing site is in range the lizard goes into an upward glide and alights delicately on the new tree using its sharp claws for purchase. Without pausing to reflect on its feats, the lizard folds its ribs against its body and frantically scrabbles up the tree seeking more prey or a mate. For such a small animal the distances achieved in these glides are remarkable. Glides as long as 60 m have been recorded, over which the animal loses only 10 m in height. When you consider that a flying lizard is only around 20 cm long, this is quite some distance.

+ There are 35 known species of flying dragon. They are among the few modern lizards able to take to the air. There are fossils of extinct reptiles with gliding surfaces formed from ribs, as in the flying dragons.
+ Flying dragons are diurnal and are active from around 8 A.M. until it gets too hot, around midday, at which point they will seek out shade. Their day resumes at around 1 P.M.
+ These lizards are insectivores and specialize in catching ants that scurry around on their trees. They are 'sit and wait' predators, clinging very still to their tree, until a hapless insect wanders past.
+ Courtship amongst these lizards is complex and colorful and is thought to take place between December and January, although in some areas the lizards may breed all year round. In many species the gular flap and the wings of the lizards, especially the males, have bright splashes of color. When an amorous male likes the look of a female he extends his flap and bobs his head drawing attention to the brightly colored bib. He will also open and close his wings in an effort to impress the female with his dazzling colors. In some species the courtship display is topped off with three, body-bobbing circuits around the female, who at this point is hopefully impressed enough to allow the male to copulate.
+ Male flying lizards are fiercely territorial, defending perhaps two or three trees, on which there might be two or three females. Trespassers are treated with disdain and the owner will chase and harangue the interloper until it leaves.
+ The only time a flying lizard ventures to the alien terrain of the ground is when a female is ready to lay her eggs. She descends the tree she is on and makes a nest hole by forcing her head into the soil at the base of the tree. She then does an about turn and lays a small number of eggs (2–5) before filling the hole and patting the soil down with her head. For the first 24 hours she is a model parent, guarding the eggs vehemently, but then, seemingly bored, she leaves and has nothing more to do with her offspring. The eggs hatch after approximately one month.
+ Throughout the course of animal evolution, several unrelated groups of animal have taken to the air. Animals that master the air have access to a whole new way of life.

Flying animals can exploit new sources of food, travel great distances and successfully evade their ground dwelling predators. The air has been conquered by (in chronological order): insects, pterosaurs (extinct), birds and bats. In all of these animals, gliding was probably an intermediate stage. Today, there are several different animals capable of gliding. The gliding frog has modified feet for gliding, while the flying geckos have flaps of skin along their body. The flying snake flattens its body to glide and the numerous gliding mammals have a membrane of skin that is stretched between their fore and hind limbs. Flying fish can glide for considerable distances on modified pectoral fins. There are even gliding squid. Gliding is very common in the forests of Southeast Asia, especially Borneo; yet there are fewer gliding species in South American or African forests. The reason for this could be the fact that the trees in the Southeast Asian forests are often taller and more widely spaced than the trees in other forest. In these Old World forests there are also fewer vines and other connective tendrils between the trees.

Further Reading: Card, W.C. *Draco volans* reproduction. *Herpetological Review* 25, (1994) 65; Hairston, N.G. Observations on the behavior of *Draco volans* in the Philippines. *Copeia* 4, (1957) 262–65; Mori, A., and Tsutoma, H. Field observations on the social behavior of the flying lizard, *Draco volans sumatranus*, in Borneo. *Copeia* 1, (1994) 124–30.

FOUR-WING FLYING FISH

Four-Wing Flying Fish—A four-wing flying fish spreads its fins to take to the air. (Mike Shanahan)

Scientific name: *Hirundichthys affinis*
Scientific classification:
    Phylum: Chordata
    Class: Actinopterygii
    Order: Beloniformes
    Family: Exocoetidae

What does it look like? A fully grown four-wing flying fish is around 30 cm in length. The pectoral fins and the pelvic fins are greatly enlarged with bold banding patterns. The tail fin is nonsymmetrical with a long lower lobe. The body is conical and slim.

Where does it live? This is a fish of the Atlantic Ocean and is found in both the East and West as far south as northern Brazil. It is also found in the north of the Gulf of Mexico and the Caribbean as well as the Arabian Sea.

A Whole New Meaning to Water-Wings

In the waters of the western Atlantic, a small shoal of dolphin fish, large, fast marine predators, pursue a shoal of smallish, pretty unremarkable looking fish. With powerful flicks of their tail fins the dolphin fish surge through the water in a final lunge at their unfortunate prey. Sensing imminent danger, the prey fish takes evasive action, developing an impressive turn of speed at the surface of the water, covering about 30 body lengths per second. With their dorsal fin and head breaking the surface they give a few more powerful sweeps of their tail and they break free of their aquatic environment. Now, this is itself is not all that unusual. Lots of fish species breach the surface, some in amazing leaps, but what is amazing about this particular fish is what it does next. In the air, the fish extends its pectoral and pelvic fins, revealing them to be huge and graceful looking wings. The flying fish can't flap these wings, but the momentum it built up beneath the water is enough to allow it to glide effortlessly for at least 50 m and occasionally much further. The pectoral fins provide the surface for gliding while the smaller pectoral fins act as stabilizers giving the fish a certain degree of control over its movements whilst airborne. Released from the impeding drag of the water the speed of these gliding fish increases greatly, up to around 60 km/h and perhaps even more. To remain airborne for as long as possible the flying fish dips the elongated lobe of its tail fin into the water and frantically thrashes it. This gives it another burst of forward momentum and the glide continues. To extend its glides still further the flying fish can also use waves to its advantage by gliding on the updrafts at their leading edge. Using the waves in such a way flying fish can glide for distances of at least 400 m. Apart from the elegantly adapted fins, flying fish also have eyes that allow them to see as well out of the water as they do in it. Light is not refracted in the air; therefore, to accurately judge distance out of the water the flying fish has eyes that are less rounded than those of other fish. Eventually, the muscles powering their tail fin will tire, and they will slip cleanly back into the water, hopefully a long way from the confused dolphin fish.

+ The ability to leave the water and predators behind is a highly effective means of survival. There are at least 70 species of flying fish found throughout the world's oceans. They are fish of tropical and subtropical waters as the cold waters in the far north and south would not be conducive to the rapid muscle activity that takes them clear of the water.
+ Some species of flying fish are only around 15 cm long, while others may be as much as 45 cm in length. The band-wing flying fish is one species with four wings. Some species have only enlarged pectoral fins and relatively normal pelvic fins. Regardless of this they are all adept at gliding.
+ It is not uncommon to find flying fish on the decks of boats after an overeager leap from the water.
+ Although the flying behavior of these fish is probably a way of avoiding predators it may also be used as means of getting from one place to another as some scientists have suggested it may be more efficient than swimming through the dense medium of water.

- Flying fish are actually predators themselves, feeding on small crustaceans, and so forth, when small and progressing onto larger prey, including other fish as adults.
- In the Caribbean, especially Barbados, the flying fish are a very popular food animal. They are used in a dish called Cou-Cou. These fish are also popular in Japan where they are normally dried and eaten. Their eggs are also used in some types of sushi.
- Some species of flying fish make regular migrations, following food or to and from their breeding grounds. In some areas, especially the Caribbean, human activities have polluted and damaged these routes, harming the populations of these fantastic fish. The damage to their habitats is compounded by commercial fishing as large numbers of these fish are taken every year for human consumption.
- Some species use the floating mats of *Sargassum* seaweed as nurseries for their young. During their early life, flying fish are incapable of the gliding feats of adults and so they lurk amongst the weed relying on camouflage for protection.

Further Reading: Davenport, J. How and why do flying fish fly? *Reviews in Fish Biology and Fisheries* 40, (1994) 182–214; Saidel, W. M., Strain, G. F., and Fornari, S. K. 2004. Characterization of the aerial escape response of the African butterfly fish, *Pantodon buchholzi* (Peters). *Environmental Biology of Fisheries* 71, (2004) 63–72.

GRANT'S GOLDEN MOLE

Grant's Golden Mole—Under the cover of sand, a Grant's golden mole ambushes an insect larva. (Mike Shanahan)

Grant's Golden Mole—An adult of this species above ground. The dense fur completely covering the eyes and the powerful forelimbs can be clearly seen. (Galen Rathbun)

Scientific name: *Eremitalpa granti*
Scientific classification:
　Phylum: Chordata
　Class: Mammalia
　Order: Afrosoricida
　Family: Chrysochloridae
What does it look like? Grant's golden mole is a small mammal. Adults are 7 to 8.5 cm in length and between 16 and 32 g in weight. They lack any outward signs of eyes or ears, giving them a rather surreal appearance. The fur is grayish yellow with a golden sheen.

Where does it live? This mammal is limited to a small part of Southwestern Africa. Its range encompasses some of South Africa and the Namib Desert in Namibia. The habitat of this very specialized animal is the pinkish/red coastal sand dunes. They are restricted to these habitats and cannot spread out inland as the sand is firmer and unsuitable for their style of tunneling.

Swimming through the Sand

In some of the desert areas of Southern Africa, shallow grooves appear in the sand as if an invisible finger is tracing a line. No invisible, giant hand is at work here and the tracks are actually caused by a small, industrious mammal—Grant's golden mole, the smallest of the golden moles, but perhaps one of the best understood. Unlike other subterranean mammals Grant's golden mole does not burrow. The sand is too fine and too loose for this so this little animal actually swims through the sand. Like all other golden moles, Grant's golden mole has become supremely adapted to a subterranean existence, losing many of the mammal characteristics that are surplus to requirements underground. Soon after birth, the eyelids fuse and thicken, to eventually be covered by thick, shimmering fur, which has a beautiful iridescence. To prevent soil and other debris from entering the nostrils these animals have a little leathery flap of skin on the end of their snout, which also helps them nose their way through the sand. The digging prowess of golden moles is made possible by their heavily muscled shoulders, which power the forelimbs through the sandy soils of their habitat. The third digit on each paw is very well developed and along with the first and second digits bears hefty, curved claws. It is with these tools the golden mole paddles through the sand, shoveling the material towards the rear where it is kicked backwards by the webbed hind feet. Although it is an excellent burrower, tunneling is an expensive way of getting around, in terms of energy and it will often move around on the surface. During its hunting forays, which take place during the cool of the evening this little mammal can cover a distance of almost 6 km whilst looking for food. That is equivalent to you or I covering a distance of 150 km in one night to look for dinner. The favored prey is termites, but they also take other desert invertebrates like beetles, moths and spiders. Unwary reptiles are also on the menu. During these forays Grant's golden mole spends a lot of time on the surface occasionally slipping beneath the sand to pinpoint the victim with its ears before 'swimming' stealthily up to the delicious morsel like some manner of miniature submarine. Unlike most other mammals, the ear openings are tiny and sound energy is actually picked up as vibrations through the sand and beneath the surface the golden mole probably picks up the astonishingly faint pitter-patter of termite feet. This feat is made possible by the massively enlarged hammer ear bone that picks up these vibrations and amplifies them.

The deserts where Grant's golden mole dwells are notoriously inhospitable. Daytime temperatures are uncomfortably high, so to seek shelter when the sun is beating down on the sand the mole will dig down to depths of around 50 cm and go into a state of torpor, switching off its temperature regulation systems thus conserving valuable energy. Water is also very scarce in these habitats, but the golden mole's very efficient kidneys mean that it never needs to drink, instead it obtains all the fluid it needs from its food.

+ There are 21 species of golden mole, and all but one of them are restricted to Southern Africa. They all have the same basic body plan and most species are conventional burrowers, forming lasting tunnels in soils and even in sphagnum moss.
+ Fossils of golden moles around 25 million years old have been found, but they are so like the living species; they tell us little about the origins of these enigmatic mammals.

+ Although golden moles may look like true moles and marsupial moles they are all unrelated. The similarity in appearance is simply due to the fact they have all evolved to live a subterranean existence.

+ There is still a great deal to learn about the life of Grant's golden mole. As it cannot form any permanent burrow or nests in the loose sand it is unknown how or where they breed. No nests have ever been found, but as they suckle their young they must have a permanent base for a while.

+ Some of the golden moles have very small geographical ranges, making them intensely vulnerable to habitat loss. Of the 21 living species at least 11 are endangered.

Further Reading: Fielden, L. Home range and movements of the Namib Desert golden mole *Eremitalpa granti namibensis* (Chrysochloridae). *Journal of Zoology* 223, (1991) 675–86; Fielden, L., Perrin, M., and Hickman, G. Feeding ecology and foraging behaviour of the Namib Desert golden mole, *Eremitalpa granti namibensis* (Chrysochloridae). *Journal of Zoology* 220, (1990) 367–89; Mason, M., and Narins, P. Seismic sensitivity in the desert golden mole (*Eremitalpa granti*). *Journal of Comparative Psychology* 116, (2002) 158–63; Perrin, M., and Fielden, L. *Eremitalpa granti*. *Mammalian Species* 629, (1999) 1–4.

LEATHERBACK TURTLE

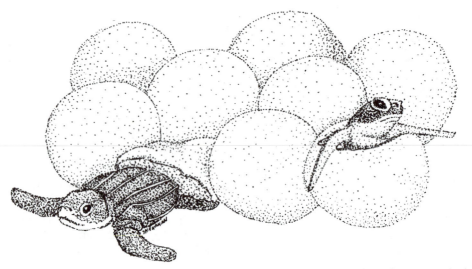

Leatherback Turtle—Leatherback turtle young hatch from their eggs to begin their life out at sea. (Mike Shanahan)

Scientific name: *Dermochelys coriacea*
Scientific classification:
   Phylum: Chordata
   Class: Reptilia
   Order: Chelonia
   Family: Dermochelyidae
What does it look like? The leatherback is the largest turtle, with a shell length of 2 m, a flipper span of 3.5 m and a weight of more than 700 kg. It lacks the distinctive, bony shell of the other turtles, instead this structure has been replaced by a lightweight, streamlined carapace that has

a series of seven ridges running along its length. The front flippers are huge and generate the forward momentum for swimming. Its dark upper surface is flecked with patches of white.

Where does it live? The leatherback turtle lives in the open ocean. It can be found throughout the world's oceans, but cannot tolerate the cold conditions in the high polar regions.

Carefree Ocean Cruising

The leatherback turtle is the largest and most peculiar of all the turtles. They begin life deep down in the sand of a favored nesting beach, where females haul their considerable bulk from the water to excavate nest holes for their eggs. For a marine animal, supremely adapted to life in the water, this is no mean feat and the female scrapes and pants and snorts until she has dug out a deep hole that will protect her eggs. This hole will be made in sand that is suitably moist, but safely beyond the reach of waves. She can lay as many as 1,000 eggs in a season, separated into batches of 50–170. The digging of a nest hole is as far as the female's maternal instincts extend and as soon as she has covered over the clutch with sand she makes for the sea in a zigzag fashion to cover her trail. The young hatch after several months and their first task is to dig themselves from the nest hole. For a turtle that is just getting used to its flippers this is not easy, but soon enough they clamber out of the hole and on to the beach. This signals the start of a long and dangerous gauntlet the young must run. Apart from some tough scales and strength of numbers, the turtles are defenseless and many beady eyes have been surveying the sand waiting for their emergence. Gulls, skuas, and crabs all descend on the turtles and make short work of them. So many young emerge at once that a few are bound to break through this barricade to the safety of the surf. The young's swimming abilities are instinctive and as soon as they enter the water their tiny, flailing flippers propel them out to sea. In the open ocean, danger is also ever present and many more turtles will be taken by predators. A fortunate few are able to avoid the gnashing jaws of other sea creatures and scour the oceans looking for food. The favored foods of these turtles are gelatinous, floating creatures such as jellyfish. They are hardly the most appetizing sea food, but they are easy to catch. There are no teeth in the turtle's jaws, but its throat is lined with backward curving spines that help it swallow its jellylike food. As the turtle grows, the unique structure of its carapace becomes ever more obvious. The thick leathery skin covers a complex arrangement of small, linked bones embedded in cartilage. Compared to a normal shell, this structure is very lightweight and enables the leatherback turtle to range for thousands of miles around the globe. Only females ever make the return journey to land to deposit their eggs, males on the other hand will live out their whole life at sea, cruising the oceans with powerful strokes of their flippers. No one is sure over what distances the leatherback turtle ranges, but an individual tagged in Surinam, in South America, turned up on the other side of the Atlantic, over 6,800 km away. Unlike the other marine turtles, the leatherback turtle is not limited to tropical and subtropical waters. A large animal loses heat more slowly than a smaller creature so the great bulk of this reptile gives it an advantage in cold water and a thick layer of fat minimizes the amount of heat that is lost through the skin. An efficient circulatory system uses the heat generated from swimming to warm the blood returning from the cool limbs. All these adaptations allow the turtle to hunt in the cold productive waters from Iceland all the way down to New Zealand.

+ There are around 300 species of living turtle. They are divided into the marine forms with legs modified into flippers, terrestrial forms with thick, pillar-like legs and semiaquatic forms (the terrapins and snapping turtles). The turtle skeleton is perhaps the most peculiar of all the vertebrates. Not only do they have a shell, composed of fused, bony plates, but the pelvic and pectoral girdles supporting the forelimbs and hind limbs are found inside the rib cage.

+ The breeding beaches used by the leatherback turtle were a mystery until relatively recently. Today, several beaches scattered throughout the tropics are known to be used by the turtle. Most are mainland sites facing deep water. They seem to avoid those beaches protected by coral reefs. Many leatherback turtles will deposit their eggs on the same beach they hatched from.

+ It is difficult to say exactly how long a leatherback turtle can live for, but such a large, slow growing creature must live for at least 100 years and possibly far longer.

+ Today, many of the turtle species are severely threatened by habitat loss and hunting. A turtle's shell can protect it from most predators, except humans. Many species are hunted for their meat, which is a delicacy in many parts of the world. The marine species must all come to land to breed and the sheltered, sandy beaches on which they depend are at the mercy of tourist and industrial developments. Negligent use of the seas has also endangered many species as they are caught and suffocated in fishing nets, injured by boat propellers and harmed by pollution. Even on protected beaches, hatchlings are often confused by city lights and end up scuttling away from the sea.

Further Reading: Perrine, D. *Sea Turtles of the World.* Voyageur Press, Osceola, WI 2003; Spotila, J.R. *Sea Turtles: A Complete Guide to Their Biology, Behavior, and Conservation.* Johns Hopkins University Press, Baltimore, MD 2004.

NORTHERN BLUEFIN TUNA

Northern Bluefin Tuna—A section through the body of a northern blue fin tuna showing the arrangement of its enormous muscles. (Mike Shanahan)

Scientific name: *Thunnus thynnus*
Scientific classification:
    Phylum: Chordata
    Class: Actinopterygii
    Order: Perciformes
    Family: Scombridae

What does it look like? The bluefin tuna is flattened from side to side but is very deep-bodied. The powerful body tapers strongly to the thin tail and sickle shaped tail fin. The pectoral fins are long and rigid and are positioned not far behind the conical head, with its large mouth and eyes. The back of the fish is deep bluish purple and the underside is a shimmering silver. The largest bluefin on record weighed more than 750 kg and was around 3 m in length.

Where does it live? This fish is found in the Atlantic Ocean and ranges throughout this expansive body of water, including temperate and tropical waters from the far north to the Mediterranean, Black sea and coasts of Brazil and West Africa.

Hydrodynamic Perfection

The bluefin tuna is, arguably, the pinnacle of fish evolution in terms of swimming ability. Its body is so well adapted to its way of life that it can cruise for thousands of miles with ease and in the pursuit of prey or escaping from danger it can accelerate to speeds more than a match for all but the fastest boats. Such speed and power is made possible by the huge muscular bulk of this fish. The whole animal is little more than a dense block of muscle. From the skeleton outwards there is layer after layer of muscle fibers all of which pull in concert to drive the sickle shaped tail fin through the water in short, powerful sweeps. Much of muscle in the tuna has a limited ability to use the oxygen present in the animal's blood; it is anaerobic, therefore most of the energy is generated in the absence of oxygen. These fast-twitch muscles contract and relax very rapidly generating a lot of power, but only for short bursts at a time as. Such activity in the absence of oxygen leads to the build up of lactic acid, eventually causing muscular fatigue. A much smaller proportion of the fish's muscle mass is dependent on the oxygen carried in the blood and it is this aerobic tissue, tapering down the sides of the body to the tail, which contracts to provide the slow, steady action required for long distance cruising.

Both of these muscle groups channel their straining effort through a pair of tendons anchored to the sickle-shaped tail fin. With each flick of the tail, in contrast to most other fish, the whole of the body remains more or less rigid, increasing the efficiency of each stroke. In bouts of fast and furious swimming, the tuna's progress through the water is further enhanced by the other fins nestling in shallow depressions on its body, reducing drag.

Not only can the tuna shoot through the water like an aquatic bullet, but an interesting arrangement of blood vessels allows it to swim in water cold enough to deter other fishes. The exertion of the animal's muscles generates a great deal of heat and this warmth is captured by a heat exchange system where it is passed to cold blood traveling to the central nervous system and eyes, keeping them responsive even in chilly, temperate waters. This is a massive advantage for the bluefin tuna as it allows the animal to hunt effectively in waters, normally the reserve of warm-blooded creatures, such as marine mammals.

    ✦   Apart from the northern bluefin tuna the world's oceans are also home to the southern bluefin and Pacific bluefin. There are nine other species of tuna, all of which have the same general appearance as the bluefin, but are considerably smaller. All the tuna and their relatives, the mackerels and bonitos are carnivores adept at using their great bursts of speed to capture prey.

+ The heat exchange system of the bluefin tuna enables it to hunt as far north as the cold waters off Newfoundland.

+ The bluefin tuna vies with a few other species for the crown of the fastest fish. There is the sailfish and the marlin, both of which are reputed to reach speeds of 100+ kmh in short bursts.

+ The tuna's gills are ventilated by the rapid flow of water around them as the animal swims. If the tuna stops swimming it will die from a lack of oxygen.

+ Bluefin tuna swim huge distances. Tagged individuals have been tracked from the East Atlantic to the West Atlantic and back again in one year and it is very likely they cover even greater distances. During their ceaseless oceanic wanderings they tend to stay quite near the surface but they dive to depths of almost 1,000 m to hunt.

+ Tuna have small deposits of magnetite in their head and it is thought these act like a built-in compass, picking up the earth's magnetic field allowing the fish to orientate itself in the vastness of the open ocean.

+ The bluefin tuna, unlike the other tuna species is slow growing. Individuals may live for at least 30 years.

+ The large size of the tuna and the quality of the flesh has made it a favorite of fishermen everywhere. The bluefin tuna, with its huge size, is the most coveted of these fish and sadly, today, it is the rarest species. Before 1970 the bluefin tuna was not held in high regard. Captured specimens were normally rendered down for use in pet food. Then, the species rapidly gained a following in Japan as something of a delicacy. A large fishery grew and today stocks are severely depleted to the extent where fine examples now command huge prices. A 200 kg specimen sold in 2001 fetched more than $173,000 dollars.

+ Catching a bluefin tuna is no mean feat. They range over thousands of square kilometers of open ocean and in many cases long-line fishing is the weapon of choice. Longlining involves paying out up to 130 km of line bristling with thousands of hooks.

Further Reading: Blank, J.M., Morrissette, J.M., Landeira-Fernandez, A. M, Blackwell, S.B., Williams, T.B., and Block, B.A. In situ cardiac performance of Pacific bluefin tuna hearts in response to acute temperature change. *Journal of Experimental Biology* 207, (2004) 881–90; Block, B.A., Teo, S.L.H., Walli, A., Boustany, A., Stokesbury, M.J.W., Farwell, C.J., Weng, K.C., Dewari, H., and Williams, T.D. Electronic tagging and population structure of Atlantic bluefin tuna. *Nature* 434, (2005) 1121–27; Carey, F.G., and Lawson, K.D. Temperature regulation in free-swimming bluefin tuna. *Comparative Biochemistry and Physiology* 44, (1973) 375–92; Stokesbury, M.J.W., Teo, S.L.H., Seitz, A., O'Dor, R.K., and Block, B.A. Movement of Atlantic bluefin tuna (*Thunnus thynnus*) as determined by satellite tagging experiments initiated off New England. *Canadian Journal of Fisheries and Aquatic Science* 61, (2004) 1976–1987.

SEA LAMPREY

Scientific name: *Petromyzon marinus*
Scientific classification:
   Phylum: Chordata
   Class: Cephalaspidomorphi
   Order: Petromyzontiformes
   Family: Petromyzontidae

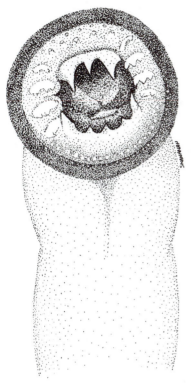

Sea Lamprey—The sea lamprey's fierce looking mouth. (Mike Shanahan)

Sea Lamprey—An adult sea lamprey. Note the gill holes and complete lack of jaws. (Horst Taraschewski)

What does it look like? The sea lamprey has a snakelike body with no paired limbs of any kind and no jaws. In the place of biting jaws there is an oral disk. The eyes are large and well developed and behind them are seven pairs of gill openings. Toward the end of the animal there are fins running along its back and underside. Color varies, but the back is normally green or gray, while the belly is white or pale gray. A fully grown sea lamprey may be as much as 1 m long.

Where does it live? The habitat of the sea lamprey is the northern Atlantic Ocean and the western Mediterranean, together with streams on the eastern coast of North America, Western Europe, and the Mediterranean basin.

A Chinless Wonder

The sea lamprey is a relic, a real blast from the past. It has many features of the first, primitive backboned animals, and it somehow manages to survive in a world dominated by higher, more evolved forms. Like many, outwardly primitive animals, the sea lamprey has a very interesting life history and some unique adaptations allowing it to thrive in a number of different areas. The tiny eggs of a female lamprey are a little over 1 mm in diameter. These shell-less, unspecialized eggs hatch into 6–10 mm, pink, wormlike larvae, which are so unlike their parents that early biologists described them as distinct animals, christening them *ammocoetes*. For a week or so after they hatch the ammocoetes stay in the nest, surviving on the remainder of the yolk stores that sustained them during their time in the egg. Soon, the time is ready for them to leave and they rise up into the flowing water to be carried downstream by the current. The rushing water deposits them in sleepy backwaters or calm banks. Here they burrow into the soft sediment and this is where they remain for the next three to seven years, safely tucked away in their muddy retreat, filtering the water for edible matter. Their oral hood projects from the burrow where it acts as a funnel to channel water through the muscular, pumping pharynx where particles of food are trapped in a layer of sticky mucus before being swallowed. When it is around 10 cm long, it is time for the larva to begin the transformation that will turn it into a young adult lamprey. This metamorphosis takes place in mid-summer and by the spring of the next year the juvenile may be ready to begin its long, arduous migration to the sea. The adult lamprey is quite a different beast to the larva. Gone are the delicate hood and the genteel, filter feeding way of life. In its place there is a carnivorous parasite with feeding apparatus straight out of a nightmare. The oral disk of the adult lamprey is wreathed with sharp, conical, inward curving teeth and in its mouth there is a rasping, muscular tongue. In the ocean, the lamprey swims with awkward undulations of its serpentine body, searching for suitable prey. When they locate a fish or perhaps even a marine mammal they latch on with their tooth-lined sucker and begin to rasp at the flesh of the prey. As they tear and rend the skin and muscle of the prey they secrete an anticoagulant that keeps the blood flowing freely. After they have gorged themselves on this highly nutritious food they let go to digest their meal leaving the prey with a large gaping wound, which can occasionally be fatal. After one year of this ravenous behavior they head away from the open ocean and back to their breeding grounds in the continental freshwater streams. Their progress is slow as they must fight the current and obstacles. Waterfalls can be overcome by using their oral sucker to cling on to the wet rocks and hoist themselves slowly through the rushing water of the cataract. As soon as they reach a suitable spot the males and female set about constructing nests in the gravelly bottoms of their breeding streams. They thrash their bodies around and use their suckers to move larger stones. After a great deal of exertion an oval depression has been formed into which the female sheds her eggs, closely followed by the fertilizing sperm of the male. This reproductive gesture is the last they make and the lampreys die soon after breeding is complete. The nests fill in with sand and silt carried by the current and the young, unseen, grow steadily, ready to begin the intriguing cycle all over again.

+ There are 41 known species of lamprey, 17 of which are found in North America. The sea lamprey is the biggest. They are found all over the world, except the high polar regions and the tropics. Fossils of lamprey ancestors are rare, but remains have been found in rocks from the early Carboniferous period, more than 320 million years old. It is very likely the history of these animals extends even further back into the mists of time.

+ Some lampreys, unlike the sea lamprey, are not parasitic. They develop as ammocoetes in the river sediment but breed straight after metamorphosis and then die. Other lamprey do not journey to the sea, but live out their parasitic existence in freshwater streams and rivers.

+ The skeleton of the lamprey is very simple and is composed of cartilage and not bone. There are vertebrae along the spine, but they are very small. The brain is small and is located above the blind ending channel of the single nostril.

+ The ravenous feeding behavior of the lampreys, particularly the larger species puts them into direct conflict with fishermen, who blame these animals for depleting fish stocks. A lamprey, latched onto a large fish will eat as much as it can at one site before releasing its grip and attaching onto a new area. The result is a fish with large parts of its body stripped almost down to the bone.

+ The lampreys, after finding their way into the great lakes through man-made canals have ravaged fish stocks and numerous measures have been employed to try and stop them, including chemical toxins and traps.

SLOTHS

Sloths—A cutaway of the two-toed sloth's limbs, showing its specialized skeleton, adapted for hanging. (Mike Shanahan)

Sloths—A captive three-toed sloth kept as a pet by a tribe in the Amazon. (Rhett Butler)

Scientific name: *Choloepus* and *Bradypus* species
Scientific classification:
   Phylum: Chordata
   Class: Mammalia
   Order: Pilosa
   Family: Megalonychidae and Bradypodidae
What do they look like? Sloths are divided into two main types: two-toed and three-toed. These names actually refer to the number of fingers they have as all sloths actually have three toes. All have rounded heads and flattened faces with a doleful look. The forelimbs are especially long and end in 8–10 cm, curved claws. The short, fine underfur of the sloth is covered by a longer coat of coarse hair, giving the animal a scruffy, grizzled appearance. Three toed sloths are 56–60 cm in length and around 4 kg in weight, while the two-toed species are 58–70 cm long and weigh between 4 and 8 kg.

Where do they live? Sloths are tree dwelling animals, preferring lowland and upland tropical forests up to an altitude of 2,100 m. They are found in Nicaragua, through Central America and as far south as northern Argentina.

Just Hanging Around

In the sixteenth century, early Spanish visitors to Central America were unimpressed by the sloth. Oviedo y Valdes wrote that he had never seen an uglier or more useless creature than the sloth. Granted, the sloth may not be the most beautiful animal, but it is a very successful, New World mammal. They have become so specialized to an arboreal, plant-eating way of life that there are few animals to compete with them and hardly any to eat them. They are nothing like those other arboreal specialists, the monkeys, who spend their days chattering and frolicking among the foliage. The sloths live life at a much slower pace. For most of their life they hang from the branches of their canopy home. Their limbs and claws are structured so that no muscular effort is needed to hold them in this position—they just hang. This slow pace of life is a consequence of the food they eat. Vegetation forms the bulk of their diet and to cope with this difficult-to-digest food the sloths have an elaborate, many chambered stomach containing cellulose digesting bacteria. The stomach is very capacious and when full it may account for about a third of the body weight of the mammal. Digestion, like everything else about the sloth is a slow process. Ingested food remains in the stomach for about a month, undergoing very slow digestion, until it passes into the short intestine. This slow process means that sloths only have to go to the toilet once a week and they do so, strangely, by leaving the safety of the canopy to deposit their feces and urine at regularly used spots at the base of trees.

The sloths have made a success on this meager diet by living a frugal life. They move slowly and only when they have to and as a result their muscle mass is only 50 percent of that of a similar-sized mammal. They don't waste energy by making lots of sounds. They sleep a lot, often for 18 hours a day and their body temperature is much lower than other mammals of a similar size.

Although the sloths have opted for a low energy life they are very successful in their forest homes. In some areas, half of the mammalian energy consumption is accounted for by sloths and two thirds of all the mammals, by weight, are sloths. In some closely studied forests, 8.5 sloths are found, on average in every hectare of forest. These successes are due not only to the sloths frugal way of life, but also its lack of predators. Only jaguars, harpy eagles and humans can catch and kill an adult sloth and the first can only get at them when they visit the ground. Humans can only shoot them, but they often hang on even when dead, making it difficult to retrieve the body. As protection against the harpy eagle, the sloth not only hangs lifeless for long periods, but is also camouflaged. The long, coarse hairs on its body have grooves inhabited by two species of cyanobacteria, giving the fur a greenish tinge, helping the sloth blend into the verdant foliage.

- There are two species of two-toed sloth: Hoffman's and Linné's and three species of three-toed sloth: brown throated, pale-throated, and maned. It is thought the sloths originated in South America about 35 million years ago. Their closest relatives are the armadillos and anteaters.
- Today there are two genera of sloth, but the prehistoric New World was home to 35 genera of sloth, many of which were exclusively ground dwelling animals. Some of

these animals were huge, as large as elephants, and constituted what is known as the American megafauna, almost all of which is now extinct. These huge sloths ranged from Alaska to parts of Antarctica.

+ The teeth of sloths do not have any enamel, making them very soft and of little use, except for chewing soft leaves.

+ Sloths give birth to a single young at a time, which has to cling to its mother for dear life. For a month or so the young is suckled by the mother, but after this time it begins eating whatever leaves it can reach. The young stays with its mother for six to nine months, but then it goes it alone and resides in a portion of territory left vacant by its mother. The type of leaves eaten by the youngster is inherited from its mother, enabling several sloths to cohabit in a similar home range.

+ Two-toed sloths have well developed canine teeth and these are often employed to good effect in disputes between males during the breeding season.

+ Only the maned sloth of south eastern Brazil is considered to be threatened due to the loss of its coastal rain-forest habitat. If destruction of the South American tropical forests continues unabated then the other species may soon be threatened.

STENUS ROVE BEETLES

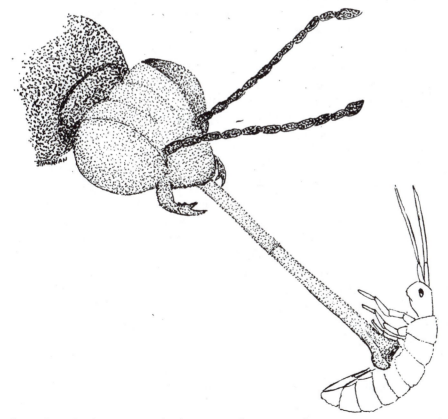

Stenus Rove Beetles—A stenus beetle uses its telescopic mouthparts to capture a springtail. (Mike Shanahan)

Scientific name: *Stenus* species
Scientific classification:
Phylum: Arthropoda
Class: Insecta
Order: Coleoptera
Family: Staphylinidae
What do they look like? These are very slender rove beetles with long legs and large, bulbous eyes. Typically, they are black, although some species have red spots on the wing-cases.
Where do they live? They are found throughout the world, often in moist habitats, or in the vicinity of water.

A Beetle with Tricks Galore

The undergrowth is home to all manner of minibeasts. It is a microcosm where danger lurks around every corner, prowled by small but fierce predators. It is an interesting place to say the least and among its most intriguing residents are the *Stenus* rove beetles. These insects, with their large eyes and thin bodies can be found during daylight moving gracefully on bare patches of ground or on aquatic vegetation looking for small invertebrates to eat. The way they catch and subdue prey is unique, because when in range of an appropriate target they don't just pounce or lunge, but maintain some distance between themselves and their quarry so they can employ their secret weapon. The lower mouthparts (the labium) of this beetle can be shot out from the head using blood pressure. The thin rod ends in a pad of bristly hairs and hooks and between these hairs are small pores, which exude an adhesive, glue-like substance that sticks the potential prey to the mouthparts of the beetle. The labium is then retracted, reeling in the food, bringing it in range of the sicklelike mandibles. This feeding strategy is particularly useful when the beetle is clambering around the vegetation and spies some prey, usually springtails. It is thought this amazing way of capturing food evolved due to the rapid reflexes of their springtail prey. The rod is so thin and can be shot with such speed that a springtail probably does not have sufficient time to react and hurl itself clear using its flexible tail.

Springtails are not the only animals with a surprise escape mechanism, as the *Stenus* beetles also have a remarkable means of eluding danger. These beetles are so small and light they will often take to water, their weight supported by the surface tension. On the water, they can scull along using their legs at a speed of 2–3 cm per second, however, at the slightest sign of danger they have a surprising turn of speed thanks to a substance called *stenusin* produced by their anal gland, a droplet of which is dabbed on the surface of the water. This compound is very hydrophobic, and it spreads with such force on the water the beetle is propelled forwards at a rate of 45–70 cm per second, which is considerable for an animal less than 5 mm long. This equates to 90–140 body lengths in one second (600–900 km/h in human terms). It is possible for the beetle to do a number of these super skims before its reserve of stenusin is exhausted.

> + There are more species of rove beetle than any other type of beetle, with perhaps the exception of the weevils. Approximately 50,000 species of rove beetle have so far been identified, but as with many animals in this book, many more are yet to be discovered and named. They live in a myriad of different ways, in a huge variety of habitats. Many species are carnivorous, while others are herbivorous. Some are scavengers and some, parasitic. There are also a few specialized species, which have taken to living with ants.

✦ Rove beetles are among the few beetles with wing cases (elytra) which do not cover their abdomens. Therefore, their abdomen is very flexible, enabling these insects to exploit microhabitats (i.e., tiny crevices in the soil) that are out of bounds to other, more rigid-bodied insects. This unusual morphology is probably the biggest reason why these insects are, arguably, the most successful beetles. Almost all of the rove beetles have wings and many are accomplished fliers, but the wings have to be folded away beneath the wing cases to prevent them from being damaged when not in use. This requires the skills of an origami expert to get the wings into the space available.

> ## 🔍 Go Look!
>
> Stenus beetles can be most easily found in waterside habitats, especially where there are bare patches of ground. Careful searching in these habitats may yield a specimen or two. The normal technique for catching land crawling insects is to use a pit-fall trap, which involves inserting a small beaker into the ground and then part filling it with a solution of a preservative, such as a mixture of water and glycerol. Animals foraging on the ground will fall into the trap. This technique can catch Stenus beetles, but they have excellent grip and will often walk down the sides of the container, take a sniff of the fluid and trot back out again. Also, any beetles falling into the solution will be killed. A dry pitfall trap can be used, but it must be checked regularly for specimens. Stenus beetles can be kept in small humid containers and fed with springtails. The way in which they catch springtails can be observed as can their ability to skim across the water if they have access to a tray of water.

✦ The short wing cases of the roves beetle leave the abdomen vulnerable to attack by predators. It is thought the lack of mechanical protection was instrumental in the evolution of an extensive range of compounds to deter potential predators. Stenusin is one such compound, and originally it may have been an odorous feeding deterrent, as it is known to be slightly toxic, but now its primary role is to provide an efficient way of evading danger or reaching land if the beetle falls into water.

Further Reading: Betz, O. A behavioural inventory of adult *Stenus* species (Coleoptera: Staphylinidae). *Journal of Natural History* 33, (1999) 1691–1712; Betz, O. Comparative studies on the predatory behaviour of *Stenus* spp. (Coleoptera: Staphylinidae): The significance of its specialized labial apparatus. *Journal of Zoology* 244, (1998) 527–44; Betz, O. Life forms and hunting behaviour of some Central European *Stenus* species (Coleoptera, Staphylinidae). *Applied Soil Ecology* 9, (1998) 69–74; Betz, O., and Fuhrmann, S. Life history traits in different life forms of predaceous *Stenus* beetles (Coleoptera, Staphylinidae), living in waterside environments. *Netherlands Journal of Zoology* 51, (2001) 371–93.

STOWAWAY FALSE SCORPION

Scientific name: *Cordylochernes scorpiodes*
Scientific classification:
  Phylum: Arthropoda
  Class: Arachnida
  Order: Pseudoscorpionida
  Family: Chernetidae
What does it look like? This is a large, dark brown false scorpion (4–5 mm long), and it has the large pincers typical of this group of animals.

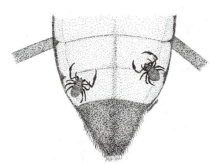

Stowaway False Scorpion—Stowaway false scorpions hitch a ride on the abdomen of a harlequin beetle. (Mike Shanahan)

Stowaway False Scorpion—Note the well-developed pincers and the short body, typical of these very interesting arachnids. (Ross Piper)

Where does it live? This animal is found in Central and South America and some of the Caribbean islands.

Why Walk When You Can Hitch

Animals have evolved a huge number of ways of getting around. Some fly, some walk and some swim. Others, however, have perfected the art of hitchhiking, relying on other species for a ride. The harlequin beetle, a large, strikingly colored insect from South America is one such ride. On the beetle's back, tucked underneath its wing covers, there are often several small false scorpions. What are these small arachnids doing in such a place? This story begins on a fig branch in the South American rain forest. Inside these branches live the larvae of the harlequin beetle, feeding on the wood. Scuttling around on the log are false scorpions feeding on small invertebrates, which they dispatch with their pincers. The population of false scorpions on the log may grow to a size where there is insufficient space and food; therefore, a new branch needs to be found, but how? These animals are very small and finding a new home in the vastness of the forest is made even more difficult by the fact they don't have wings. Fortunately, the solution to their problem is beneath their feet as the harlequin beetle larvae have completed their development and they are ready to emerge as adults from the branch. As soon as the adult beetle is fully out of its tunnel, the false scorpions gather around its rear end and nip its abdomen with their pincers. Annoyed by these pinches, the beetle flexes it rear, making enough space beneath its wing cases and wings for the little arachnids to jump aboard. A large male false scorpion will prevent other males from boarding the beetle, but will happily allow females on, in the hope of fathering their young when they reach their destination. In total, there may be as more than 30 false scorpion passengers on one beetle. The beetle takes off from the branch that nurtured it, along with its arachnid stowaways. Eventually, the beetle will land on a suitable branch and the passengers will disembark. The bullying male who shepherded the females will not father all of the young the females produce as they stored sperm from previous matings with the males who were left behind on their old branch.

 + False scorpions are amazing animals, but due to their small size they are often overlooked. They are rarely more than 8 mm in length and many species are much smaller. Over 2,000 species have been described throughout the world, but due to their size and secretive habits, it is highly likely that many more species are yet to be identified. Although they are rarely seen, false scorpions are actually very common and a single square meter of ground can be home to several hundred individuals.

+ They have the name false scorpion or *pseudoscorpion* because of their similarity to their relatives, the true scorpions. Unlike the true scorpions, however, these animals lack a sting-bearing tail. Like the true scorpions, they do have a well-formed pair of pincers, from which long sensory hairs project. Venom is produced by glands in the pincers and is used for subduing the prey of these animals (all pseudoscorpions are carnivorous). Silk can also be produced by the pincers.

+ Like other arachnids, pseudoscorpions must shed their skin in order to grow. To do this, some species construct a small silken igloo, which keeps them safe during this very vulnerable time.

+ Reproduction in these animals depends on the male producing a small sperm-filled stalk (spermatophore), which he sticks to the ground. He must then 'dance' with the female and lead her over his stalk. A substance in the female's genital tract makes the sperm mass swell up, helping the male's gametes into the body of his mate.

+ Using their pincers to get a grip, false scorpions use a whole range of large flies, wasps and beetles to get from A to B.

Further Reading: Weygoldt, P. *The Biology of Pseudoscorpions.* Harvard University Press, Cambridge, MA 1969; Zeh, J.A., Zeh, D.W., and Bonilla, M.M. Phylogeography of the harlequin beetle-riding pseudoscorpion and the rise of the Isthmus of Panama. *Molecular Ecology* 2, (2003) 2759; Zeh, D.W., and Zeh, J.A. On the function of harlequin beetle-riding in the pseudoscorpion, *Cordylochernes scorpioides* (Pseudoscorpionida: Chernetidae). *Journal of Arachnology* 20, (1992) 47–51.

TOKAY GECKO

Scientific name: *Gekko gecko*
Scientific classification:
Phylum: Chordata
Class: Reptilia
Order: Squamata
Family: Gekkonidae

What does it look like? The tokay gecko is a glorious specimen, and it is one of the largest living geckos. Males, the larger of the two sexes, can be up to 40 cm long and weigh as much 300 g.

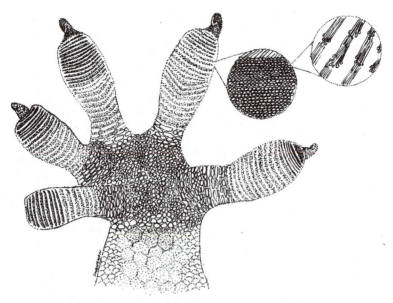

Tokay Gecko—The foot of the tokay gecko with insets showing the microscopic arrangement of bristles and their feathery ends. (Mike Shanahan)

Their body is flattened and the head is rather large with striking eyes. The big mouth is normally agape in what can only be described as a grin. Not only are they large geckos, but tokays are also very boldly patterned. They have small red/orange blotches all over their body against a background or green/blue.

Where does it live? The native range of these animals extends from India to New Guinea, including much of Southeast Asia. They are forest dwellers and can be found scrambling up the trunks of rain forest trees or scrambling around on rock faces.

Get a Grip!

The tokay gecko, with its large size, fierce temperament and distinctive call is one of the most familiar of the all the geckos. Like many geckos it has an amazing ability. With apparent ease it can clamber around on vertical surfaces or even upside down. No animal can really touch them for sure-footedness and for many, many years, the secret of their death defying stunts was a complete mystery. All that people could see for sure were what looked like tiny treads on the underside of the lizard's digits. Some people thought these little flaps were acting like little suction cups, enabling the gecko to stick to even the most slippery of surfaces. This was disproved when it was shown that even in a vacuum the lizard remains firmly attached. If the treads on the digits were sucking the animal on to a vertical pane of glass they would quickly become useless in a vacuum. Other people thought the gecko's secret was some adhesive produced by glands in the feet, in the same way a fly appears to mock the laws of physics by scuttling with poise on a ceiling. It was reasoned an adhesive would spread through the tread-like structures giving a large surface area of stickiness when the foot was planted. This too, turned out to be incorrect. The feet of geckos have no specialized glands for the production of an adhesive. Finally, in the year 2000, some scientists finally unraveled the gecko's mystery. Using very powerful microscopes they were able to look in minute detail at the underside of a gecko foot. What they saw was row-upon-row of tiny, flat hairs—2 million of them on a single toe. Cranking the electron microscope up to near the limits of its abilities, they were able to zoom in on the tips of these hairs. The end of ear hair branched out into hundreds of thousands of microscopic spoon-shaped structures, which the scientists likened to a florets of broccoli. These tiny structures are applied to the surface on which the gecko is walking and being so small they can be pressed right up against the fissures, lumps, and holes that are invisible to the naked eye of a human. Their contact with the surface is so intimate that can actually latch onto the molecular binding forces present in all the substances they walk on. The force they are exploiting is known as van der Waals force and although it is very weak and essentially invisible to us, there are enough tiny little spatulas on the gecko's feet to provide an excellent grip. The grip is so strong the reptile can hold all of its weight on one toe, but when it wants to lift its foot away from the surface all it has to do is roll its digits upward, thus breaking each of the microscopic attractions. Even more impressive is the fact this stickiness never grows weak. The hairs on the gecko's feet are self cleaning. Any dirt on the hairs is more strongly attracted to the surface the lizard is walking on rather than its feet. This incredible ability has given some people inspiration for a whole new generation of adhesives that are very strong and can be used over and over again.

- There more than 800 species of gecko. The largest species is only known from one stuffed specimen found in a museum basement in Marseille, France. It is known as Delcourt's gecko and an adult specimen was at least 60 cm long. It was native to New Zealand, but like much of the native fauna of this island it became extinct at the end of the nineteenth century. The smallest gecko is a mere 16 mm long, small enough to curl up on a dime. This species, the Jaragua Sphaero, was only discovered in 2001 on a small island off the coast of the Dominican Republic.
- The name *tokay* comes from the distinctive call this gecko makes.
- Although the tokay gecko is an impressive beast it doesn't make good pet. It is the most aggressive gecko and any attempt to handle them is usually met with its snapping jaws. Their bite can be painful and once attached they have a reluctance to relinquish their grip.
- All geckos are carnivorous, although some species supplement their diet with fruit, nectar and sap.
- As many species of gecko are naturally found on rocky outcrops, where their amazing climbing ability comes in very useful, they are more than at home around human dwellings as they climb around the walls and ceilings. Several species of gecko have become house specialists as their favored food, insects, are drawn to the lights around human dwellings.
- There are at least 12 species of gecko that reproduce without the need for a male. These parthenogenetic species are represented by females only as sperm isn't required to fertilize the eggs.
- The clutch size of geckos is small, the typical number of eggs being two. The small species, possibly due to their tiny bodies are restricted to one egg at a time.

Further Reading: Autumn, K., Liang, Y., Hsich, T., Zwaxh, W., Chan, W.P., Kenny, T., Fearing, R., and Full, R.J. Adhesive force of a single gecko foot-hair. *Nature* 405, (2000) 681–85; Autumn, K., Sitti, M., Liang, Y., Peattie, A.M., Hansen, W.R., Sponberg, S., Kenny, T., Fearing, R., Israelchvili, J.N., and

Full, R.J. Evidence for van der Waals adhesion in gecko setae. *Proceedings of the National Academy of Sciences* 99, (2002) 12252–56.

WHITE WORM LIZARD

White Worm Lizard—A white worm lizard raises its tail and head off the ground to confuse predators as to which end is its head. (Mike Shanahan)

White Worm Lizard—A white worm lizard removed from its subterranean lair. The grooves, encircling its body and giving it the appearance of a large worm, are clearly visible. (Laurie J. Vitt)

Scientific name: *Amphisbaena alba*
Scientific classification:
    Phylum: Chordata
    Class: Reptilia
    Order: Squamata
    Family: Amphisbaenidae
What does it look like? Amongst amphisbaenians, the white worm lizard is one of the largest species reaching a length of approximately 72 cm. The animal is commonly yellowish-white, although a great deal of variation in color exists.
Where does it live? This species has a wide distribution across central and northern South America and occurs on some islands in the Caribbean, such as Trinidad. It is found both in forested (e.g., Amazon rain forest) and open (e.g., Cerrado of Brazil and Llanos of Venezuela) areas and is often associated with leaf-cutter ants.

The Subterranean Life of a Legless Lizard

The white worm lizard spends almost all of its time below ground and on the island of Trinidad, they have occasionally been found in the deep galleries of leaf-cutter ant nests, particularly the garbage tips, where the ants deposit their waste and their dead. What are the reptiles doing in these subterranean rubbish tips? They are looking for food. This reptile seems to be particularly partial to the larvae of a large beetle (*Coelosis biloba*), which feeds on waste and detritus in the ant's rubbish pile safe from most predators, but the worm lizard with its long, thin body and tunneling abilities can easily find these plump grubs. The relationship between the worm lizard and beetle also involves a small parasitic crustacean, known as a tongue worm. These live in the respiratory tract of the worm lizard, its definitive host. To infect the reptile it must use another animal, the intermediate host, in this case the beetle grub. The tongue worm's eggs are found in the feces of the worm lizard, which the beetle grub inadvertently ingests. The worm lizard then devours the infected beetle larva reinfecting itself with the tongue worm parasite and the cycle is complete!

The white worm lizard is a large animal, so before long it may exhaust the supply of beetle larvae in one leaf-cutter ant nest and have to search for more larvae in other nests. To do this it leaves its familiar subterranean habitat and heads for the surface. Due to its fossorial nature, the animal's eyesight is almost nonexistent, so it must employ an alternative sense to find another ant nest. It uses its acute sense of smell to detect the pheromone trails left by worker leaf-cutter ants as they forage along the forest floor. These trails are markers for their nest mates to follow and they indicate the way back to the nest. The reptile finds and follows one of these trails to its source. It enters the nest and makes for one of refuse tips deep underground to continue its search for the fat beetle grubs.

- Approximately 140 species of worm lizards are known, distributed across tropical and subtropical regions of the globe. Consequently, most are in the southern hemisphere. A single species occurs in the United States and only one occurs in Europe. They range in size from 9–12 cm to the 70 cm white worm lizard and they are all superbly adapted to a fossorial way of life, rarely venturing to the surface, although rainfall and flooding will force one of these animals from its burrow. Due to their secretive way of life, very little is known about them. It is likely that many more species remain to be discovered.

- The body of the worm lizard displays a number of amazing features, some of which are adaptations to a subterranean lifestyle. As in other, fossorial animals, many of the bones of the skull are fused, making it particularly strong for ramming a tunnel through the soil. Some species have a spade shaped head for compacting tunnel walls by moving their head up and down, while others have a keel shaped head used to compact their tunnels with a side-to-side motion. Many species have a bullet shaped head used to forge a tunnel through the soil. Also, the skin of a worm lizard is very loosely attached to its body, enabling the animal to 'slide' through its integument. This helps the worm lizard move backward just as well as it can move forwards, which is important in the confines of a tunnel.

- The Latin name *Amphisbaena* means to "move in both ways" reflecting the animal's ability to move forward and backward equally well. Also, the term *amphisbaena* is used to describe a venomous creature of myth and legend, said to have a head at both ends of its body. Interestingly, when a worm lizard is on the surface and finds itself threatened it will raise both ends of its body toward the sky. This defensive posture confuses the predator as to which is the real head. The head end with its powerful jaws is capable of inflicting a powerful and painful bite.

- The vast majority of worm lizards are completely legless. However, one peculiar group has retained its front legs, which are strong and molelike, an obvious adaptation for burrowing.

- The benign appearance of the worm lizard belies its voracious, predatory ways. They will eat any animal they find in their tunnels and will even make forays to the surface to forage for suitable prey. Their dentition is unique, with two large teeth in the upper jaw and one in the lower jaw, meshing together to form an effective pair of pincers.

◆ The reproductive habits of worm lizards, like many other aspects of their lives, are very poorly known, but some species lay eggs, while others give birth to live young. The white worm lizard lays 8–16 eggs, primarily during the tropical dry season. Nothing is known of their population densities as they are so difficult to find. The worm lizards are, without doubt, among the most enigmatic land animals.

Further Reading: Colli, G.R., and Zamboni, D.S. Ecology of the worm-lizard *Amphisbaena alba* in the Cerrado of Central Brazil. *Copeia* (1999) 733–42; Riley, J., Winch, J.M., Stimson, A.F., and Pope, R.D. The association of *Amphisbaena alba* (Reptilia: Amphisbaenia) with the leaf-cutting ant *Atta cephalotes* in Trinidad. *Journal of Natural History* 20, (1986) 459–70.

5

LOOKING OUT FOR THE NEXT GENERATION

BEE WOLF

Bee Wolf—A female bee wolf carrying a honeybee at the entrance to her elaborate nest. (Mike Shanahan)

Bee Wolf—An exceptional, close-up image of an adult female bee wolf. (Roger Key)

Scientific name: *Philanthus triangulum*
Scientific classification:
  Phylum: Arthropoda
  Class: Insecta
  Order: Hymenoptera
  Family: Sphecidae
What does it look like? Superficially, the bee wolf resembles a common wasp as it has the same color pattern (black and yellow) and is approximately the same size.
Where does it live? The bee wolf is found in habitats with sparse, short vegetation where the soil is sufficiently sandy to allow digging. The climate is also a crucial prerequisite as

bee wolves need comparatively dry summers and occur mostly in places with less than 700 mm rain per year. This insect is distributed from Spain in the west to Afghanistan in the east, and from Norway in the north to South Africa in the south. This is about the same distribution as their prey, the (western) honeybee.

The Lengths a Mother Will Go To

Many of the bees and wasps go to great lengths to make sure their young have the best possible start in life. The bee wolf is a perfect example of this as the female goes to great lengths to provision her nest with food for the next generation. The adult female feeds on nectar and pollen to provide her with the energy needed for flight, but to provide for her young, she must stalk and capture unfortunate honeybees. Her prey is often distracted gathering nectar and pollen from flowers, and upon sighting suitable quarry, she flies straight toward the honeybee and knocks it from its flower. On the way toward the ground, the bee wolf stings the honeybee in the weak membrane joining the plates of its thorax, injecting a paralyzing venom. While the honeybee is tangled in the vegetation or wrestling on the ground, the venom acts quickly, and when the prey is safely immobilized, the bee wolf grabs it with her middle legs and takes off back to her nest.

The nest of the bee wolf is a lavish affair. The subterranean lair is dug in sandy soil by the female using her front legs and mandibles. The tunnel can be as much as 1 m long, a considerable distance for an animal that is no more than 16 mm in length. The first part of the tunnel slopes downward at an angle of 30°, after which there is a long horizontal section. Up to 34 lateral tunnels ending in a brood chamber are excavated horizontally on each side of the main tunnel. The female bee wolf stocks each brood chamber with one to six paralyzed honeybees and a single egg that hatches into a hungry larva. The bees are smeared with a secretion from glands on the female's head, and only recently has the reason for this behavior been discovered. The nest of the bee wolf is warm and humid—a great place for the growth of fungi and bacteria—and without some means of preservation, the larva of the bee wolf would be left with little more than a mound of rotting insects. The secretions of the bee wolf act like embalming fluid and smooth all the tiny nooks and crannies on the bee's body that offer a great place for the condensation of water. Without these minute condensation nuclei, droplets of water cannot form on the bee and its body stays dry—protected from the ravages of bacteria and fungi, which require moisture to survive. Not only this, but the female bee wolf also secretes a whitish substance from its antennae that contains symbiotic bacteria. These bacteria are taken up by the larvae and spun into the walls of the cocoon when it is preparing to pupate. The bacteria protect the pupa from potentially harmful fungi, especially during the long period of hibernation.

With the brood chamber fully stocked the female daubs the white bacteria paste from her antennae on the ceiling of the chamber. The chamber is then sealed. The white microorganism marker left by the female indicates where her fully developed offspring should dig to escape from the brood chamber. This is important, as the entrance of the chamber was filled in by the female, as was the outer entrance. The fully developed wasp cannot depend on light to orientate itself, so the marker prevents it from wasting time looking for the nest exit in the wrong place.

> ♦ The bee wolf larvae hatching from the eggs have a voracious appetite. In less than two weeks, they consume all of the food left by their mother. Ready to pupate, they spin themselves a bottle-shaped silken cocoon, which is attached at its thin end to the wall of the chamber.

+ A single bee wolf female can stock its nest with well over 100 honeybees. In many areas of their range, they are the only animal preying specifically on the honeybee.
+ The substances found in the head gland secretions of the bee wolf are potent anti-microbial agents, and they are being actively investigated as starting points for new types of antifungal and antibacterial compounds. It is possible that many wasp species produce substances like the ones produced by the bee wolf. Since they often store food in nests for long periods of time, the prevention of decay is very important.
+ There are close relatives of the bee wolf that specialize in preying on other insects. One interesting species hunts weevils on the ground or in short vegetation, while another hunts flies, which are paralyzed with venom before being flown back to the nest after being impaled on the female wasp's stinger.
+ In the United Kingdom, this species was once thought to be very rare and was known to inhabit only a handful of sites in the south of the country. In recent years, however, the range of the animal has exploded, and it can now be found in suitable sites in all of southern England, which is bad news for honeybees! It is thought that this range expansion is direct evidence of global warming. The vast majority of insects are warmth loving and will therefore respond to even slight increases in temperature.

Further Reading: Kaltenpoth, M., Göttler, W., Herzner, G., and Strohm, E. Symbiotic bacteria protect wasp larvae from fungal infestation. *Current Biology* 15, (2005) 475–79; Strohm, E., and Linsenmair, K. E. Females of the European beewolf preserve their honeybee prey against competing fungi. *Ecological Entomology* 26, (2001) 198–203.

BLUE WHALE

Blue Whale—An adult blue whale compared in size to a pick-up truck. (Mike Shanahan)

Scientific name: *Balaenoptera musculus*
Scientific classification:
   Phylum: Chordata
   Class: Mammalia
   Order: Cetacea
   Family: Balaenopteridae
What does it look like? This is a huge animal with an elegant, tapering body topped off with a snoutlike rostrum and ending in a large tail fluke. It is a mottled blue color with a yellowish

underside. The massive mouth can be made even more cavernous thanks to a series of long pleats along the throat.

Where does it live? The blue whale is found in all the world's oceans. Outside the breeding season, they are found in the high latitudes of their feeding grounds, but in the breeding season, the females swim to tropical and subtropical regions in order to give birth.

Milk Makes You Big and Strong

The blue whale has to be included in this book for the simple reason that it is so massive. There are frequent claims from fossil hunters that the dinosaurs and their kin could have attained huge proportions, but it is very doubtful that any animal has ever exceeded the unbelievable bulk of this mammal. The blue whale starts its life as a calf, 9 m in length and a hefty 3 tonnes in weight—the same weight as a fully grown hippopotamus. Even from birth the race is on to amass the great weight of the adult, and so the calf begins suckling in earnest, swallowing around 200 L of milk per day. The term *milk* is used loosely as the substance produced by the mammary glands of the female is more like runny cheese and contains 35–50 percent fat. On this highly nutritious diet, the calf grows faster than any other young animal. Every day, its weight increases by 90 kg (4 kg every hour), and its length by 4 cm. The female cannot sustain this level of guzzling indefinitely as her weight plummets by about 50 tons during the suckling period. At seven months old, the calf is weaned off the milk onto the adult diet of the small shrimp known as krill. Even at the tender age of seven months, the calf is 16 m long and weighs 23 tons. This is also the time when the females and calves part company, but both head back to the feeding grounds in the northern or southern latitudes where the red krill can be found in immense swarms.

Blue whale feeding is far from subtle. They simply swim through these crustacean clouds and open their mouth as they near the surface, gulping up thousand of liters of water and millions of the small crustaceans. To maintain its huge bulk, the whale needs gigantic helpings. It has been calculated that an adult blue whale requires at least 1.5 million calories a day (a human needs 2,500), but as they only feed intensively for part of the year, this figure is more like 3 million calories a day. This energy requirement translates into 40 million little crustaceans per day, or something on the order of 4 tons. At any one time, the whale's stomach can hold between 1 and 2 tons of krill!

Even after its initial growth spurt, the blue whale continues to grow rapidly. A female reaches sexual maturity at the age of five, at which point she is around 24 m long. Fully grown, the females are around 2 m longer than the males, and the size they can attain is a contentious subject. The most reliable measurement is for a female caught by Japanese whalers in the 1946–47 season. She was 29.9 m long, but it is almost certain that the biggest specimens were caught a long time before this. A length of well over 30 m is certainly within the realms of possibility. Weight is even more difficult to estimate. It is exceptionally difficult to weigh a marine animal of this size. The only measurements come from the days of whaling when they were butchered on shore and each piece was weighed separately, giving weights of 160–190 tons—the same as about 26 fully grown African elephants. These numbers did not take into account the blood and other fluids that were lost when the animal was cut up. It is likely that the largest blue whales weighed more than 200 tons—considerably more than a fully laden Boeing 737.

- ♦ The family of whales to which the blue whale belongs includes five other species: fin whale, sei whale, humpback whale, Bryde's whale, and minke whale. They all have the long throat pleats that make the mouth roomier during feeding gulps. There is still a great deal to learn about these animals because, although they are so huge, these whales swim fast and range over thousands of square kilometers of ocean, making it difficult to find and study them.

+ All of these whales feed in the same way, trapping huge quantities of food-laden water in their bulging mouth and using their massive, elephant-sized tongue to squeeze the water through hanging fringes of baleen plates—made from keratin, the same material found in mammalian nails and hair. The baleen acts like a sieve and traps the food as the water is squeezed out.

+ The vital statistics of the blue whale are mind-boggling. Much of its great weight is blubber and muscle. The skeleton is actually quite simple and light. The heart is as big as small car, pumping blood around lungs weighing more than 1 ton. The lung capacity is probably on the order of 2,000 L; compare this to the 4–5 L for an average human. Used air is expelled, as a high-pressure cloud of carbon dioxide and water vapor, over 9 m into the air when they surface to breathe through a nostril wide enough for a toddler to crawl through. The blue whale, like other large marine mammals is a very efficient breather. Each time it breathes, 80–90 percent of the air in its lungs is exchanged, compared to only 10–15 percent for mammals like humans.

+ Not only is the blue whale the largest animal, it is also the loudest. The sounds they produce are very low-frequency and beyond the range of human hearing, but they have been measured at around 190 decibels. Sound travels very well through water, and these very loud moans are probably audible for thousands of miles. It is possible that populations of blue whales separated by thousands of miles of ocean can communicate with one another. These sounds were only detected in the 1950s, when military hydrophones were installed beneath the waves to listen for enemy submarines. It was only in the 1960s that these sounds were found to be coming from blue and fin whales. The complete function of these sounds is not fully understood.

+ During the early years of commercial whaling, the blue whales were nothing but a majestic sight for the hunters with their sailing ships and hand-held harpoons. They were too fast to be caught, and even if they could be caught and killed, their corpses would sink. It was only toward the end of the nineteenth century when various enterprising Scandinavians devised the means of pursuing, killing, and keeping hold of these gigantic animals that the whalers could turn their attention to the blue whale. With the tools at their disposal, the whalers set about hunting the blue whales remorselessly. The slaughter is startling and peaked in the season of 1930–31 when more than 28,000 blue whales were killed. In total, well over 350,000 were killed and butchered in the whaling years, and 90 percent of these were taken in Antarctic waters. Today, a few thousand blue whales survive. The huge population of blue whales in Antarctic waters probably numbered 250,000 animals, but by 1963, it was found to number no more than 2,000 individuals. Fortunately, this plight ensured their protection, and today, the populations in some areas are increasing, but other small groups have been lost forever. As whaling targeted the largest animals, the individuals in the oceans today are far below the maximum attainable size of this whale, and it will be many, many decades before the oceans are once again graced by truly immense blue whales.

Further Reading: Calambokidis, J., and Steiger, G. *Blue Whales.* Voyageur Press, Osceola, WI 1997.

BURYING BEETLES

Burying Beetles—A pair of burying beetles tend to their young on the small mammal corpse they have smoothed to a ball. (Mike Shanahan)

Scientific name: *Nicrophorus* species
Scientific classification:
  Phylum: Arthropoda
  Class: Insecta
  Order: Coleoptera
  Family: Silphidae
What do they look like? These medium to large beetles are 15–30 mm in length. They are black, and many have orange markings on the wing cases, which do not cover the full length of the abdomen. The head is large, with bulbous eyes and a pair of clubbed antennae. The legs are long and sturdy.
Where do they live? These beetles are found in various undisturbed habitats, including coniferous and deciduous woodlands, open grasslands, and marshes. They are animals of temperate and alpine areas, and there are also some species in North Africa.

Nature's Grave Diggers

The burying beetles are accomplished scavengers, and they are some of the most dedicated parents in the insect world. To breed, they must locate the body of a dead animal, such as a small bird or mammal, which they accomplish by scent. They are attracted to these corpses by the smell of decay, which they sense using their antenna. These odors can be detected over long distances, up to 8 km, and as soon as the beetle gets a whiff of them, it takes to the air and flies to the source. Should a male beetle locate a corpse and find a female of his species already there, the two will mate, but if there is no female, he will adopt a posture on top of the carcass and emit a sex pheromone to attract a mate. Other burying beetles will also be attracted to the carcass, but a newly mated couple will have to repel them. The pair must work quickly to hide the body before other animals are attracted to it. They eagerly burrow around and beneath the body, excavating a small pit, which the animal eventually drops in to, disappearing from view and sometimes descending up to 60 cm in to the ground. With the carcass hidden in the confines of the burial chamber, the female deposits her eggs in the soil just above the dead body. Both beetles climb all over the small dead body, removing fur or feathers and smoothing it until it looks like a featureless lump of matter. Before the eggs hatch, the prospective parents nibble a hole in the corpse and coat the resulting pit with a substance from their anal glands. The larvae emerge from their eggs and are attracted by the odor of the small dead animal and head straight for the main burial chamber, where they climb on to the carcass to begin feeding in the little pit excavated by the male and female. The parents feed the larvae with chewed up flesh, a behavior that is a necessity in some species as the larvae can't survive otherwise. In other species, the larvae are able to feed by themselves with a little help from their parents. The male and female will also kill some of their offspring if there is not enough food to go around. The larvae feed voraciously and grow rapidly until the carcass provisioned by the parents is just a pile of bones. Fully grown, the larvae leave the remains of their larval food and burrow into the surrounding soil where they pupate, to emerge the next season as adults, thus completing the cycle.

- There are approximately 75 species of *Nicrophorus* burying beetles worldwide. Along with their close relatives, they specialize in locating and using the dead bodies of small animals. They are thought to be closely related to the rove beetles, many of which are also attracted to the bodies of dead animals, but only to feed on the other insects they find there.
- As with animal dung, animal carcasses are precious resources, representing an abundance of organic matter that can be used by a bewildering variety of organisms. The supply of small corpses is not predictable; therefore, the animals that use them must be able to detect the faintest sign of one and get to it as quickly as possible. The burying beetles will not only have to compete with others of their own kind but also with other beetles, flies, bacteria, and fungi. Other beetles can be chased away, the eggs of flies can be removed from the carcass, but bacteria and fungi present a different problem. They are too small to be chased off or removed, but the beetles have a secret weapon to deal with these microorganisms. It is thought the anal secretion of these beetles contains substances that prevent or stop the growth of bacteria and fungi. This is essential to ensure that the flesh of the corpse is not spoiled before the burying beetle's offspring get a chance to feed.
- These beetles often have very bold, orange markings on their wing cases. This is known as *aposematic coloration* and is a warning to predators that the animal displaying the bright pattern either tastes disgusting or has a painful sting. Many insects broadcast their defenses in such a way. The burying beetle's defense is its distastefulness.
- Like many large, flying insects, burying beetles are often used for transport by mites. These mites cling to the flanks or underside of the beetle, shunning the light, and will

Go Look!

These beetles are rarely seen, but they are common, and there are several ways in which they can be found. The first method is a light trap, which is typically used to attract moths. As the burying beetles are mainly active after dark and probably orientate themselves by the moon, a light trap can be used to lure them. The trap, consisting of a special, bright bulb and collecting box, can be set up in various habitats, such as woodlands, open areas, or scrub, and left running throughout the night. In the morning, the trap can be opened, and at the bottom, there may be some burying beetles. If you pick the beetles up, you will notice their very pungent odor, which potential predators must find unpleasant. Look closely at the beetle, and you will see the mites that scuttle all over its body. Burying beetles are active animals and strong fliers, and as soon as they are released, they will take flight.

Another way to catch burying beetles is by using a small pitfall trap baited with meat, such as chicken or fish. The beetle will smell the decaying meat and fly to its source, eventually ending up in the plastic container you have partially buried in the ground. The trap should be checked regularly so that any captured animals can be released.

move over the beetle as one large group, resembling an orange patch sliding over the insect. Exactly what these mites do is unknown. They are obviously associated with the burying beetle, as all individuals carry them. It is possible they live in the nest of the burying beetle feeding on the scraps left by the larvae and get into a new nest by hitching a ride.

- Few insects demonstrate the level of parental care shown by the burying beetles. Even more remarkable is the involvement of the male. The males of almost all insects have nothing to do with their young.

Further Reading: Milne, L. J., and Milne, M. The social behavior of the burying beetles. *Scientific American* 235, (1976) 84–89; Scott, M. P. Competition with flies promotes communal breeding in the burying beetle, *Nicrophorus tomentosus*. *Behavioural Ecology and Sociobiology* 34, (1994) 367–73; Scott, M. S. The ecology and behavior of burying beetles. *Annual Review of Entomology* 43, (1998) 595–618; Trumbo, S. T. Reproductive success, phenology, and biogeography of burying beetles (Silphidae, Nicrophorus). *American Midland Naturalist* 124, (1990) 1–11.

FIG WASPS

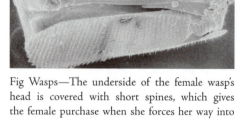

Fig Wasps—A sequence showing a female fig wasp on the outside of the fig, squeezing down the fig's entrance and inside, pollinating the flowers. (Mike Shanahan)

Fig Wasps—The underside of the female wasp's head is covered with short spines, which gives the female purchase when she forces her way into the fig. (Steve Compton)

Scientific name: Agaonids
Scientific classification:
  Phylum: Arthropoda
  Class: Insecta
  Order: Hymenoptera
  Family: Agaonidae
What do they look like? Fig wasps are small insects. The females are winged with flat heads, while the males are wingless with pale, relatively soft bodies.
Where do they live? These insects are found in tropical and subtropical areas, always in association with fig trees.

Close Ties

Figs are a spectacularly successful group of flowering plants, numbering about 900 species, many of which dominate tropical terrestrial ecosystems. Fundamental to the success of this group of plants is the unique way in which they are pollinated. A flower is an advertisement for pollinators, typically insects. The shape of the flower normally gives it away, but with the figs, the flowers develop inside the growths coveted by humans and other animals as food. What could be the pollinator in this very odd arrangement? Over millions of years, the life of the fig tree and the life of a number of small wasps have become inextricably entwined. The majority of fig trees have their own dedicated species of tiny pollinating wasp that fertilize the bounteous fruit of a fig tree (a single fig tree may produce up to 1 million figs). These wasps are so small they could easily crawl through the eye of a small needle. The development of the figs on one particular tree is often synchronized, and when they are young and small, they give off a particular odor that attracts females of its pollinator wasp from far and wide. The wasp, after following this scent to its source, alights and enters the fig through a small hole surrounded by scales—a wasp turnstile that admits only the right pollinator species. It is a tight squeeze, and the female tears off her wings and antennae as she squeezes along the tunnel, inching her way along with backward pointing teeth on her head and limbs. This is often a one-way ticket—she may never leave the fig once she has fought her way inside. The fig is actually a thick-walled chamber, the sides of which sprout tiny flowers. In some fig species, there may be only female flowers in the chamber, whereas other species always have male and female flowers in the same fig. Many female fig wasps have a pouch on their underbelly that stores pollen taken from the fig they developed in. The female pollinates the female flowers, diligently sprinkling a few grains on each flower visited and laying eggs in some of them, threading her egg-laying tube into the thin neck of the bloom. This is her final act, and she dies soon after. A month passes, and some of the fig flowers are ruptured by the escape of the female wasp's first offspring. These are all males, and they are pallid and wingless. They crawl around the fig chamber with heavy heads and well-developed limbs. Their first act is to mate with their sisters that are still developing in some of the other flowers of the fig. The male wasps then use their well-developed mandibles to chop a tunnel from the chamber toward the surface of the fig. Sadly for the males, their life in the dim half-light of the fig chamber is short, and when the excavations are complete, they grow weary and die. Shortly after, the female wasps emerge, with their own young already developing inside them. They work their way around the chamber, either carefully collecting pollen from the male flowers that have sprouted or being accidentally dusted by the pollen. Using the hole chewed by the males to escape, they then fly away from their birth fig. The heady tropical air will be thick with all manner of odors, yet only one will grab the attention of the freshly emerged females, and that is the scent given off by one species of developing fig as it signals its readiness to its dedicated pollinator.

+ The relationship between the wasp and the fig tree has been at least 80 million years in the making. It is one of the closest in nature. Neither the fig nor the wasp can live without one another.

+ The finer points of this complex relationship are still to be understood. It appears that some species of fig tree are pollinated by more than one species of fig wasp.

+ Any self-respecting lazy wasp would take advantage of the refuge for its young provided by the fig without returning the favor and pollinating the plant. After all, the wasp has to go to a lot of trouble to actively pollinate the fig. What stops the wasp from cheating? It has been found that unfertilized figs can be more likely to fall from the tree than fertilized ones and that this would spell the end for the offspring of any cheating gall wasps. Even if the unpollinated figs do not fall, it is likely they will produce fewer wasp offspring.

+ These tiny insects and their role as the sole pollinators of fig trees make them a key component of tropical terrestrial ecosystems. Countless animals rely on fig trees, not only for their nutritious fruit but also for the habitats they provide. After the female fig wasps have emerged, the seeds develop and the whole fig ripens, becoming attractive to birds, monkeys, and a raft of other tropical animals.

+ As with any other relationship on the planet, the fig/fig wasp symbiosis is exploited by various other animals. There are many other species of wasp that prey on the fig wasps. The females of these parasitic wasps have long, needle-like ovipositors, which they use to introduce their eggs into the developing larvae of the fig wasps. They perch on the outside of the fig, and with the accuracy of a sniper, pick out the fig wasp larvae in the multitude of fig flowers using the sensitive tip of their ovipositor.

Further Reading: Cook, J. M., Bean, D., and Power, S. Fatal fighting in fig wasps—GBH in time and space. *Trends in Ecology and Evolution* 14, (1999) 257–59; Cook, J. M., and Lopez-Vaamonde, C. Figs and fig wasps: evolution in a microcosm. *Biologist* 48, (2001) 105–9; Cook, J. M., and West, S. A. Figs and fig wasps. *Current Biology* 15, (2005) 978–80; Ronsted, N., Weiblen, G. D, and Cook, J. M. 60 million years of co-divergence in the fig-wasp symbiosis. *Proceedings of the Royal Society (Series B)* 272, (2005) 2593–99.

KING COBRA

Scientific name: *Ophiophagus hannah*
Scientific classification:
  Phylum: Chordata
  Class: Reptilia
  Order: Squamata
  Family: Elapidae
What does it look like? Fully grown king cobras can reach almost 6 m in length, although the typical size is more like 3–4 m. Their color varies, and adults can be yellow, brown, green, or black. The adult's underside is typically light in color and sometimes sports dark bars. The youngsters are black with light-colored bars on their head and body.

King Cobra—A female king cobra rearing up to warn intruders away from her nest. (Mike Shanahan)

King Cobra—Here, the world's largest venomous snake adopts a threatening posture by rearing up and flattening its neck into the distinctive hood. (Rhett Butler)

Where does it live? The king cobra is found throughout much of India, and its range extends into southern China, Malaysia, and the Philippines. It is essentially a forest animal, preferring tree cover, bamboo thickets, and mangrove swamps. It has been found in mountainous regions at altitudes in excess of 2,000 m. It is often found near water.

A Dedicated and Venomous Nest Guardian

As its name implies, the king cobra is a very majestic beast. It is the largest venomous snake by far, and because of its size and fearsome appearance, it has always been surrounded with superstition and is rather legendary in its native range. Although the king cobra does not have the most potent venom of all the snakes, it more than makes up for that with the huge volumes of venom it can inject in a single strike. The sheer quantity of venom an adult king cobra injects through its 10 mm fangs is enough to bring down a fully grown Asian elephant, or to put it another way, 20–30 adult humans. The toxin affects the nervous system of the victim, quickly causing respiratory failure. Even with its nasty nip, the king cobra is not an aggressive snake. Only when it is feeding or feeling threatened will it resort to using its venom. At other times, it slips quietly into the undergrowth out of the way of approaching trouble. The only time a king cobra will stand its ground is when it is guarding its nest, as females in particular are dedicated parents. The nest is quite an impressive structure especially for an animal without limbs of any kind. Using the coils of her long body, the female scoops up leaf litter to make a mound into which she deposits a clutch of 20–40 eggs. The mound of rotting plant material acts like an incubator speeding the development of the embryos within their leathery shells. During this time, the female is a very attentive mother, staying on or near the mound for the whole incubation period. Any trespasser is treated to an impressive defensive display. The female rears up until her head may be almost 2 m off the ground. Her neck flattens into a hood, and she emits a high-pitched hiss.

The male may also help guard the nest, but this is rare and only extends to him loitering in the vicinity. The female may be a tenacious guard of the nest, a behavior almost unknown in snakes, but as soon as the young writhe from their shells, they have to fend for themselves in the big, wide world with its many dangers. For the first part of their lives, the young snakes consume any prey they can get their elastic jaws around, but as they grow, so do their prey items. The favored prey of an adult king cobra is other snakes. Rat snakes are often taken, but pythons up to around 3 m in length may be subdued and eaten.

- The king cobra injects an impressive amount of venom, but it is less than that injected by the gaboon viper, a thickset snake of sub-Saharan Africa. The gaboon viper also has the longest fangs of any snake (up to 5 cm!)
- Snake venom is no less than a wonder of the animal kingdom. It is produced by modified salivary glands, which, over the eons, have evolved to produce a cocktail of different chemicals, the effects of which depend on the snake species. The venom can be used to subdue prey or as a defensive weapon, and it is squeezed from the glands along the jaw and out through modified teeth called fangs. The fangs are hollow, and in some species, such as the vipers, they are so big that they have to be hinged. Some snakes produce venom that attacks the nervous system of prey, leading to rapid death, while others produce venom that degrades and kills whatever tissue it comes into contact with. These substances start digesting the prey from the inside out, even before the snake begins the elaborate process of swallowing its meal.
- Snake venom is so diverse and so potent because snakes are sadly lacking when it comes to other weapons. They don't have any limbs with sharp claws. The lack of limbs also makes snakes quite slow-moving animals, so their poisons have to be fast acting enough to kill prey before it has a chance to get a good distance from the reptile.
- The king cobra is an active predator and will go out looking for its prey instead of using the ambush tactic. To help it locate its prey, the king cobra has very good eyesight and can spot its prey from distances of around 100 m.
- Traditionally, the king cobra was revered in India and held in such high respect that villages were often vacated if a nesting female had taken up residence. Times change, however, and the king cobra is increasingly regarded as a menace, but it can still be encountered fairly commonly in this subcontinent.

Further Readings: Mehrtens, J. M. *Living Snakes of the World.* Blandford Press, Dorset 1987; Seigel, R. A., and Collins, J. T. *Snakes: Ecology and Behavior.* McGraw-Hill, New York 1993.

MALLEEFOWL

Scientific name: *Leipoa ocellata*
Scientific classification:
   Phylum: Chordata
   Class: Aves
   Order: Galliformes
   Family: Megapodiidae
What does it look like? Adult malleefowl are large birds, about the same size as a domestic chicken. They are around 60 cm long and can be 2.5 kg in weight. The head is small with a small bill, and the plumage is in muted tones with white, black, and grey barring.
Where does it live? The malleefowl is only found in Australia. Its range extends from southwestern Australia to central New South Wales. In this range, it is restricted to semiarid scrub and woodland dominated by the mallee eucalyptus.

Malleefowl—A section through a malleefowl nest mound also showing the male bird scraping away soil and leaf litter to regulate the mound's temperature. (Mike Shanahan)

Build Your Own Incubator

The malleefowl is a shy, retiring bird of the Australian outback. During winter in the Southern Hemisphere, changing day length gets the malleefowl in the mood for courtship. Males survey their territory for a suitable site to build a nest. Unlike the rickety twig nest of many birds, the malleefowl's nest is a real feat of avian engineering. In a suitable site where the soil is deep, the male scrapes at the ground with his beak and feet. These birds are not really built for digging, nonetheless they are tireless, and before long, they have excavated a substantial pit, which can be 3 m across and 1 m deep. When the pit is deep enough, the male begins the laborious task of dragging material into the pit. Depending on the location, this may be leaf litter, sand, or a mixture of the two. He drags and scrapes until he has formed a mound more than half a meter high. Winter in this location is the time for rain, and following a decent downpour, he will root among the nest mound to mix it. Turning the material over encourages decomposition, and he constructs a next chamber within the mound. His mate, which up until this point has been rather lazy, comes to inspect his work and deposits a clutch of eggs in the mound chamber. Over the next few days, the female lays eggs in more chambers eagerly prepared by the male until she has laid between 15 and 24, although a fecund female may produce 30. The eggs are large, and when she has finished laying, she may have produced more than 2.5 times her own body weight in eggs.

With the eggs in place in the mound, the expectant parents have constructed themselves a large, organic incubator. The decaying plant matter in the mound and the heat of the sun warm the nest, nurturing the developing brood. The male is a devoted father, tending to the nest mound regularly, constantly making sure the temperature is correct for the eggs. If the nest gets too hot, he will open it up, and if it gets too cool, he will add more material. After the male has tended the giant compost heap for anywhere between 50 and 100 days, the young are ready to hatch. They break out of the thin-shelled eggs using their powerful legs, but they are also confronted with the difficult task of digging their way out of the nest. In 5–10 minute bursts, they scrabble frantically at the walls of their chamber, lying on their backs and kicking with their well-developed legs. Each of these digging bouts is separated by an hour's rest. Depending on where they dig from, the journey to the surface of the mound may take anywhere between 2 and 15 hours. Eventually, they emerge from their improvised incubator with beaks and eyes tightly closed. They take a look at their surroundings for first time and suck in a deep breath. Then, perhaps gathering their energy or listening for signs of predators, they stay completely motionless for 20 minutes or so.

By the time they have emerged, their parents have long since abandoned them to forage amongst the scrub. They leave their hole and fall or stagger to the bottom of the mound. Completely independent, they scamper off into the scrub to begin their life outside the warm, cosseted confines of their compost heap.

- The malleefowl is just one of a group of birds known as *megapodes* (Greek for "big feet"). There are 19 species of megapodes, or incubator birds as they are also known.
- They are all found in the Australasian region, including Australia itself (three species), New Guinea, some of the islands in the western Pacific and Indonesia, and the Andaman and Nicobar Islands in the Indian Ocean.
- The breeding season of these birds may extend for up to 11 months, and the males and females normally form long-lasting bonds. These established pairs occupy the same territory, centered around a mound, from one year to the next.
- Only in his fourth year does a male malleefowl look for a suitable location to construct his own mound. The mounds of young males are nowhere near as impressive as those constructed by older, more experienced males.
- Mallefowl can fly and will take to the trees if particularly alarmed, but they normally rely on their tremendous camouflage and ability to remain motionless for long periods to keep out of danger.
- Until Europeans colonized Australia, the malleefowl was a successful bird, stretching over huge swathes of Australia. Colonists, with their agriculture and menagerie of introduced animals, soon had an impact on these fascinating birds. Much of the malleefowl's habitat was given over to the cultivation of crops or the rearing of livestock, and animals like foxes made short work of these birds, which are unwilling to use their power of flight.
- Bones and old tales are the clues to two species of giant megapode that lived on certain islands in the Pacific Ocean. Compared with today's megapodes, these birds were very large and could have easily weighed 30 kg. A 2.5 kg malleefowl can construct quite an impressive mound, but the mounds produced by these birds must have been huge. Both species are sadly extinct, undoubtedly a result of human colonization of these islands. Exactly when they became extinct is a mystery. Some sources suggest populations could have survived until modern times, providing a conceivable inspiration for the almost mythical bird known as the *du*.

Further Reading: Brickhill, J. Malleefowl: A remarkable bird with an uncertain future. *Australian Natural History* 21, (1987) 147–51; Frith, H. J. Breeding of the mallee fowl, *Leipoa ocellata*. *CSIRO Wildlife Research* 4, (1959) 31–60.

MARBLE GALL WASP

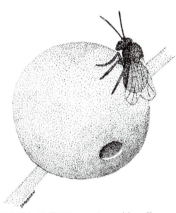

Marble Gall Wasp—A marble gall wasp resting on the gall she developed in as a larva and pupa. (Mike Shanahan)

Marble Gall Wasp—This picture shows a winged female that emerges from the spherical galls on oak trees. (György Csóka)

Scientific name: *Andricus kollari*
Scientific classification:

   Phylum: Arthropoda
   Class: Insecta
   Order: Hymenoptera
   Family: Cynipidae

What do they look like? There are several stages in the life of this insect, but the most obvious is the heavy-bodied female. She has a huge, globular abdomen, a humped thorax, and small head. The entire body is an attractive, chestnut brown color.

Where do they live? The marble gall wasp is found throughout temperate Europe wherever its host, the oak tree, can be found; however, it is more common in less-disturbed habitats.

Invasion of the Bud Snatchers

Over millions of years, insects and flowering plants have a developed a very close relationship with one another. In many cases, it is the insects that live at the expense of the plants, nibbling their leaves, chewing holes in their flowers and seeds, and generally making a nuisance of themselves. In the case of gall wasps, this relationship has grown to be one of the most intricate relationships in any terrestrial ecosystem.

The complex story of the marble gall wasp begins with a pregnant female wasp using her sharp, little egg-laying syringe to introduce an egg into the rapidly growing tissue of a leaf bud. Due to reasons that are still poorly understood, the egg and perhaps the substances injected with it trigger frantic cell growth in the plant tissue, forming a gall. The effect is very similar to what happens when the cells of an animal go out of control to form a cancerous tumor. However, in the gall, the cell growth is controlled by something associated with the developing wasp. The gall expands, forming a spherical, woody structure. This little ball not only protects the growing wasp but also provides abundant, nutritious plant matter on which the larva feasts. Somehow, the growing wasps manipulate the cellular machinery of the plant to divert resources away from the where they are needed (i.e., the leaves, stems, and reproductive structures) to its nursery sphere.

If this weren't complicated enough, the entire life cycle of the wasp is even more elaborate, involving a generation of males and females, and an asexual generation consisting entirely of female wasps. In the autumn, female wasps emerge that have no need of a male to breed. These *parthenogenetic* females seek out a suitably sheltered spot in which to see out the winter. With the arrival of the spring, the wasps leave their overwintering lairs and lay eggs into the developing leaf buds of oak trees. If the oak happens to be a common oak, new marble galls develop, like the ones described above. However, if the eggs are deposited into the developing leaves of a turkey oak, small galls form containing a sexual generation of males and females.

- The gall wasps are diverse animals. There are 1,300 known species, and all of them make unique galls on a wide variety of plants. Other species undoubtedly await discovery. The galls of some species are tiny disks on the underside of leaves, whereas others are huge, spongy wasp mansions.
- Gall wasps are not the only animals that are able to hijack the inner workings of a plant to meet their own ends. Mites can also form galls, as can nematodes. Galls can even be formed by bacteria and other microorganisms.
- A gall gives the young insect a safe refuge during the most vulnerable stage of its life, but as with any protective measure in the living world, it is far from invulnerable. The evolutionary arms race existing between all predators and prey means that regardless of how an animal evolves to protect itself, it is only a matter of time before a predator evolves new ways of bypassing its defenses. Several types of parasitic wasp exploit galls by using their long ovipositors to lay an egg on or in the occupant of the gall.
- Apart from the various parasitoids that punish gall-forming insects, there are numerous animals that take advantage of the gall and the food and shelter it offers. These animals are known as *inquilines*. Some live in comparative harmony with the gall former, while others are greedier and compete with the occupant for food, sometimes causing its death
- In some species of gall wasp, the life cycle is far more complicated than that of the marble gall wasp. One such species is the small wasp that forms the large galls on oaks known as oak apples. Each oak apple contains only male or females. The females mate and descend to the ground to lay their eggs on the fine roots of the host tree. These eggs form small galls that give rise to an asexual generation of parthenogenetic females.
- Gall-forming insects have been around for many millions of years, and they probably evolved from an ancestor that burrowed into the host plant, seeking food.

+ The science of studying galls is known as *cecidology* and involves understanding what animal forms a gall and exactly how the animal does so. The *how* is complicated, as a feeding insect inside one of these galls will produce a range of compounds, some of which are in its saliva, while other are in its feces, and still others exude from its cuticle. Exactly which of these is involved in manipulating the plant is difficult to study. There could be a huge number of interactions taking place, all of which play their own small part in the formation of a gall.
+ Although the late Alfred Kinsey is famous for his *Kinsey Reports*, he was first and foremost a zoologist who specialized in the study of gall wasps.

Further Reading: Schönrogge, K., Walker, P., and Crawley, M. J. Complex life-cycles in Andricus kollari (Hymenoptera, Cynipidae) and their impact on associated parasitoid and inquiline species. *Oikos* 84, (1999) 293–301.

PLATYPUS

Platypus—A female platypus nurturing her newly hatched young in her burrow. (Mike Shanahan)

Scientific name: *Ornithorhynchus anatinus*
Scientific classification:
　Phylum: Chordata
　Class: Mammalia
　Order: Monotremata
　Family: Ornithorhynchidae
What does it look like? The platypus is a bizarre-looking mammal with a ducklike bill, black, beady eyes, dense fur, large webbed feet with claws, and a flattened tail like that of a beaver. The males are far larger than females, ranging from 45 to 60 cm and tipping the scales at 1–2.4 kg

Where does it live? The platypus is found only in Australia, and even there it is restricted to the eastern part of the country. It has a liking for freshwater and is found in rivers and streams.

Egg Laying, but Not by a Bird

The duck-billed platypus is right up there with the naked mole rat in the odd mammal stakes. It has the face of a duck, but has little in common with birds. Like other mammals it can regulate its own body temperature, it has fur and mammary glands, yet it lays eggs like a reptile. It is also one of the few venomous mammals.

In the late-eighteenth century when the dried skin of this animal was first seen in England it caused uproar among the zoological fraternity. There were cries of "fake" and "sham" as many experts of the time claimed it to be nothing more than an ungodly creation of a mischievous taxidermist. Alas, it was not a taxidermist's trick, cobbled together in a back street from the body parts of various animals. No, it was a living, breathing animal, completely unlike any other mammal known to science at the time.

So then, just what is the platypus? Well, although it looks like a mishmash of different animals, the platypus is undoubtedly a mammal, but a very primitive one at that. Internally, the male platypus's reproductive organs resemble other mammals, with a pair of testicles (which never descend from the body cavity). The female's internal reproductive anatomy, however, is not the mammal norm. She has paired ovaries resembling those found in birds and reptiles. Only the left ovary is functional, the other is small, underdeveloped, and never produces eggs. After mating, the female platypus digs a burrow, which can be as much as 30 m long and is blocked with plugs along its length to protect against flooding. This tunnel is lined with wet vegetation dragged in by the female under her curved tail. The nest chamber is lined with fallen leaves and reeds, and it is here, 2–3 weeks after mating, that she lays two small eggs (occasionally a female lays one or three eggs). The eggs are small (about 11 mm in diameter), round, and leathery like a reptile's. The female incubates the eggs for around 10 days until they hatch and the tiny, naked young crawl out and cling onto their mother. The female doesn't have teats, so the youngsters lap the milk that soaks the fur and collects in grooves on the female's abdomen near the openings to the mammary glands. Every time the female leaves the brood chamber, she walls up the entrance. The young leave the burrow after 3–4 months, but they continue to take milk from their mother. Needless to say, the male plays no part in rearing the young.

Another interesting feature of the platypus, which may be linked to breeding, is the male's ability to produce and use venom. On each hind leg of the male, on the back of the ankle, there is a short, curved, hollow spur that sits in a fold of skin and can be erected at will. This spur is connected to a venom gland behind the knee of the animal. When manhandled or grabbed, the male platypus will forcefully jab this spur into the transgressor with immediate results. A platypus sting is said to cause agonizing pain in humans and will easily kill a dog or small domestic animal. In humans, the affected limb swells up rapidly and can hurt for days or even months. The venom produced by the platypus is unlike other venom. Its unique ingredients seem to be intended to cause intense pain and immobilization, but not death, in humans at least. Although the platypus can use its spur and venom to defend itself, it is probably primarily important in the disputes occurring between males in the breeding season when they are defending their territories, but it is not known if the venom is lethal to the platypus' own kind.

- The closest living relatives of the platypus are the echidnas, or spiny anteaters, of which there are two species. It is unclear how these animals are related to the placental mammals and the marsupial mammals. The reproductive organs of these animals have a lot in common with the marsupials. The eggs of marsupials, like those of the platypus, are encased in shells for at least some of their gestation.

+ The body temperature of the platypus is around 6°C lower than that of other mammals (32°C instead of 38°C).

+ The bill of the platypus is actually a very sensitive organ used for locating the small, bottom-dwelling invertebrates on which it feeds. The bill is not like a bird's beak, and the mouth of the platypus is actually beneath this fleshy protuberance. Nerves in the rubbery bill not only sense touch but also the electrical field of their prey. When searching for food in the gravelly bottom of a stream, the platypus will move its head from side to side, searching for the tell-tale signs of its prey. Scientists are still trying to understand exactly how the platypus's electrosense works.

+ The teeth of the platypus drop out when they are very young, so they grind their food between horny plates at the back of their jaws.

+ The platypus used to be hunted for its very dense pelage, which was valued in the fur trade.

+ Mammals typically have two sex chromosomes: X and Y. Interestingly the platypus has 10, and some of these are similar to the sex chromosomes found in birds.

+ When they are slumbering, mammals and birds all experience varying durations of rapid eye movement (REM) sleep, a phenomenon that in humans is associated with dreaming. It was thought that since the platypus is considered primitive, it would not have REM sleep, but it turns out that the platypus has eight hours of REM sleep a night, which is considerably more than other mammals and birds. Exactly why the platypus should have so much REM sleep is still unclear.

Further Reading: Augee, M. L. *Platypus and Echidnas.* The Royal Zoological Society of New South Wales, Mosman, New South Wales 1992; Griffiths, M. *The Biology of Monotremes.* Academic Press, New York 1978.

RED-AND-BLUE POISON-ARROW FROG

Scientific name: *Dendrobates pumilio*
Scientific classification:
    Phylum: Chordata
    Class: Amphibia
    Order: Anura
    Family: Dendrobatidae

What does it look like? This is a tiny, jewel-like frog measuring only 25 mm from the tip of its snout to its rear end. The body of this minute amphibian is a fabulous red, while its limbs and small marks on it back are a shimmering blue. The eyes are large and dark, and beneath the head there is an inflatable throat sac.

Where does it live? This species of poison-arrow frog inhabits the perpetually wet rain forests of Costa Rica in Central America. It can be found on the forest floor and also up in the trees, where it breeds.

Parenting in a Poisonous Amphibian

A red-and-blue poison-arrow frog hops and scrambles along the rain forest floor, providing a lively splash of color in the brown carpet of decaying vegetation. It reaches the base of a tree and surmounts a small mound of moss. Here, it perches for a few moments and emits a shrill call

Red-and-Blue Poison-Arrow Frog—A female red-and-blue poison-arrow frog taking the tadpole on her back to a bromeliad nursery. (Mike Shanahan)

Red-and-Blue Poison-Arrow Frog—An adult specimen of this tiny frog, photographed on the forest floor in Costa Rica. (Marcus Bartelds)

by inflating its throat sac. Another flash of color in the undergrowth nearby reveals this frog's amorous intentions. He is a male, and he has spotted a female that he is trying to impress with his calling. His voice will not be enough to impress the female, and he gives chase with the hope of catching her and engaging in some romantic wrestling. Impressed by the calling, chasing, and wrestling abilities of this suitor, she lays 2–16 eggs on the ground in a secluded spot. The male, taking his cue, fertilizes all the eggs with his sperm and moistens them with fluids from his skin. The frogs are model parents, and they guard their eggs religiously until they hatch. Soon, the waiting is over and young tadpoles break free from their eggs and instinctively wriggle upon to the back of their parents, who must now take them to water. The journey may take several hours, so the tadpoles are securely glued to the parents with sticky mucus that also keeps them moist. Ponds and streams in the rain forest are crawling with all sorts of animals that would make short work of a tiny tadpole; therefore, the devoted parents head for the tree tops. Small pads on their feet afford them an excellent grip, and before long, they are in amongst the branches. Trees are not the sort of places you would expect to find good amphibian breeding spots, but lo and behold, growing on the branches are epiphytic plants, like bromeliads (related to pineapples), which, at their center have a bowl that collects condensation and rainwater. The parents deposit up to three of their young in every suitable pool they find. In the pool, there will be various insect larvae and other invertebrates that the tadpoles will feed on. To supplement the diet of her young, the female frog will also deposit unfertilized eggs in the bromeliad bowls, which will be greedily eaten by the tadpoles. In this sheltered microhabitat, the young develop rapidly and eventually metamorphose into adult frogs.

After the careful nurturing provided by their parents, the froglets must fend for themselves. They descend to the forest floor and selectively hunt ants and other small animals among the leaf litter. Trace quantities of toxins in the frog's diet are assimilated by the frog into its own defensive systems. The chemicals are stored in skin glands, and should any animal eat or even lick one of these frogs, it will be dead in no time at all. The amazing colors of the frog are not to brighten up the drab forest floor, but are a warning for other animals to give it a wide berth.

+ There are approximately 150 species of poison-arrow frog, but their small size and habitat means that new species are being found all the time. They are only found in South and Central America. The red-and-blue poison-arrow frog is one of the more northerly species, but poison-arrow frogs as a whole are found all the way down to southern Brazil. Of the 150 or so species in the family, only around half are brightly colored like the species mentioned here.

+ It is quite common for small amphibians in the forest to use bromeliads and other epiphytic plants as breeding pools. In a large patch of forest, there will be thousands of these pools, adding another level of complexity to the already bewildering array of microhabitats that can be found in pristine equatorial forests.

+ A large brood of red-and-blue poison-arrow frog tadpoles may require repeated trips to the forest floor by their parents to transport them to their nursery pools.

+ The poison-arrow frogs, although remarkable for the lengths they go to rear their young, are also amazing for the toxicity of the substances produced by their skin glands. The red-and-blue poison-arrow frog is one of the less toxic species, but its relative, *Phyllobates terribilis*, contains enough poison to kill 100 fully grown men. The sheer potency of this poison in such a tiny animal is astonishing, but in the lushness of the rain forests, where danger lurks everywhere, it makes sense to be well defended.

+ When these frogs are maintained in captivity and fed on small flies and crickets, they gradually lose their toxicity, suggesting that their poisonous secretions are derived from the foods they eat, such as ants and beetles. It is thought that the same toxic compounds in the insects are taken in the by the frog and concentrated in the skin glands.

+ Advertising the production of a toxin with bright colors is common in the animal world. Such colorful displays are known as *aposematism*. It makes no sense for a poisonous animal to be drab and instantly forgettable. Bright, bold colors ensure that a predator with no experience will remember an animal that makes it sick, thus ensuring it will give the same species a wide berth in future encounters.

+ American Indians use the toxins secreted by these frogs to coat the points of their blow darts. Some of the more virulent species are held down with a stick, and the toxin is simply wiped on to the dart. Other, less-dangerous species are not so lucky. They are caught, impaled on sticks, and gently toasted over a fire to yield their poison, which is then concentrated and smeared over the dart points. Today, these rapidly acting nerve toxins are used for hunting game in the forest, but they were probably once used in tribal warfare as well.

+ Although these frogs are very poisonous, there are some animals that appear to be immune to the toxins produced by these frogs. A snake, *Leimadophis epinephelus*, is a specialist predator of these frogs and is an example of the constant evolutionary arms race that takes places between all predators and prey.

Further Reading: Prohl, H., and Hodl, W. Parental investment, potential reproductive rates, and mating system in the strawberry dart-poison frog, *Dendrobates pumilio*. *Behavioral Ecology and Sociobiology* 46, (1999) 215–20; Saporito, R. A., Garraffo, H. M., Donnelly, M. A., Edwards, A. L., Longino, J. T., and Daly, J. W. Formicine ants: An arthropod source for the pumiliotoxin alkaloids of dendrobatid poison frogs. *Proceedings of the National Academy of Sciences USA* 101, (2004) 8045–50.

SAND TIGER SHARK

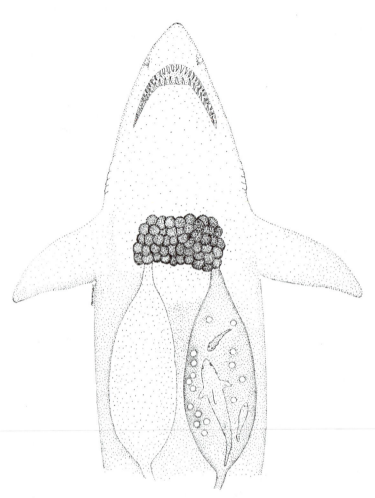

Sand Tiger Shark—Cutaway of the female sand tiger shark showing her wombs and the developing, cannibalistic young within. (Mike Shanahan)

Scientific name: *Carcharias taurus*
Scientific classification:
    Phylum: Chordata
    Class: Chondrichthyes
    Order: Lamniformes
    Family: Odontaspididae
What does it look like? The sand tiger shark can grow to lengths of 3.5 m, though there are records of specimens over 5 m in length, and reach weights of 160 kg.. Adults range in color from grayish brown to bronze with a whitish underside. Youngsters have small dark spots, which fade as they grow older. The head of the shark is flattened, and the wide mouth is adorned with many fearsome-looking teeth. The pectoral fins are fleshy and immobile, and the first dorsal fin is positioned a long way back on the body. The tail fin is asymmetrical, with an upper lobe that is far longer than the lower one.

Where does it live? The sand tiger shark is found in all warm seas except the eastern Pacific. They can be found at a range of depths, down to a maximum of around 200 m.

Cannibalism in the Womb

Sharks are one of the most misunderstood groups of animals. Although human perceptions of these animals are slowly changing, long-standing beliefs have painted these animals in a bad light, making them out to be savage, unflinching killers of the deep. The facts are far removed from the fiction, as these ancient marine animals are perfectly attuned to their watery world with a host of adaptations that contribute to their success. These adaptations are perfectly exemplified by the huge variety of ways in which they nurture their young, and of these strategies, none is more remarkable than the gestation of the sand tiger shark.

A female sand tiger shark will only give birth to two young every two years, which is relatively slow for an animal of this size. The sand tiger's secret to success is quality not quantity, as the few young that the female gives birth to are very well developed and are capable of looking after themselves as soon as they are born. To understand how the female can produce such well-formed young, we must learn a little about the environment inside her, where the young undergo their early growth. Inside the female, there is a pair of separate, capacious tubes. Each of these is a uterus and is similar, in some ways, to the womb found in female mammals. Eggs produced by the female leave the ovaries and descend into these tubes where they divide and grow, eventually forming baby sharks. The baby sharks are nourished by the store of yolk in their flimsy-shelled egg, and the first embryo in each uterus to reach around 6 cm in length hatches from its egg and actively moves through the fluid contained in the so-called nurseries. This is normal enough, as many other animals, such as marsupials give birth to tiny, poorly-formed, but active babies. However, what the baby sharks do next sets them apart. The young swim around inside the female and start feeding on their developing brothers and sisters. Even at this very early stage, they have a well-developed set of teeth and can make short work of their soft siblings. This baby predator grows and eats until it is alone in the uterus, all of its brothers and sisters consumed. To provide further nourishment for her cannibalistic babies, the mother continues to produce eggs that descend into the uteruses to be gobbled up by the two remaining young. After two years, the young are fully developed, and the female gives birth to them in a lengthy labor, which is unsurprising as each shark is around 1 m long at birth. With their predatory instincts well honed by their time in the womb, the young swim off, perfectly capable of fending for themselves and big enough to be vulnerable to only the largest predators.

+ There are more than 360 species of shark, with new species coming to light as the oceans are explored. Most sharks are active predators, but there are a few species that strain the water for plankton, such as the whale shark and basking shark. They are extremely ancient creatures with a heritage that goes back hundreds of millions of years. The fossil record shows that the ancestors of sharks patrolled the seas over 400 million years ago. Their basic anatomy is thought to be quite primitive, but some of their bodily systems, particularly their senses, have evolved to levels that cannot really be matched by any other living vertebrate.
+ The senses of the shark are very refined, and they have a whole battery at their disposal to sense their environment. Their sense of smell, which is really their ability to detect chemicals in the water, is unparalleled. They can detect a specific chemical (i.e., a blood protein) in the water even when it is as dilute as 1 part in 10 billion.

Around their head and along the sides of their body are the tubes and pores of the lateralis system. This sensory system detects minute changes in water pressure, picking up the tiny vibrations made by the thrashings of a struggling fish. Also around the head are small pits that detect the electrical fields of other animals. This is very important for species that snuffle around in the sand and mud searching for prey. Even when the prey is buried, its faint electrical field can be detected by the shark. Finally, they also have very sensitive vision. A special layer of light-reflecting crystals at the back of the eye improves their vision in dimly lit waters. As sharks are so well adapted to their aquatic environment and can occur in high densities, they are very important in the maintenance of marine ecosystems the world over.

+ Their bodies are a model of hydrodynamic perfection allowing them to cruise through the water with a minimum of effort. To provide buoyancy, many shark species have a huge, oily liver. However, the sand tiger is the only species of shark that sucks air into its stomach to act as a buoyancy aid.

+ The manner in which young sand tiger sharks are nurtured inside their mother is one of many strategies. Many species lay large, leathery eggs. Washed ashore, these empty egg cases are commonly called mermaid's purses. Tendrils on each corner of these eggs anchor them to seaweed and rocks. Many sharks give birth to live young, and inside their mother, they are nourished by yolk, eggs ovulated by the female, or a milklike substance secreted by the uterus wall. The young of others are hooked up to a structure very similar to the placenta of mammals.

+ The predatory nature of sand tiger embryos was discovered in the late 1940s by a marine biologist who was internally examining a landed female. The young shark mistook the researcher's probing fingers for food and gave him a sharp nip.

+ Although sharks are feared by humans, they are the ones under threat. Every year, more than 100 million sharks are killed for sport, for food, and for the other products they yield, such as leather and oil. With their slow rate of reproduction, sharks cannot sustain this unchecked level of exploitation, and many species are already critically endangered. Without complete protection in the very near future, it is highly likely that several shark species could be lost forever.

Further Reading: Breder, C. M., and Rosen, D. E. *Modes of reproduction in fishes.* T.F.H. Publications, Neptune City, NJ 1966; Gilmore, R. G. Reproductive biology of lamnoid sharks. *Environmental Biology of Fishes* 38, (1993) 95–114; Gilmore, R. G., Dodril, J. W., and Linley, P. A. Reproduction and embryonic development of the sand tiger shark, *Odontaspis taurus* (Rafinesque). *Fish Bulletin* 81, (1983) 201–25; Joung, S., and Hsu, H. Reproduction and embryonic development of the short-fin mako, *Isurus oxyrinchus* Rafinesque, 1810, in the northwestern Pacific. *Zoological studies* 44, (2005) 487–96.

SHIP TIMBER BEETLE

Scientific name: *Hylecoetus dermestoides*
Scientific classification:
   Phylum: Arthropoda
   Class: Insecta
   Order: Coleopetra
   Family: Lymexylidae

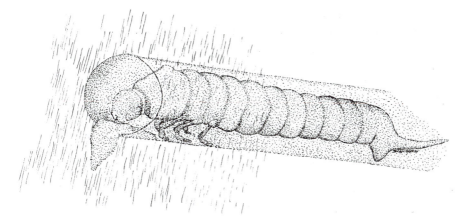

Ship Timber Beetle—A ship timber beetle larva keeping its tunnel clean to provide good growing conditions for the fungi on which it feeds. (Mike Shanahan)

What does it look like? This elongate, cylindrical beetle can be up to 2 cm long, although the males are far smaller. The females are orange to light red with a very long abdomen. The males are also thinner bodied than the females, with a darker front end and a pair of feathery palps on the underside of their head.

Where does it live? This beetle is found in woodlands and areas where wood is cut and stored. The natural habitat is deciduous woodlands, particularly wood edges, clearings, and sheltered, open areas. It is found throughout Europe and into the Russian part of Asia.

Fungi for the Future Generation

Female ship timber beetles, after they have mated, have the difficult task of finding a suitable place to lay their eggs. They are quite selective in their choice of egg-laying site. They hardly ever use healthy trees, but look for those trees that are diseased or are already dead. They will also use timber that has been felled. The female deposits eggs in crevices and cracks in the wood using her telescopic egg-laying tube. As each egg is laid, the female coats it with spores from a little pouch near her ovipositor. Secreting the egg in a safe place and coating it with fungal spores is the extent of her maternal care, after which she takes to the air in a labored fashion to find more suitable nooks and crannies on other trees. After a few days, the egg hatches, and a small, whitish larva wriggles out into the world and loiters around its empty egg shell for a while to pick up some of the fungal spores that its mother left for it. After a while, it begins to tunnel into the wood using its powerful mandibles, carrying some of the fungal spores with it. Initially, the tunnel is very narrow as the larva is small, but as it grows, the tunnel must also increase in width to accommodate the grub. The tunnel may run for over 30 cm, snaking into the wood, but it is not the wood the larva is eating. The tunnel is, in fact, a sheltered fungus farm. The fungal spores provisioned by the female and carried by the larva, infect the wood, until the tunnel is carpeted in a white layer of fungus. It is this fungus that the beetle larva eats. The larva takes excellent care of its fungus garden so that it will have enough food to complete its development, doing everything it can to keep the conditions just right for the fungi. The fungus requires oxygen to thrive, so the larva must rid the tunnel of any debris to maintain a good flow of air. The larva shuffles along the confines of its tunnel pushing any wood dust and waste to the outside, where it falls to the base of the tree. By the winter, the larva will not be fully grown, so it will have to retreat to the deepest reaches of its burrow and enter a resting state to survive the cold, harsh conditions. With the spring, the larva reawakens and continues to feed on whatever fungi survived the winter. Soon, it is ready to pupate, and it wriggles to the tunnel entrance and enlarges the width of its burrow to form a chamber for its imminent metamorphosis.

+ The ship timber beetles all depend on wood in which to breed. They are regarded as forestry pests because of the damage they do to timber, affecting high-quality wood that today is destined for construction but was once used for shipbuilding. In reality, their effects on commercial timber operations are minimal, as they tend to go for diseased or dead trees.

+ The fungi carried by the female and eaten by the larvae is a yeastlike fungus (*Endomyces hylecoeti*) that has struck up a symbiotic relationship with the beetle. The fungus gets access to wood in the safety of tunnels, and in return, the larvae get something to eat.

+ As you can imagine, wood hardly makes for an appetizing diet. It is tough, and the nutrients are mostly bound up within the molecule known as cellulose. No animal can produce the enzyme, cellulase, to digest this material; yet many animals, especially insects, feed on wood. Just how do they do this? Most wood-munching insects depend on another organism to break down the cellulose for them. The ship timber beetle uses the services of a fungus, while other insects have gone one step further and actually harbor the wood-digesting organisms inside themselves. Wood-feeding termites have cellulase-producing bacteria in their gut, which break down the cellulose, releasing the simple sugars that can be used by the insect. Other insects have not yet developed this sophisticated trick and instead must feed for a very long time on this poor diet to complete their development.

+ Fungi are very proficient at breaking down and digesting wood, and this is a massively important component of the energy recycling that takes place in most land-based ecosystems. The fungi are the recyclers, taking the energy locked up in the woody tissue and making it available to other organisms in the system.

+ The ship timber beetle makes its tunnels in many types of wood, but it will avoid pines, larches, and hornbeams. Birch is a favorite, and a large, diseased birch with the correct conditions (warm and no less than 30–40 percent moisture content) will be home to many larvae, which give their presence away by the sawdust that accumulates at the base of the tree after it has been ejected from their holes.

Further Reading: Crowson, R. A. *The biology of the Coleoptera.* Academic Press, London 1981; Francke-Grosmann, H. "Ectosymbiosis in wood-inhabiting insects." In Henry, S. M. (ed.) *Symbiosis (Volume 2).* Academic Press, New York 1967.

6

LIVING AT THE EXPENSE
OF OTHERS: PARASITISM

ALCON BLUE BUTTERFLY

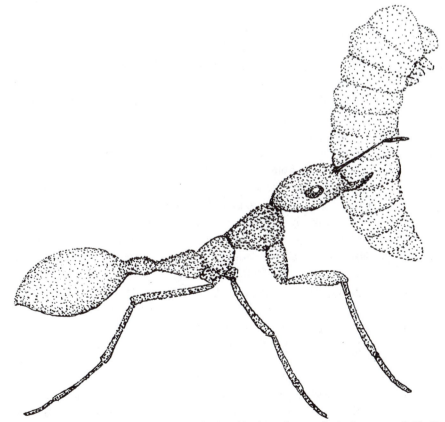

Alcon Blue Butterfly—A worker ant carrying the alcon blue butterfly caterpillar back to its nest. (Mike Shanahan)

Scientific name: *Maculinea alcon*

Scientific classification:

Phylum: Arthropoda

Class: Insecta

Order: Lepidoptera

Family: Lycaenidae

What does it look like? The blue butterflies are beautiful insects, and the alcon blue is no exception. The upper surface of the wings is a shimmering, metallic blue framed with a darker border and a narrow fringe of silvery white scales. The underside of the wings is paler, sometimes with a brownish hue and bears small dark spots rimmed with silver. The head has a large pair of dark compound eyes and thin antennae ending in small clubs. The sexes are similar, but the females are larger with fatter abdomens, and the upper surface of their wings may be adorned with small dark spots.

Where does it live? The alcon blue is found in many European countries, and its range stretches into Asia. It prefers warm/sheltered habitats such as flower meadows where it may find the flowers of the marsh gentian and certain species of red ant.

The Crafty Caterpillar

Ants, with their very ordered and productive societies, make nests that are coveted by many different animals. These subterranean strongholds provide excellent protection and a bounteous supply of food for any animal that is brave or smart enough to inveigle its way inside. The alcon blue butterfly is one such interloper that has opted for intelligent gate crashing. In the European summer, around the end of July, female alcon blue butterflies have mated and flutter earnestly around their home meadows looking for the closed flowers of the marsh gentian. The females alight and deposit their white eggs on the flowers. A few days later, each tiny caterpillar hatches out of the bottom of its egg and munches a tunnel straight into the closed flower of the gentian. The small caterpillar remains within the safety of the flower for around two weeks, eating some of the flower tissue and developing seeds, but growing slowly. In the flower, the caterpillar sheds its skin three times, and after the third molt, it is ready to take its leave of this safe house. In the early morning or evening, it chews its way from the base of the flower and shuffles along the petals to the apex of the bloom. Apparently tired of its nursery, it lets go of its grip and falls to the ground on a silken thread and waits. This is the riskiest time of its short life. Predators abound amongst the short turf, and all would make short work of a tiny, plump caterpillar, but a small, foraging red ant smells the caterpillar and goes for a closer look. The ant, seemingly intrigued and mesmerized by the caterpillar, strokes it all over with its quivering antennae. Result! If it could, the caterpillar would breathe a huge sigh of relief as this is exactly what it was waiting for. To express its relief, the caterpillar produces a drop of sweet fluid from its rear end that the ant immediately starts suckling. This can go on for some time, until the caterpillar flattens the middle or rear of its body. This act is enough to completely fool the ant into mistaking the caterpillar for a grub from its own nest that has somehow gone walkabout. It tenderly picks the caterpillar up in its jaws and makes for the nest. The caterpillar is deposited in the nursery of the ant's nest alongside the countless young of the colony. Here, it blends right in. It smells right, and smell is all important to ants. The ants feed the caterpillar by regurgitating nutritious fluid, and somehow these tiny tricksters persuade the ants to give them preferential treatment so that they receive more attention and food than the ant grubs. To add insult to injury, the caterpillars also supplement their liquid diet by scoffing the odd ant grub or two. On such a nutritious diet, the caterpillar grows quickly, increasing its weight by about 100 times during its first month in the nest. The caterpillar, well looked after and safe in the ant's nest, stays put for the summer, autumn, winter, and spring, and only when the following summer arrives does it begin the transformation that will turn it into a fine butterfly. When it emerges from its chrysalis as an adult butterfly deep underground, the tricks used to deceive the ants have long since worn out. It must make for the surface as quickly as possible to escape the wrath of the ants. Fortunately, it has one last trick to avoid being held to account. The angry ants try and bite the fleeing charlatan, but all they come away with is a mouthful of scales. The whole body of the newly emerged butterfly is densely covered in loose scales, and the ants cannot get a grip on it. Eventually, the butterfly manages to leave the nest and makes for a perch among the lofty vegetation of the meadow. Here, its wings will inflate and harden in the summer sun until it is ready to flutter off with an air of innocence.

- There are several species of blue butterfly distributed throughout Europe and Asia. All depend to varying degrees on ants to provide food and shelter for their developing young.

- Ant societies are founded on pheromones. Nest mates recognize one another by the way they smell. These distinctive odors have been hijacked by a multitude of different animals, mostly other insects. An ant will not challenge an invader if it smells like one of its sisters. An ant's nest is not only a safe place, but it is stuffed full of resources. There are huge numbers of defenseless grubs and pupae and all sorts of nutritious goodies to be found in the nest refuse piles. Some types of insects, especially beetles have become so inextricably linked with ants and their nests that they are found nowhere else. Some have modified antennae or bulbous glands on their bodies that produce fluids the ants find alluring—very useful if an appeasement is needed should the host smell a rat.

Further Reading: Damm Als, T., Nash, D. R., and Boomsma, J. J. Adoption of parasitic *Maculinea alcon* caterpillars (Lepidoptera: Lycaenidae) by three *Myrmica* ant species. *Animal Behaviour* 62, (2001) 99–106; Elmes, G. W., Thomas, J. A., and Wardlaw, J. C. Larvae of *Maculinea rebeli*, a large-blue butterfly, and their *Myrmica* host ants: wild adoption and behaviour in ant-nests. *Journal of Zoology* 223, (1991) 447–60; Elmes, G. W., Wardlaw, J. C., and Thomas, J. A. Larvae of *Maculinea rebeli*, a large-blue butterfly and their *Myrmica* host ants: patterns of caterpillar growth and survival *Journal of Zoology* 224, (1991) 79–92.

ANT-DECAPITATING FLIES

Scientific name: *Apocephalus* species
Scientific classification:
  Phylum: Arthropoda
  Class: Insecta
  Order: Diptera
  Family: Phoridae
What does it look like? Ant-decapitating flies are tiny, measuring between 1 and 5 mm in length. They are typically humpbacked and drab colored in black, browns, and yellows, with large, delicate wings, long legs, and bristly heads.

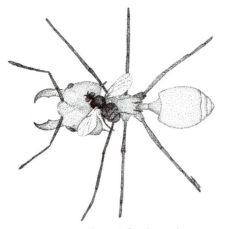

Ant-Decapitating Flies—A female ant-decapitating fly preparing to lay an egg behind the head of a leaf-cutter ant worker. (Mike Shanahan)

Ant-Decapitating Flies—A female ant-decapitating fly swooping in to lay an egg behind the head of a worker leaf-cutter ant. (Sanford Porter)

Where does it live? These flies are found in any habitat where there are ants. They are particularly diverse in the tropics. In the United States, they can be found from Alaska to southern Texas, although the largest number of species is to be found in the southwestern states.

Don't Lose Your Head

Ants in their structured communities are normally found in dense aggregations. For the most part, there is safety in numbers—their nests are heavily fortified, and there are the snapping jaws of countless colony members. All in all, ants do not appear to be the easiest pickings around. Be that as it may, many types of animal have become specialist predators of ants, and one of these is the ant-decapitating fly, an insect that terrorizes ants in a particularly gruesome way. It is in the tropics that the ant-decapitating flies are at their most devastatingly diverse, and it is the leaf-cutter ants—those tireless laborers of the verdant neotropical forests—that are singled out for special attention. Weaving their way along the forest floor to a suitable foraging spot, worker leaf-cutter ants are blissfully unaware of the danger hovering silently above their heads: a female ant-decapitating fly that is ready to lay eggs. Spying a suitable target, she makes a darting flight downward and lands delicately on the back of the ant, which is many times her size. She is very picky about the ant species she selects. Some ant-decapitating flies will only prey on one species of ant, whereas others may use a handful of species. The fly probes around the ant using her sharp egg-laying tube, and with the dexterity of a seamstress, she pierces the thin membrane joining the various plates of the ant's exoskeleton. Depending on the fly species in question, the egg is laid in the head, thorax, or abdomen of the ant. With a single egg deposited, the females takes off to search for more potential hosts. In the meantime, a larva eventually hatches from the egg, and if deposited in the abdomen or thorax, it will wriggle and squirm its way up to the ant's head. Once in the snug head capsule, the larva settles down to feed, gorging itself on the muscles and other tissues that pack out the head of the unfortunate host. In the heat and humidity of the tropics, the development of the larva is rapid, and soon enough, the head has been emptied of all edible matter. The head capsule drops off, and the fly larva completes the rest of its development in the safety of this small shell. In an animal like a mammal, where all the massively important bits of the central nervous system are seated in the brain, such a fate would undoubtedly end in death. However, the nervous system of an ant has several small brains along its length. These ganglia can control walking and other activities, and as long as the ant has sufficient food reserves, it may soldier on for a while until it keels over. In other cases, the head capsule may remain attached to the ant, but it is completely empty save for a well-developed grub or pupa of the ant-decapitating fly.

+ The phorid flies, the family to which the ant decapitators belong, is a diverse and very successful group of insects. There are more than 3,000 identified species, but as they are so small and so numerous in the tropics, the true number of species must be far, far higher. These flies are commonly known as scuttle flies due to their propensity for running instead of flying. More than 100 species of ant decapitator are known so far, but new species are being discovered on a regular basis.
+ The egg-laying tube of the female ant-decapitating fly is very different from species to species. Some species have a fearsome-looking right-angled hook, whereas others have a stubby, piercing trident.
+ The full extent of the relationship between these flies and ants is only slowly being revealed. In ant species, like the leaf-cutters of Central and South America, workers take special precautions to avoid being parasitized. For example, during their foraging trips, worker leaf-cutter ants are vulnerable to attack from these flies, especially if the ants are carrying sizeable slabs of leaves in their jaws and are not able to fend off the flies. To counter this, a tiny worker often rides on a big worker's cargo and doggedly defends its bigger sister from the attentions of these meddlesome flies.
+ The fire ant has become a real problem in the United States after its accidental introduction in the 1930s. Fortunately, help may be at hand in the form of ant decapitators. The fire ant was accidentally imported with cargo from Brazil through the port of Mobile in Alabama. In the United States, this ant has no parasitoids to keep its numbers in check; however, in South America, certain ant-decapitating flies prey specifically on the ant. The appropriate flies have been introduced to the Southeast from their native home in South America. Some introductions have not been successful, but more time is needed to see if these fascinating little flies can stem the plague of the fire ants.
+ Other ant-decapitating flies in the neotropics have become specialist parasitoids of stingless bees. These bees may play host to more than one larva, and some individuals have been found to contain 12 developing fly larvae.
+ Although very small, phorid flies are strong fliers, able to cover distances of at least 10 km in a 24-hour period.

Further Reading: Erthal, M., Jr., and Tonhasca, A., Jr. Biology and oviposition behavior or the phorid *Apocephalus* and the response of its host, the leaf-cutting ant *Atta laevigata*. *Entomologia Experimentalis et Applicata* 95, (2000) 71–75.

BIRD FLUKE

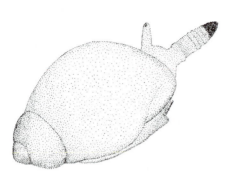

Bird Fluke—The sporocyst of the bird fluke bulging into the eye stalk of its snail intermediate host. (Mike Shanahan)

Bird Fluke sporocyst—This picture shows the stage of the parasite that can be found in the host snail. The sporocyst is filled with many young flukes. (Oldrich Nedved)

Scientific name: *Leucochloridium paradoxum*

Scientific classification:

  Phylum: Platyhelminthes

  Class: Trematoda

  Order: Strigeata

  Family: Leucochloridiidae

What does it look like? The bird fluke is a small animal. Adults are only a few millimeters long, flattened, and spiny. On their lower surface, they have suckers. The preceding stages of the fluke are microscopic and take on a variety of forms, including elongate and transparent mobile forms to larger, saclike ones lacking any discernible features.

Where does it live? This is an animal of the temperate woodlands of Europe and North America, but as it is a parasite, its actual habitat is the interior of snails and birds.

Subduing a Snail from the Inside

This small, unassuming animal typifies the bewildering complexity of some parasite life cycles. To complete its life cycle, the worm depends on two very different animals: a definitive host, in which the parasite multiplies, and an intermediate host, which harbors the growing parasite. This is quite a common setup for a parasite, but it results in one huge complication. How does the young freeloader get from its intermediate host to the definitive host? The solution used by the fluke is also complex, and it deserves an award for its ingenuity.

The adult flukes live within the intestines of woodland birds, such as crows, jays, sparrows, and finches, gripping onto the intestinal wall with their powerful suckers and feeding on the tissue and mucus of the bird's insides, perhaps even ingesting some of the intestinal contents. They are hermaphrodites and have the necessary apparatus to produce tiny, oval fertilized eggs. These eggs leave the bird's body in its feces, hopefully finding their way to the moist woodland floor. This is the beginning of a daunting journey in which the baby parasites must eventually find their way back to the definitive bird host.

With a lot of luck, the eggs are swallowed by an unknowing and unfortunate snail, while it is munching on vegetation that happens to have been soiled from above by an infected bird. The egg hatches into a clear, transparent larva, which creeps around the body of the snail before undergoing a transformation and taking on a saclike appearance. This sporocyst, as it is known, grows through the snail from the central body in the mollusk's digestive gland and forms a brood sac in the head and muscular foot of the snail, extending into the host's tentacles. It is in the central body that the parasite replicates itself, producing a multitude of embryos, which move to the brood sac via a connecting tube. The embryos mature into what represents the juvenile stage of the bird-inhabiting adult. The development of the fluke has almost gone full circle, but these juveniles must still get from the snail into a bird. The sporocysts in the eye stalks of the snail swell, change color, bearing bands of green or brown, and begin to pulsate rapidly, clearly visible through the taut, thin skin of the snail. With its eye stalks flashing like beacons, the snail wanders aimlessly into the open, unusual behavior for a retiring snail, which is influenced by the parasite. The throbbing tentacles look for all intents and purposes like small caterpillars, and they soon attract the attention of a hungry bird. The bird hops up to the snail, yanks off one of the succulent caterpillars, and swallows it. In doing so, the bird has aided the young flukes in their quest to complete their life cycle. Soon after being swallowed, the flukes mature and take up residence, as adults, in the bird's gut.

- The group of trematodes to which the bird fluke belongs is a very successful group of wholly parasitic organisms. To date, more than 6,000 species have been identified, but as the majority of vertebrates appear to be infected by at least one species, there are probably many more yet to be identified. They range in size from 0.2 mm to over 7 cm, and because many can have drastic effects on the health of humans and livestock, they have been studied in great detail by scientists. The life cycles of some of these animals are so complex that many have taken centuries to unravel. Species such as the liver fluke were known about for centuries, but it took a very long time indeed to find out how the adults actually infected cows, sheep, and humans.
- In most species, the intermediate host is a snail. Their affinity for these shelled mollusks is not fully understood. Perhaps it is because snails are common in the moist habitats where the eggs and immature stages of these parasites are normally deposited. Snails also fit the bill because they are eaten by many other animals. This does beg the question, however, of why the parasite bothers with a definitive host like a bird or a mammal at all. One possibility for this relationship is that vertebrates often range over great distances and can therefore spread the eggs of these parasites far and wide.
- Insects and rodents vie for the title of champion reproducer, but flukes must surely be among the contenders, if not the undisputed kings. Typically their life cycle involves the following stages: egg—miracidium—sporocyst—redia—cercaria—metcercaria—adult. The sporocyst gives rise to many redia, and these, in turn, yield large numbers of cercaria. One egg infecting the intermediate host can therefore give rise to many thousands of infective cercaria, each of which can develop into an egg-laying adult. Of course, nature is not extravagant, and these flukes do not do this just because they can. Their reproductive abilities say an awful lot about how slim the chances are of a single egg reaching adulthood. They have to pull out all the stops and be able to produce huge numbers of progeny in the hope that one will reach the definitive host to reproduce.

◆ The blood flukes cause a debilitating disease in humans called *schistosomiasis* (bilharzia). This is thought to infect over 200 million people worldwide, and the species responsible (mostly *Schistosoma haematobium*, *S. mansoni*, and *S. japonicum*) are some of the only flukes in which there is a male and a female. Another species of importance to humans is the Chinese liver fluke, which infects at least 30 million people in East Asia.

Further Reading: Poulin, R. "Adaptive" changes in the behavior of parasitized animals: A critical review. *International Journal of Parasitology* 25, (1995) 1371–83; Rennie, J. Trends in parasitology: Living together. *Scientific American* (1992) 123–33.

COD WORM

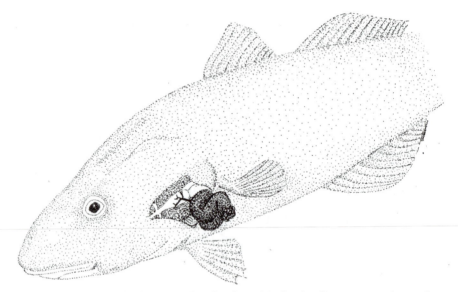

Cod Worm—A cod worm clinging to the gills of a cod, its head end growing into the circulatory system of the fish. (Mike Shanahan)

Scientific name: *Lernaeocera branchialis*
Scientific classification:
    Phylum: Arthropoda
    Class: Maxillopoda
    Order: Siphonostomatoida
    Family: Pennellidae
What does it look like? As juveniles, cod worms are tiny, well-protected animals. Covering their back is a large tear-shaped shell, and from underneath this, several pairs of fringed swimming legs can be seen. During its juvenile stages, the animal is only a couple of millimeters long. The adult animal is quite different, with a wormlike, S-shaped body.
Where does it live? The cod worm is an exclusively marine animal that makes its home in the waters of the Atlantic, and more specifically in certain fishes that live in this ocean.

Go Straight for the Heart

Cod and haddock have a very raw deal, not only are they pursued relentlessly by humans as food, but they also suffer from numerous, amazingly grotesque parasites. Perhaps the most interesting of these is the cod worm, an animal with a fascinating, albeit grisly way of life. These crustaceans start life as small, free-swimming larvae that paddle through the water with manic intensity. This first, short-lived stage goes through the first of several transformations and ends up as a slightly more elongate animal with a pair of grasping hooks on the front of its carapace. These appendages are use to good effect to gain a good purchase on the first of its hosts, the sluggish, rotund fish known as the lumpsucker. Using its clawed limbs, the cod worm grips onto the host's flanks and plugs a thin filament into the flesh of the fish to extract blood. During its time on the lumpsucker, the young cod worm goes through several stages, patiently waiting for a member of the opposite sex to arrive. When a male and female do eventually find one another, they mate, and the female, with her eggs fertilized, goes through yet another transformation into a longer form. Her long abdomen contains her eggs, and with these rapidly maturing, she takes her leave of the lumpfish and adopts a free-swimming existence once more—this time in search of her final host. She will seek out cod and their relatives, such as haddock. She uses a multitude of tiny sensory cells to sniff out one of these fish, tasting the water for the slightest hint of the host's odor. When a cod or a haddock has been detected, the tiny parasitic crustacean makes straight for its gills, using its hooked appendages to grip the delicate tissue of these respiratory organs. Firmly fixed to her host, the female goes through one final transformation. She changes into a creature that has to be seen to be believed. She looks nothing like a crustacean or any other animal for that matter. Her plump, wormlike body is held in an S shape, and nestled against the rear of her body is a coiled mass of egg strings. It is difficult to compare the adult cod worm to anything. In some ways, she looks like an organ that has fallen out of the fish. The front part of her body penetrates the body of the fish and enters the rear bulb of the host's heart. Firmly rooted in the host's circulatory system, the front part of the female parasite grows like the branches of a small tree, reaching down into the main artery of the fish. All the food the parasite needs is absorbed directly from the fish's blood, and there she can remain beneath the safety of the host's gill covers, releasing a new generation of cod worms into the sea.

- There are at least 1,000 species of parasitic copepod, of which the cod worm is but one. They infect most forms of marine animals, from small, burrowing worms to the largest whales. Their commitment to a parasitic way of life is almost unparalleled. As adults, many of the structures and organs necessary for a free-living way of life are completely redundant and have been lost. The cod worm is just one of a huge number of species in which the adults look like little more than an unusual adornment on the host's body.
- The fact that the larvae of these marine parasites find their hosts at all is mind-boggling. Imagine a tiny creature, scarcely bigger than the period at the end of this sentence. At the will of the ocean's current, it must find its host, which is little more than a vanishingly small speck in the vastness of the ocean. It is staggering that any of these animals manage to find their hosts at all.
- The adults of these parasites are so unusual that for centuries zoologists had no idea what they actually were. Some scientists believed they were a form of worm, whereas

others thought they were related to the mollusks. It was only when the life cycle of these animals was observed did it become clear they were actually a type of crustacean.

+ As the cod worm is such an invasive parasite, it is not surprising that its host suffers considerable damage. Not only does the parasite cause a great deal of structural damage to the host's circulatory system, but the very process of breaching the fish's skin and muscle opens a path to bacteria, fungi, and viruses. The parasite also extracts a lot of nutrients from the fish's blood. With this constant drain on its resources, an infected fish can lose almost a third of its weight, and the oil content of its liver can be reduced by half. The host's blood also loses about half of its all-important oxygen-containing hemoglobin.

+ One of the most unusual copepod parasites lives in a cyst in the muscle or body cavity of certain fish. The female is hardly more than a reproductive bag, absorbing nutrients from the blood of the host. The male, in what must be one of the most uncomfortable relationships, is squashed against the side of the cyst by the corpulent bulk of his much larger partner.

Further Reading: Rohde, K. *Ecology of Marine Parasites: An Introduction to Marine Parasitology*, 2nd ed. Redwood Books, Trowbridge, UK 1993.

CRICKET FLY

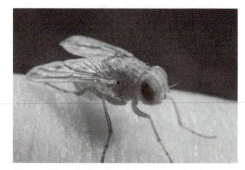

Cricket Fly—After locating her host with her ears, the female cricket fly prepares to lay an egg on it. (Mike Shanahan)

Cricket Fly—A female of this species bred in captivity. (Andrew Mason)

Scientific name: *Ormia ochracea*
Scientific classification:
  Phylum: Arthropoda
  Class: Insecta
  Order: Diptera
  Family: Tachinidae
What does it look like? Adult cricket flies are slightly less than 1 cm in length. They are a yellowish orange in color with red eyes.
Where does it live? The cricket fly is found throughout the southern United States and has even been found as far north as Ontario, Canada. They are usually encountered in open habitats, such as grassland.

Betrayed by a Love Song

To attract a mate, male animals go to great lengths. Many of them have very bold and bright colors to appeal to females, others have a range of ornaments to use in breeding disputes with rival males, and some build things to get a female's attention. These visual signs of a male's virility are all well and good, but the downside is that they make him stand out like a beacon. No self-respecting predator or parasitoid could miss such a show. Because of the disadvantages of broadcasting their suitability as a mate visually, some male animals hide themselves away and use sound. One such animal is the field cricket. In the safety of a burrow, it sings out to the females in the vicinity. The night air can be full of these subterranean stridulations, and for the most part, the crickets that make them are quite safe; however, over millions of years, a species of fly has developed the ability to listen in on these songs to locate its host.

Ears are very unusual in flies, as almost all of them rely heavily on their sense of sight and smell to locate their food and mates. Of course, they do have other senses, but hearing is very rarely one of them. On the bulbous thorax of the female cricket fly, there is a pair of very thin membranes, which function like the tight skin of a drum, similar to the eardrum found in the ears of mammals and other vertebrates. These fly ear drums are connected by a small bridge formed from the exoskeleton of the fly, and each is wired up to the insect's central nervous system. Not only are these structures possibly the smallest ears in the animal kingdom, but they are also among the most sensitive. Depending on where the sound of the singing male cricket is emanating from, the tiny ear drums will reverberate at slightly different frequencies. This difference may be as little as 50 billionths of a second, but it is enough to allow the fly to home in directly on a singing male cricket. It doesn't have to stop and cup its ears; it just homes in unerringly on the source of the noise. Even if the cricket stops singing, midhoming, the fly can approximate its position from the last sound it made.

Once it has been discovered, the cricket is powerless to stop the fly from completing its task. She walks over her host, and deposits lots of wriggling maggots on its body. A lucky maggot will find a weak spot in the armor of the cricket and wheedle its way into the body cavity of the hapless host. Once inside the host, it will grow rapidly, gorging itself on the cricket's organs. After 6 to 10 days, the maggot is ready to leave its host. In preparing to leave, it empties its gut of all of the waste material accumulated during its feast. If the cricket doesn't die from it insides being flooded with effluent, its time is definitely up when the maggot breaks free from its body. Within a short time after leaving the host, the skin of the maggot hardens and takes on the dark brown hue of the puparium. In the confines of its barrel-shaped puparium, the larva's tissues break down and rearrange themselves into the structures of the adult fly. After 10 to 12 days, an adult fly emerges, perhaps a female, her ears twitching to the sounds of singing crickets.

- The cricket fly is a type of tachinid fly. This is a large family of flies, comprising 8,200 identified species worldwide, but as with all obscure families of insects, there are many more species yet to be identified. In the United States alone, there are thought to be 1,300 species. All tachinids are parasitoids of other invertebrates, mostly insects. They commonly parasitize the larval stages of butterflies and moths, beetles, and their larvae. The larvae develop in the body of the host. Tachinid parasitism always results in the death of the host, and for this reason, some have been used as biological control agents for troublesome crop pests.

+ There are other flies, unrelated to the cricket fly, which parasitize singing cicadas. Again, they have small, but very sensitive ears that allow them to pinpoint the location of their host. One species, *Emblemasoma auditrix*, locates a cicada and clambers onto its body, edging under its wing in reverse. It uses the tough tip of its ovipositor to make a small gash in the large, sound producing organ of the cicada. It lays an egg in this tear, and the larva goes on to develop inside the body cavity of the host. A parasitized cicada is unable to sing because of its damaged sound organs. The mute host will no longer attract the attention of the parasitic flies, and the developing larva will not have to compete with others for the succulent insides of its host.

+ It might be expected that crickets and cicadas would evolve the ability to stop singing when parasitic flies are in the vicinity, but this doesn't seem to be the case. The benefit of singing and attracting a mate obviously outweighs the risk of being parasitized and meeting a nasty end.

Further Reading: Adamo, S. A., and Hoy, R. D. Effects of a tachinid parasitoid, *Ormia ochracea*, on the behaviour and reproduction of its male and female field cricket hosts (*Gryllus* spp.). *Journal of Insect Physiology* 41, (1995) 269–77; Lakes-Harlan, R., Stölting, H., and Strumpner, A. Convergent evolution of insect hearing organs from a preadaptive structure. *Proceedings of the Royal Society of London* (*Series B*) 266, (1999) 1161–67; Mason, A. C., Oshinsky, M. L., and Hoy, R. R. Hyperacute directional hearing in a microscale auditory system. *Nature* 410, (2001) 686–90.

GIANT ROUNDWORM

Scientific name: *Ascaris lumbicoides*
Scientific classification:
  Phylum: Nematodes
  Class: Secernentea
  Order Ascaridida
  Family: Ascarididae
What does it look like? Large nematode females are between 20 and 49 cm long and 3–6 mm wide, whereas males are 15–31 cm long and 2–4 mm wide. At the front end, surrounding the mouth, are three lips. The back end of the male is hook shaped.
Where does it live? They are found all over the world, wherever there are people. As adults, they live in the small intestine of humans.

Wandering Worms

This roundworm is one of the most prolific human parasites with a staggeringly complex life cycle that depends heavily on sheer numbers and coincidence. It is thought that at least 1 billion people around the world are infected with this worm, which spends its entire adult life in the human small intestine. Like many other nematodes, the giant roundworm is superbly adapted to a parasitic way of life. Their environment is the small intestine, and consequently they aren't dependent on the array of sense organs that other, free-living animals need to locate food or detect danger. Since they live amongst the host's digested food, evolution has eliminated the need for a complex gut, thus freeing up a lot of space inside the worm for prolific egg production. Much of the female's body is devoted to the production of eggs, and she does so in astounding numbers. At any given time, the ovaries of this worm can contain up to 27 million eggs, with 200,000 being

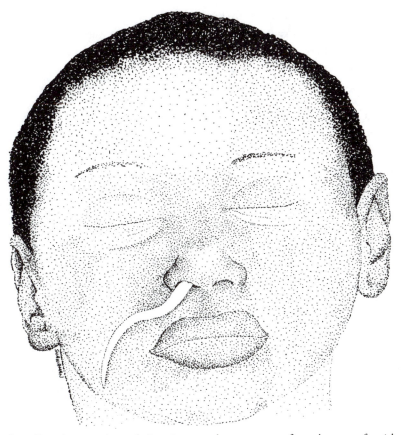

Giant Roundworm—A wandering giant roundworm emerges from the nose of a sick infant. (Mike Shanahan)

laid every day. The eggs pass from the host's body in the feces, and if they are fortunate, will be inadvertently swallowed with food or water.

Once inside the host, the eggs make their way through the stomach, protected by their tough shell, and end up in the in the duodenum where they hatch into microscopic larvae. Not content with staying put just yet, the larvae burrow through the intestinal lining into the circulatory system where they are spirited away in the blood. Eventually, they reach the right side of the heart and move into the blood stream that will take them to the lungs. In the lungs, they break out of the capillaries and into the tiny air sacs, through which oxygen and carbon dioxide diffuse in and out of the blood.

During this complex migration, many larvae get lost and end up in every organ of the body. The larvae that reach the lungs remain there for 10 days, shedding their skin twice and growing. This causes lung irritation, resulting in a characteristic cough, and some larvae may be coughed up and swallowed. Some of these larvae will not be prepared for the harsh conditions of the stomach, but those that are pass through the acidic environment of the stomach and make for the intestine, where they mature.

It is unknown why the larvae embark on this migration only to end up where they began. However, the migration route is similar to other nematodes that infect their host by penetrating the skin. These other nematodes must migrate to the lungs in order to ultimately reach the gut, so perhaps the behavior of the giant roundworm larvae is a vestige of some ancestral necessity.

The worm's normal activities in the human host go unnoticed; however, they sometimes occur in such large numbers that they can cause some bizarre conditions due to their wanderings. The worms may head downstream and find themselves in the appendix, which can be damaged, or they carry on down and emerge from the person's anus. The worms also wander upstream and may block the bile and pancreatic ducts with fatal consequences, or cause damage to the liver. Their wanderings may also take them to the stomach, an environment that is not to their liking, causing them to writhe and thrash, making the infected person feel nauseous. Often, the person may vomit one of these huge worms. Not only would this be pretty shocking, but breaking the worm as it is being thrown up can trigger a fatal allergic reaction. These upward wanderings are made easier when the host is sleeping and the worms don't have to tackle gravity. In the sleeping, horizontal host, the worm will reach the throat and may head for the lungs or will crawl even further up and in to the ear and nose, causing a lot of damage. Sometimes, a slumbering person infected with these worms may awake to find an adult giant roundworm popping out of their nose or mouth.

- At least 16,000 species of nematode have been identified, but there are many more yet to be identified. It is difficult to estimate how many species there are, but it is not in the realm of fiction to suggest that there may be a million different species. Nematodes are found everywhere, from the deepest ocean trench to the highest mountain, and from the miniscule spaces between the cells of a plant leaf to the gut of a grasshopper. Everywhere! Not only are they found in every conceivable habitat, but they are also found in profusion. In one rotting apple, there may be as many as 90,000 nematodes belonging to a number of species, and in 1 sq. m of mud from the seabed, there may be well over 4 million nematodes. Since they are normally so small, they are often forgotten about, but they are instrumental in the functioning of entire ecosystems.
- Although there are many free-living nematodes, they have turned parasitism into an art form. No animal or plant is safe from the depredations of parasitic nematodes. They display the whole range of parasitic interactions, from those species that are parasitic for only a small part of their life cycle to those species that have lived a parasitic way of life for so long they require a number of hosts and complex transmission routes to complete their development. They are extraordinary!
- The human giant roundworm is very closely related to the pig giant roundworm, but it is still unclear in which host this worm evolved. It is highly likely that the nematode was originally a parasite of pigs, which adapted to humans when pigs were first domesticated.
- The eggs of the nematode have a very tough, sculpted shell that is resistant to low temperatures, dry conditions, and strong chemicals. This protective coat enables the young worms to survive in unsuitable conditions for 10 years or more, until they are swallowed by a human. As the eggs can survive for so long, it is very difficult to eradicate this parasite. Reinfection is common, especially among small children who are always putting objects in their mouth. In some areas, human feces are sometimes used as a fertilizer (night soil), which can be a source of infection. Uncooked vegetables can be another source, as can cockroaches and the wind, both of which transport the eggs. The eggs of this worm have even been found on banknotes.

+ The wandering behavior of the worms is sometimes the result of a female unsuccessfully looking for a mate. The male nematode has a hooked tail, and the female must wriggle through this to bring their genitals into contact. Crawling through tight spaces possibly fulfills this instinct.
+ The female must produce such huge quantities of eggs because so many will never be swallowed and will perish. The production of huge numbers of eggs is very common among intestinal parasites. Their eggs are simply scattered, and more often than not, chance dictates whether they will find a host.
+ Occasionally, the worms in the intestine of an infected person will knot into a writhing mass, completely blocking the intestine with dire consequences. It is thought that certain drugs used to treat other intestinal parasites can aggravate the giant roundworm, causing them to bunch together.

GORDIAN WORMS

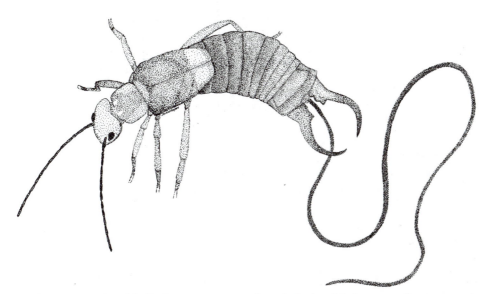

Gordian Worms—An adult Gordian worm emerges from the back end its host, an unfortunate earwig. (Mike Shanahan)

Scientific name: Nematomorphs
Scientific classification:
   Phylum: Nematomorpha
   Class: Gordioida
   Order: Chordodea and Gordioidea
   Family: Chordodidae, Gordiidae and Gordioideaincertae
What do they look like? The Gordian worms are long, threadlike worms that can reach lengths of more than 100 cm. Typically they are between 5 and 10 cm in length. Although long, they are very thin, with a diameter of around 1–2 mm. When alive, Gordian worms range from black to white with a shimmering iridescence. There are no distinct features on

the outer surface of the animal. The head end is usually a lighter color than the rest of the body. Adults are often found coiled in what looks like a loose knot.

Where do they live? Gordian worms are found all over the world in aquatic habitats, including lakes, streams, rivers, ditches, and even in the large, open water tanks that farm animals drink from. A few species are terrestrial and can be found in moist soil.

Overcoming a Dislike of Water

As adults, Gordian worms are very short lived, only seen fleetingly in the summer months, propelling themselves through freshwater by whiplike thrashings of their long, sinuous bodies. Their odd, seemingly spontaneous appearance has always intrigued people, who conjured up stories to explain what these animals were and where they came from. The adult, threadlike form is just one stage in the life of these interesting invertebrates. The life cycle begins with a male finding a female and entwining his body around hers before depositing sperm from his anus around the female's genital opening. The sperm fertilize the eggs inside the body of the female, and soon she lays many thousands of eggs in gelatinous strings that she attaches to aquatic plants. The eggs develop over a period of 15 to 80 days, eventually hatching to release small, highly active larvae. These larvae are very different from the adults. They are short, chunky grublike animals with a distinct head end bearing hooks and a proboscis armed with spines that can be projected and withdrawn like a tongue. The mobile larva must seek out and enter a host, such as a beetle, a grasshopper, a millipede, or even a leech. Exactly how it does this is uncertain; the actual act has never been seen. It is possible that the larva could latch onto a potential host and find a suitably weak point in the host's armor, puncturing it with its sharp proboscis. On the other hand, the larva could opt for a lazier strategy, forming a dormant cyst and waiting to be swallowed by a suitable, unlucky creature. Should the young worm find itself in the gut of a host, it tunnels its way out into the body cavity using its proboscis. Within the confines of the host, the worm metamorphoses into the nondescript adult form. As the worm's mouthparts are rudimentary, it cannot ingest any food; instead it absorbs all the nourishment it needs from the bodily fluids of its host. The worm grows steadily for several weeks or many months, shedding its skin several times, until it is a fully formed adult, which must leave the host that has inadvertently nurtured it. The adult worm is an aquatic animal, and to ensure that it can pass seamlessly from the sheltered environment of its host's insides to the safety of freshwater, it uses mind control. Somehow the worm hijacks the host's behavior and makes it head for water, even though it may be an animal that never normally has any need to visit bodies of open water. Sometimes, the overpowering mind-controlling abilities of the worm are so great that the host will make a headlong dive into an open tank of water

⚲ Go Look!

Gordian worms can be seen in bodies of freshwater, even those as small as water troughs. To the untrained eye, the worm does have more than a passing resemblance to an animated, knotted horse's hair. Finding an adults means that there are probably other individuals in the vicinity that are still in their host. Look for insects such as grasshoppers or beetles that appear to be paying more than a passing interest in the water. The insect, in a moment of worm-controlled madness, may leap with abandon into the water before the worm bursts out of its abdomen. The worm may also break out before the host has reached water. Earwigs are commonly infected. These are naturally creatures of the night, so look for those wandering around aimlessly in broad daylight. They will search, sometimes fruitlessly, for water. The worm, writhing with despair in the host, may simply break free even though water may be nowhere nearby.

or even a swimming pool. With its natural habitat tangibly close, the worm bursts out of the unfortunate host. This is by no means an easy or minor process for the host. The huge worm emerging from its rear end will be 10 times longer than itself, at the very least. After a long struggle, the worm frees itself from its so-called vehicle and swims off to search for a mate. The floundering host, reeling from the traumatic exit of its passenger, soon dies.

+ There are approximately 320 species of Gordian worm. The types mentioned here are all denizens of freshwater, but there are also marine species. As adults, they are free living, but all need a host in which to complete their development. The unlucky hosts of marine species are crabs and shrimps.

+ There is some interesting folklore surrounding these animals. In some places, they are known as horsehair worms in the belief that horse hairs falling into drinking troughs spontaneously give rise to them. As these worms have a liking for tying themselves in knots, they are called Gordian worms after Gordius, king of Phrygia, who apparently decreed that whoever could untie his intricate knot should rule his kingdom. Alexander the Great, ever the practical sort, cut the Gordian knot with his sword and added the king's empire to his own.

+ Very rarely, Gordian worms infect humans where they are found in the gut or in the urethra.

+ Some species of Gordian worm will go through more than one host, and some may only burst out in the autumn, an unsuitable time for breeding. To wait out the harsh winter months, they will form cysts on waterside vegetation, which they leave in the spring. In these cases, complete development may take as long as 15 months.

Further Readings: Poinar, G. O., Jr. "Nematoda and Nematomorpha." In Thorpe, J. H., and Govich, A. P. (eds.) *Ecology and Classification of North American Freshwater Invertebrates*. Academic Press, New York 1991; Poulin, R. "Adaptive" changes in the behavior of parasitized animals: A critical review. *International Journal of Parasitology* 25, (1995) 1371–83; Poulin, R. Evolution and phylogeny of behavioural manipulation of insect hosts by parasites. *Parasitology* 116, (1998) 3–11; Poulin, R. Observations on the free-living adult stage of *Gordius dimorphus* (Nematomorpha: Gordioidea). *Journal of Parasitology* 82, (1996) 845–46; Schmidt-Rhaesa, A. The life cycle of horsehair worms (Nematomorpha). *Acta Parasitologica* 46, (2001) 151–58; Thomas, F., Schmidt-Rhaesa, A., Martin, G., Manu, C., Durand, P., and Renaud, F. Do hairworms (Nematomorpha) manipulate the water seeking behaviour of their terrestrial hosts? *Journal of Evolutionary Biology* 15, (2005) 356–61; Thomas, F., Ulitsky, P., Augier, R., Dusticier, N., Samuel, D., Strambi, C., Biron, D. G, and Cayre, M. Biochemical and histological changes in the brain of the cricket *Nemobius sylvestris* infected by the manipulative parasite *Paragordius tricuspidatus* (Nematomorpha). *International Journal of Parasitology* 33, (2003) 435–43.

GUINEA WORM

Scientific name: *Dracunculus medinensis*
Scientific classification:
    Phylum: Nematoda
    Class: Secernentea
    Order: Spirurida
    Family: Dracunculidae
What does it look like? An adult female guinea worm is the only stage of this animal that most people are likely to see. They are long (up to 80 cm, although 120 cm is occasionally

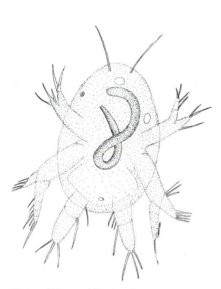

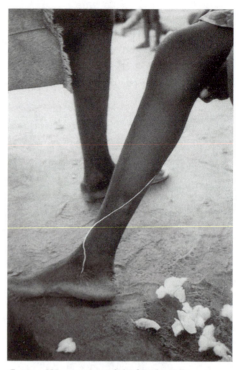

Guinea Worm—The larval guinea worm is clearly visible inside the body of this tiny, freshwater crustacean. (Mike Shanahan)

Guinea Worm—An adult female Guinea worm being drawn from an incision in the right calf of its human host. (The Carter Center/P. Emerson)

given as the longest length), thin, and pale and look for all intents and purposes like a length of thin spaghetti.

Where does it live? The Guinea worm is found in Africa, but its range is not as extensive as it once was, as it used to be endemic in the Middle East, India, and Pakistan, but has been eradicated from those countries. Its habitat in the larval stage is freshwater, while the habitat of the adults is the mammalian body, particularly humans.

Getting Under the Skin

The Guinea worm is probably one of the best documented parasites of humans, with tales of its behavior reaching as far back as the second century B.C. in accounts penned by ancient Greek chroniclers. People who have been afflicted with an infection of this worm would argue that it is a deeply offensive creature, and it is true to say that many international organizations have waged a long war to rid the earth of this worm. However, unsavory affects aside, the life cycle of this much maligned nematode is fascinating: an elaborate, but elegant route from a freshwater pool to the body cavity of a human.

The story begins with a microscopic larva drifting in the plankton of a small lake. With luck, this young nematode will be gobbled up by a passing crustacean, hopefully one of the little, fast swimming copepod species that are fixtures of pond life everywhere. The crustacean is the nematode's intermediate host, and it spends up to two weeks in the little crustacean going through the first two stages of its development. Stacking the odds even higher, the developing Guinea worm must now somehow get from the copepod to an unsuspecting human. Again, it seems to depend on good fortune, swallowed unknowingly by a human in a mouthful of water, still inside its little crustacean vehicle. Even when the young worm finds itself inside the definitive host, the journey is far from over, and the most difficult obstacles are still to come. In the stomach of the unsuspecting person, the copepod and the nematode part company. This is not through any choice of their own, but because the acids kill and digest the copepod, leaving the nematode to go it alone. Somehow, the young guinea worm is invulnerable to the potent chemical cocktail of the stomach, and it passes on through into the intestine where it burrows out into the body cavity. In the warm and moist safety of a large mammal, the worm quickly migrates to the body's musculature, where it can grow. If there is a worm of the opposite sex present, the worm can mate and start to produce eggs. All this lounging around absorbing nutrients cannot last forever, and after 10–14 months the female must embark on the last leg of her journey. She goes wandering, and in 9 out 10 cases she ends up in or near the foot. In most cases, the female worm causes a large blister to develop on the skin, just above herself. Eventually, this blister ruptures exposing a loop of the female worm. The ulcer causes a very painful burning sensation, and to relieve this the unfortunate person often bathes their feet in a pond or open well. This is the moment the female has been waiting for, and with no further invitation, she bursts, releasing thousands upon thousands of tiny larvae that find their way into the water, helped by muscular contractions of the female's uterus, which can force out more than half a million young in a single push. Over the next few days, the female is capable of releasing more larvae, until her whole body is used up. These larvae drift and wriggle off to find themselves a copepod, so continuing the cycle.

+ There are several species of nematode in the same family as the Guinea worm. All of them inhabit the tissues of vertebrates as adults.
+ The exit of the worm from the skin is not only painful, but also leaves the victim susceptible to bacterial infections, which can be life threatening in areas where hygiene is poor and antibiotics are in short supply. These lesions, particularly if a secondary infection sets in, may be painful enough to stop the person from using or even moving the affected limb, causing crippling and locked joints. Infections may also occur at times when work needs to be done, such as crop harvesting, severely affecting victims' livelihoods.
+ Once a person is affected with Guinea worm, there is little that can be done. There are no known drugs that are effective against the infection. The traditional treatment is to wait until the female emerges from the blister and slowly wind her out on a small stick. Completely removing the very long female can take weeks or even months, and winding must be conducted with extreme caution. Breaking the worm can release fluids into the body of the infected person causing anaphylactic shock or other serious reactions.
+ Several international organizations are waging an eradication campaign against the Guinea worm. Asia and the Middle East are now free of the disease, as are several African countries, including Kenya, Senegal, Cameroon, Chad, and the Central African Republic. In 1986, there were 3.5 million cases of Guinea worm infection worldwide, but in 2004, this had fallen to just over 16,000.
+ The symbol of the medical profession, the Rod of Asclepius, depicting a serpent coiled around a staff is thought to have been based on the guinea worm being wound around a stick. The fiery serpent on the end of Moses's staff is also thought to be a representation of the Guinea worm, such was its importance to people of the ancient Middle East.

- Preventing Guinea worm is straightforward. When in areas of Guinea worm infection, always filter dinking water, especially if it originates from an open source. Using the material of a shirt is more than adequate. This will strain out the tiny crustaceans that harbor the larval parasite. Taking water from a covered well or other enclosed source is another easy way to prevent infection.

Further Reading: Muller, R. "Dracunculus and dracunculiasis." In Dawes, B. (ed.) *Advances in Parasitology.* Academic Press, New York 1971.

HUMAN BOTFLY

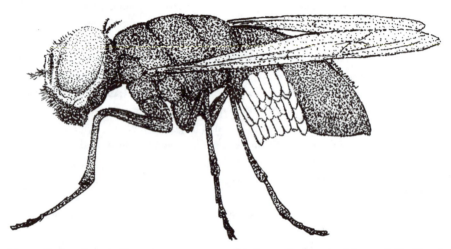

Human Botfly—A courier fly carries numerous eggs of the human botfly on its abdomen. (Mike Shanahan)

Scientific name: *Dermatobia hominis*
Scientific classification:
  Phylum: Arthropoda
  Class: Insecta
  Order: Diptera
  Family: Oestridae
What does it look like? The human botfly is a large fly that resembles a blue bottle. It is around 2 cm long with a metallic blue luster on its thorax and abdomen. From the side, it has a rather hunched appearance. The legs are reddish in color, as are the large eyes. The fly has no functional mouthparts.
Where does it live? The botfly is a forest creature of Central and South America. Its range extends as far north as Mexico and as far south as northern Argentina.

Parasitism by Courier

The forests of Central and South America are home to a great raft of insects that are desperate for a piece of a large, warm-blooded creature. Mammals fit the bill handsomely, and they are plagued by these biting and egg-laying parasites. Perhaps the most devious of all is the human botfly. Although it is known as the human botfly, it will parasitize any largish, warm-blooded animals, including birds. Livestock are a particular favorite as they are large, docile, and partially unaware of the insects buzzing around them. The female botfly, heavily laden with eggs, must deposit her developing young on a suitable host, such as a grass-chewing cow or an unlucky tourist. To do this herself would be tantamount to suicide. She is a large insect, and she makes a terrible din when she is flying. She could easily get accidentally swatted by an ungulate's swishing tail or purposefully squashed by an agitated tourist. She needs a courier to do her work for her, and so she goes off to find one. A suitable courier is a fly or a tick that also spends its live lapping at the secretions or sucking the blood of warm-blooded animals. With a suitable menial identified, she tackles it in midair, and both go cartwheeling into the undergrowth. The botfly is a powerful insect, and it holds the smaller fly down with its powerful legs before curling its abdomen around and depositing an egg on the rear end of the unknowing courier. The courier feels that it is a little bit heavier than normal and gives itself a good grooming to free whatever is sticking to its body. It can groom all it wants, but the egg is stuck fast, and no amount of tweaking and stroking will remove it. Apparently unperturbed by its tussle in the undergrowth, the fly takes to the air, but its bad luck isn't over yet.

Where there is one female botfly, there are bound to be others, all waiting for a suitable fly to carry their eggs to an unsuspecting host. Before the little carrier fly gets anywhere near a potential host, it may be lugging 12 or more eggs from different females. When it alights on the host to lick its sweat, the heat rising from the relatively massive body triggers the eggs of the botfly to hatch. Dangling from their eggs, tantalizingly close to safety and food, the botfly maggots drop onto the host and endeavor to get under its skin as quickly as possible. Safely concealed beneath the host's skin, the larvae start eating, head-down in its tissues, forming themselves a cavity that is enlarged as they grow. They breathe using small holes at the back end of their body, which face the entrance hole to the cavity. After several months of feeding, the larvae are fully grown and are visible as large, pimple-like lumps beneath the animal's skin. When they have had their fill of the host, they wriggle out, drop to the ground, and pupate in the soil.

- There are many species of flies that develop and feed in this way, although there are some very interesting deviations from the general theme. For example, the larvae of the sheep nostril fly develop in the nasal cavities of sheep and goats, rasping at the mucus membranes with their mouth hooks and ingesting the blood. Stranger still is the horse-stomach fly whose larvae live attached to the stomach lining of their equine hosts. They are mostly found in hot countries, although certain species are native to cooler, temperate countries.
- *Dermatobia hominis* is unusual as it routinely develops under the skin of humans. Infections can often involve many larvae, and many areas of the body can be affected, including the groin, the legs, and the head. Travelers to Central and South America can sometimes return to their home countries unknowingly carrying a cargo of botfly maggots beneath their skin. At first the lesions above the larvae look like small pimples, but there are sudden shooting pains caused by the maggot's feeding activity. To rid the unfortunate victim of the infection, the lesion can be smeared with petroleum jelly or animal fat, suffocating the larva. In this situation, there is a possibility the larva may die and rot in its feeding chamber, causing a serious infection. An alternative method involves the surgical removal of the maggot with forceps, taking care not rupture its body.

- The human botfly has been shown to use over 40 species of fly and one species of tick as carriers for its eggs.
- Botflys cause considerable damage to livestock in the tropics. They damage the meat, ruin the hide, and leave the host vulnerable to infection with their feeding activities; but this is because humans are encroaching on their habitat, and the fly is just doing what it does best.
- The lesions caused by the botfly in cattle can become infected by a bacteria that leads to a serious disease, known as lechiguana, characterized by rapid growing, hard lumps beneath the skin of the animal. Antibiotics are the animal's only hope, and without them it will die within 3–11 months.

Further Reading: Cogley, T. P., and Cogley, M. C. Morphology of the eggs of the human bot fly, Dermatobia hominis (L. Jr.) (Diptera, Cuterebridae) and their adherence to the transport carrier. *International Journal of Insect Morphology & Embryology* 18, (1989) 239–48; Mourier, H., and Banegas, A.D. Observations on the oviposition and the ecology of the eggs of *Dermatobia hominis* (Diptera: Cuterebridae). *Vidensk Meddel Dansk Naturhist* 133, (1970) 59–68; Tsuda, S., Nagai, J., Kurose, K., Miyasoto, M., Sasai, Y., and Yoneda, Y. Furuncular cutaneous myiasis caused by *Dermatobia hominis* larvae following travel to Brazil. *International Journal of Dermatology* 35, (1996) 121–23; Zumpt, F. "Diptera parasitic on vertebrates in Africa south of the Sahara and in South America and their medical significance." In Meggers, B. T., Ayensy, E. S., and Duckworth, W. D. (eds.) *Tropical Forest Ecosystems in Africa and South America: A Comparative Review*. Smithsonian Institution Press, Washington, D.C. 1973.

LEAF WASPS

Scientific name: Trigonalids
Scientific classification:
   Phylum: Arthropoda
   Class: Insecta
   Order: Hymenoptera
   Family: Trigonalidae

What do they look like? By parasitic wasp standards, most trigonalids are quite large (5–15 mm long), although some species are as small as 3 mm. Body shape varies considerably. Some are long and thin, resembling the ichneumonid wasps, while others are heavier bodied. All have a characteristic hole-punch ovipositor.

Where do they live? The trigonalids are found in a wide variety of habitats all over the world, apart from the high polar regions. They appear to be most abundant in the tropics, and only one species is found in Europe.

The Certainty of Chance

Reproduction in all animals is a haphazard affair involving a considerable amount of luck to ensure the continuation of the species. Nowhere is this more apparent than in the trigonalid wasps. The gauntlet they must negotiate from egg to adulthood is bewilderingly complex and relies heavily on chance. Trigonalids are parasitoids, and like the rest of their kind, they feed on other animals as larvae. In most cases, a female parasitic wasp deposits her eggs on or in the host, but the female trigonalid does neither—she deposits her eggs on the leaves of trees. Using her short ovipositor, she makes a small incision in the outer surface of the leaf to form a small pocket. Into this pocket, she deposits an egg. She repeats this procedure over and over again, scattering her

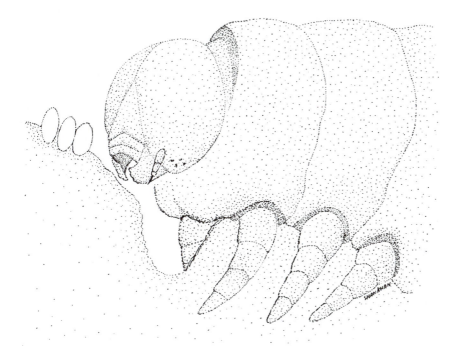

Leaf Wasps—A hungry caterpillar edges closer to the eggs of a leaf wasp. (Mike Shanahan)

eggs over a considerable area of foliage. The plants she lays her eggs on aren't just a random assortment of species, but the preferred food plant of the trigonalid's primary host. The primary host is an insect that is often the sucker for the cruel intentions of parasitic wasps everywhere, the humble caterpillar, although the caterpillar-like larvae of sawfly are also targeted. The larvae, with their ceaselessly munching jaws, chomp their way through a large number of leaves. Embedded beneath the outer surface of some of these leaves may be the developing egg of a trigonalid. The eggs are small and tough, and in the blink of an eye, they have been swallowed by the ravenous caterpillar. With the odds stacked against them, the unhatched trigonalid larvae have succeeded in the first part of their difficult mission.

Once inside the gut of the caterpillar, the enzymes and other gastric secretions act as a cue for the larva to hatch from its tough capsule. This process may even be triggered as soon as the egg is taken into the mouth of the host, where it is bathed in saliva and chewed. Experiments have shown that eggs that are chewed on are much more likely to hatch. Whatever the trigger, the larva wriggles from its egg and makes for the gut wall of its host. The environment in the gut is harsh and alkaline, and few things can survive there. Using means not completely understood, but perhaps a combination of erosive secretions and mechanical rasping, the larva burrows through the gut wall and into the body cavity of the caterpillar. This space is filled with hemolymph, the insect equivalent of blood, and through this medium it swims with grim intent, apparently searching for something. What the small trigonalid larva is looking for is the maggot of a parasitic fly or wasp, which has been growing in the caterpillar, feeding in safety on the nonessential tissues of the caterpillar and absorbing nutrients from the blood. Only a small percentage of caterpillars will be infected with the right parasite; therefore, it is not only a massive long-shot for the trigonalid larva to get into a caterpillar but also to be swallowed by one containing the right parasitoid. In this unusual, diminutive aquatic habitat, the original parasitoid larva is soon to get a taste of its own medicine. It can't leave the host until it is fully developed; therefore, it is a sitting duck, waiting for the trigonalid larva to sniff it out. This is one elaborate way in which the trigonalids complete their development. The survival of other trigonalid species also hinges on coincidence, but the supporting cast is slightly different. They still depend on a caterpillar or sawfly larvae, but this time, the caterpillar must be captured by a social vespid wasp, butchered, and fed to one of its grubs back at the nest. In the flesh of the dead caterpillar are the eggs of the trigonalid, and once swallowed by the wasp grub, they can complete their development.

+ The trigonalids are a small group of parasitic wasps. Around 90 species are known, but they are rare animals, and it is very likely many more species are yet to identified, especially in the tropics.

+ As trigonalids have such an unusual, scatter-gun approach to host infection, many eggs are needed. A single female trigonalid can produce as many as 10,000 eggs in her short lifetime.

+ Although many trigonalids are what is known as *hyperparasites*, requiring a parasite already within a host, some species parasitize the caterpillar or sawfly larva directly.

+ The distribution of some of the trigonalids suggests that within the parasitic wasps they are an ancient group. They have probably been living out their extraordinary lives for many millions of years. The oldest trigonalid known was found entombed in a lump of amber from the mid-Cretaceous, making it around 100 million years old.

+ The best way to catch an adult trigonalid is to use a malaise trap, which looks like a tent made from very fine gauze. Insects like wasps and flies will be intercepted by the upright wall of the trap and walk up to its apex, where they will enter a collecting vessel filled with preserving fluid. Such a trap can be left in place for a number of days or even weeks.

Further Reading: Carmean, D. Biology of the Trigonalyidae (Hymenoptera), with notes on one vespine parasitoid *Bareogonalos canadensis*. *New Zealand Journal of Zoology* 18, (1991) 209–14; Clausen, C. P. Biological notes on the Trigonalidae (Hymenoptera). *Proceedings of the Entomological Society of Washington* 33, (1931) 72–81; Weinstein, P., and Austin, A. D. Primary parasitism, development and adult biology in the wasp *Taeniogonalos venatoria* Riek (Hymenoptera: Trigonalyidae). *Australian Journal of Zoology* 43, (1995) 541–55.

RABBIT FLEA

Scientific name: *Spilopsyllus cuniculi*
Scientific classification:
　　Phylum: Arthropoda
　　Class: Insecta
　　Order: Siphonaptera
　　Family: Pulicidae
What does it look like? Adult rabbit fleas are approximately 1 mm long and dark brown in color. They are compressed from side to side with no wings. Carried beneath the head are the sharp, piercing mouthparts shielded by a spiky comb, reminiscent of a moustache.
Where does it live? The rabbit flea is found wherever there are European rabbits or any suitable hosts, regardless of whether they are wild or not. Naturally, rabbits have a wide

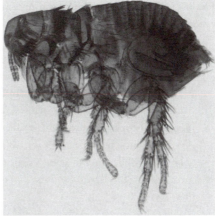

Rabbit Flea—A preserved, adult rabbit flea on a microscope slide. Note the numerous spines and bristles that help keep it in place on its host. (David Merritt)

Rabbit Flea—Rabbit fleas jump from a mother rabbit to her young. (Mike Shanahan)

distribution, but this has been increased considerably by the export of rabbits for food and the pet trade.

From Mother to Baby

The rabbit flea is a skin-dwelling parasite of the European rabbit and will take up residence on other rabbit species, hares, and occasionally dogs and cats. Like all fleas, they are proficient parasites with a number of adaptations allowing them to find and stick to a host. Their powerful back legs enable them to jump huge distances for such a small animal. Their flattened bodies allow them to lie close to their host's skin, making them difficult to dislodge, and their sharp mouthparts can pierce the tough skin of their host to suck the blood flowing beneath. The rabbit flea has a further adaptation, refining its parasitic ways still further. Its relationship with the rabbit is a long one, and during this time, it has tuned in to its host's cycles and rhythms to the extent that its own reproduction is dictated by the hormones of the rabbit.

The rabbit flea does not move around much and spends most of its time attached to the ears of its host, keeping its head down and sucking blood. However, 10 days before the female host gives birth, the female fleas change. They start to reach sexual maturity. Their ovaries develop and start producing eggs of their own. It doesn't need a calendar to know when the time is right; instead it tastes the changing levels of cortisol and corticosterone, hormones in the rabbit's blood, indicating that the arrival of the host's young is imminent. The fleas sit out these 10 days with their mouthparts plugged into the bunny, tasting the changing chemistry of its blood. As soon as the babies are born, they must extract their sucking stylets, leave the lofty perch of the ears, and make for the face of the rabbit, accompanied by the male fleas. Once at the face, the insects can easily hop onto the newborn baby rabbits as they are nuzzled by their flea-bitten mother. On these new, smaller hosts, the fleas feed voraciously, mate, and lay their eggs—all triggered by increasing levels of growth hormones in the blood of the baby rabbits. After 12 days of feeding, mating, and egg laying, the fleas leave the virgin territory of the newborn rabbits and return to the mother rabbit. As a female rabbit can give birth to a litter of young every 31 days on average, the fleas make this miniature migration on a regular basis, successfully parasitizing each new generation of rabbits.

+ There are around 2,500 species of fleas, and every single one of them parasitizes mammals or birds. Unlike many of the insect parasites of vertebrates, the fleas have a larval stage that can only develop in a sheltered environment. This limits them to hosts with nests or regularly used burrows. A perfect example of this is the human flea. The homes of people make an ideal breeding ground for flea larvae. No other primates have fixed dwellings; therefore, fleas can never make a living on monkeys or apes.
+ The larvae of fleas are maggoty creatures that wriggle around in the nests of their host, feeding on detritus and the digested blood defecated by the adult fleas on the nest owner.
+ All fleas are wingless. Wings would be a burden for a parasite of birds and mammals as scrabbling through fur and feathers would quickly damage them. Normally, fleas crawl clumsily around their host, using their piercing mouthparts as anchors, but when they first hatch from their pupae, or if they are dislodged, they use their back legs to make huge leaps. Their prodigious jumping abilities are not due to massive muscles, but the elastic properties of a protein called *resilin*. This material is found in the wings and thorax of many insects, and its elastic abilities make a bouncy rubber ball look a bit flat. These stretchy qualities are used to their maximum in the fleas. Preparing for a leap, the flea notches back its hind legs to a catching point, compressing the resilin like a tiny, but powerful spring. Poised, the flea releases the catch and is catapulted into the air. In human terms, this is equivalent to a person jumping to a height of over 200 m and covering a distance of 140 m. Measure for measure this is the greatest jump of any animal.
+ During these jumps, a flea reaches an acceleration of 140G in a little more than a millisecond, a force that would tear a larger animal to pieces. Muscles would never be able to simulate this feat as they contract too slowly and don't perform very well in low temperatures.
+ Although fleas are very interesting animals, their reputation is tarnished because of the numerous diseases they transmit. Plague, probably the single most important disease in human history, is transmitted by the oriental rat flea. The pandemic of plague in the fourteenth century killed more than a quarter of all Europeans (25 million people). The bacteria that causes plague is found in rodents and is sucked up by the fleas when they are feeding on blood. Inside the flea, the bacteria reproduce to the extent that they can block the insect's throat, and when the flea bites a human, which it inevitably will, the bacteria are regurgitated into the wound, eventually causing the nasty symptoms of plague. This disease goes through cycles of epidemic and remission, and at the moment, it appears to be in remission. However, a huge reservoir of bacteria can be found in rodents, dogs, and fleas, and it is only a matter of time until there is another large outbreak.
+ Myxomatosis is an infamous virus transmitted by the rabbit flea, amongst other biting insects. It originated in South America, where rabbits are quite resistant to it and was introduced intentionally into Australia to control the burgeoning rabbit population. In the 1950s and 1960s, huge numbers of rabbits were killed across Europe and in Australia.

Further Reading: Rothschild, M. Fleas. *Scientific American* 313, (1965) 44–53; Rothschild, M. and Ford, B. Breeding of the rabbit flea (*Spiloptyllus cumculi* (Dale)) controlled by the reproductive hormones of the host. *Nature* 201, (1964) 103–4.

RED-TAILED WASP

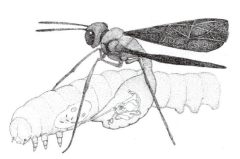 

Red-Tailed Wasp—A red-tailed wasp injects its eggs and viral particles into the body of a caterpillar. (Mike Shanahan)

Red-Tailed Wasp—A female red-tailed wasp injecting an egg in her host, a tobacco bud worm caterpillar. (Andrei Sourakov and Cosuelo De Moraes)

Scientific name: *Cardiochiles nigriceps*
Scientific classification:
    Phylum: Arthropoda
    Class: Insecta
    Order: Hymenoptera
    Family: Braconidae

What does it look like? This braconid is a small wasp, about 9 mm long. Its wings and the front part of its body are black, while its abdomen and its hind legs are red. The first few segments of the abdomen are rather narrow, giving it a slim waist. The antennae are long and black and are formed from many minute segments.

Where does it live? This wasp is native to the eastern and southwestern United States, but it can also be found in California. It is found in association with tobacco plants, on which it finds its host. It has been purposefully introduced to other parts of the world where tobacco is grown commercially.

Biological Weapons

Today, tobacco is grown all over the world to provide millions of people with the nicotine fix they require every day. Although this crop is valuable to humans, many other animals like to nibble it. One of these is the tobacco budworm, a dingy little caterpillar that likes nothing more than the succulent parts of this valuable plant. These budworms do untold damage as hungry hordes of them chew their way into the flower buds and blossoms of the tobacco plant. Fortunately for smokers everywhere, these voracious little caterpillars do not have free reign of the tobacco harvest. Flying delicately through the foliage of the crop, small, glossy female red-tailed wasps search for the tell-tale signs of the caterpillar's feeding activity. The wasp tastes the air for the distinctive scent given off by the tobacco plant when it is being eaten and perhaps even the odorous scent of the caterpillar's waste products. However they find their prey, they are very good at it, and as soon as they are in the vicinity of a caterpillar, they alight and do the last stage of the hunt on foot. Normally the caterpillar is oblivious to most things apart from eating, but it will detect the presence of the wasp and may convulse wildly in attempt to scare the parasite away. Carefully, so as not to alarm the caterpillar, the wasp rears up and twists its abdomen through its legs. The wasp's syringe-like ovipositor is inched toward the host and is delicately used to puncture the skin of the plump caterpillar. All of this happens in the blink of an eye, and before the caterpillar knows what has happened, it is playing host to the tiny egg of the female wasp. In itself this is not that special. Thousands of smallish wasps parasitize their hosts in the same way, but what sets this animal and its relatives apart from the majority of other parasitic wasps is the other things it injects into the host along with the egg.

Surrounding the egg like microscopic bodyguards are tiny particles that bear more than a passing resemblance to viruses. These viruses appear to be unique to these wasps and are created in an organ inside the female called the *calyx*. The DNA of the wasp actually contains portions that are the templates for the components of the viral particles. The role of these viral particles gives a whole new meaning to the term *biological warfare*. Somehow, they trick the immune system of the host, and the wasp's egg develops unmolested. They cloak the egg, making it appear to the host's immune cells as part of caterpillar's body and therefore safe. Completely free from the attentions of the host's last line of defense, the egg hatches and the larva feeds on the soft, juicy insides of the caterpillar. The host continues as normal, but when it has shed its skin for the fourth or fifth time, it pupates prematurely and dies, a result of the growing larva controlling its development and feeding on its essential organs.

+ Braconids and their close relatives, the ichneumonids, are among the most numerous types of insect in terms of species. Over 12,000 species of braconid have been identified so far, but it is very likely that more than four times this number exist.
+ Their great diversity is a reflection of the variety of insect life as the hosts of many of these wasps are the larvae of beetles, flies, butterflies and moths, and the various stages of true bugs, such as aphids. Many braconids parasitize only one host, although others parasitize a range of species. Fossils suggest that these wasps first appeared in the Creataceous period, more than 65 million years ago. This was when the flowering plants first appeared. The appearance of many new types of food led to the evolution of new species of plant-feeding insects, which in turn provided food for an increasing diversity of predators and parasitoids.
+ Any familiar insect you see will be the host of at least one species of parasitic wasp. For every butterfly you see, many others will have perished as caterpillars and pupae at the hands of these amazing little parasitoids.
+ The red-tailed wasp and its predilection for the tobacco bud worm is an example of how farmers can protect their crops without having to rely on toxic insecticides. All the animals in an ecosystem live in balance with one another. The parasitoids cannot kill all of their hosts as it would lead to their own extinction. In a situation free from human intervention, the parasitoids regulate the populations of their hosts. However, when farmers spray their crops with poisonous chemicals, they not only kill the pests,

but also the predators and parasitoids of these pests. The latter often struggle to recover from these dousings, while the pests bounce back rapidly. As they are no longer being eaten and parasitized, these pest populations reach plague proportions. This is the inevitable long-term result of intensive agriculture. Harnessing the populations of natural predators and parasitoids allows farmers to protect their crops from pests without resorting to pesticides.

Further Reading: Fleming, J. G. W. Polydnaviruses: mutualists and pathogens. *Annual Review of Entomology* 37, (1992) 401–25; Schmidt, O., and Schumchmann-Feddersen, I. Role of virus-like particles in parasitoid-host interaction of insects. *Subcellular Biochemistry* 15, (1989) 91–119; Stoltz, D. B., and Whitfield, J. B. Viruses and virus-like entities in the parasitic Hymenoptera. *Journal of Hymenoptera Research* 1, (1992) 125–39.

SABRE WASP

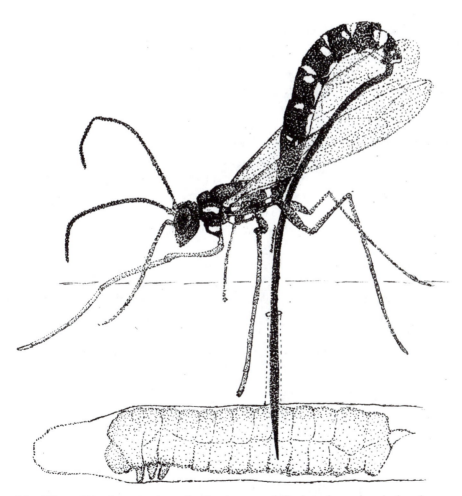

Sabre Wasp—Using her ovipositor a female sabre wasp drills through wood to the host larva on which her young will develop. (Mike Shanahan)

Scientific name: *Rhyssa persuasoria*
Scientific classification:
    Phylum: Arthropoda
    Class: Insecta
    Order: Hymenoptera
    Family: Ichneumonidae

What does it look like? This is one of the largest ichneumon wasps. The body of the insect can be around 4 cm long, with another 4 cm in the form a long, thin egg-laying tube (ovipositor). This splendid insect is glossy black with white/yellow markings. The legs are normally of an orange hue. A long pair of constantly flickering antennae issue from the head of the wasp.

Where does it live? The sabre wasp is a denizen of conifer forests in the Northern Hemisphere all the way down to the latitudes of North Africa. Within these forests, open patches such as clearings and paths are preferred because of the sheltered, warm conditions they offer.

Out of Sight, but Not Out of Mind

During the summer months in conifer forests, the parasitic sabre wasp can often be seen flying purposefully through clearings and along sunlit paths. As it flies, it senses the air for tell-tale signs of its host. Exactly what it smells is unclear, but it is thought to be the feces and feeding debris of large, wood-boring wasp larvae that live in tunnels deep in the conifer wood. As soon as the wasp detects the distinctive odor, it follows the source to a large, fallen conifer. Alighting on the trunk of the tree, it moves jerkily back and forth, its antennae constantly twitching. It senses that somewhere beneath its feet is a wasp larva greedily munching a tunnel through the wood, but it is impossible to tell the exact position of its prey from smell alone. By flicking its antennae on the wood, the wasp listens for minute differences in sound that may pinpoint the position of the host. It is also possible that the wasp can actually hear the host larva rasping at the wood with its powerful mandibles several centimeters below the bark. Confident it has identified the correct location, the wasp begins to drill. It doesn't do this by chewing the wood or frantically scrabbling with its feet, but by arching its abdomen high into the air and levering the long needle-like ovipositor until the tip of it engages the wood. The ovipositor is composed of two parts that can slide past one another, enabling the tip to be forced through the wood, like some sort of mechanical drill. The tip of the ovipositor steadily edges through the wood, and after approximately 30 minutes, it may have breached the tunnel wall of the host larva, which is blissfully unaware of the piercing egg layer that is heading straight for it. The very flexible ovipositor delivers a paralyzing sting to the host larva and then deposits an egg in the tunnel near the doomed grub. After a few days, the parasite egg hatches, and there in front of it is the only food it will ever need as a juvenile. It makes straight for the paralyzed grub and starts tucking in straight away, being careful not to nibble or chomp any crucial bits that could kill the host. The parasite wants the host to remain alive for as long as possible so the flesh doesn't spoil, but when it is nearly fully grown, it will deliver the coup de grâce and kill the hapless host. Safe in the tunnel, the sabre wasp larva pupates in a cocoon that it spins from silk extruded by glands in its mouth. In the cocoon, the tissues of the larvae are reordered into the adult form, and it may remain in this state in the tunnel for the winter, emerging as an adult the following summer.

+ The ichneumonids and their close relatives, the braconids, are thought to be represented by 80,000 species, at the very least, all of which are parasitic. Many species of ichneumon will tuck their egg inside the host using the pointed ovipositor; however, some species, like the sabre wasp, go after larvae concealed in various plants. In these tunnels, the eggs can be deposited on or near the host.

+ For a long time, scientists have cogitated on how the sabre wasp drills through wood with an ovipositor composed of chitin. Surely, over time the drilling process would wear the ovipositor down to a nub. It was discovered recently that the cuticle of the wasp, especially in crucial places like the ovipositor and mandibles, is impregnated with metals, including zinc and manganese. These elements afford the ovipositor added strength for penetrating the wood. Not only do these metals assist in the drilling process, but they also come in useful when the adult insect has hatched from its pupa and needs to chew its way out of the tree.

+ In North America, there is a species of ichneumon very similar to the sabre wasp, but it is much larger. It is known as *Megarhyssa* and is brown with yellow and black markings.

+ When the wasp is laying an egg onto its host, the egg has to be forced through the very narrow tube of the ovipositor; therefore, it is deformed into a sausage shape and pushed through the narrow channel.

+ Sabre wasps and their hosts, the large wood wasps, are often found inside newly built homes. They don't fly in through windows, but emerge from the timbers used in the construction of the house.

> ### ⚲ Go Look!
>
> Ichneumon wasps and their close relatives, the braconids, can be commonly encountered in the summer months. They are predominantly black with bold markings and constantly twitching antenna often bearing white flashes. They are typically to be seen feeding on nectar from flowers or scudding low amongst the vegetation. The nectar and pollen they collect from flowers fuels their flight, and both male and females will visit floral blooms. When you observe one of these wasps flying with intent low among the herbage, you are witnessing a female looking for a host in which her offspring can develop. Caterpillars are often the targets of these searches. It is rare to catch an ichneumon in the act of slipping its egg into the plump body of a living caterpillar as the whole act is over very quickly, with the host barely looking up from the leaf it is earnestly munching. An excellent way to watch the development of an ichneumon is to collect a few caterpillars from the wild, either by carefully looking on vegetation, or by using a beating tray (a sheet of material stretched over a frame used to catch insects dislodged by tapping a branch or bush with a stick). The caterpillars can be taken home with their food plant and observed until they pupate. Instead of a moth or butterfly, what pops its head from the chrysalis may be an adult ichneumon. Parasitic wasps, like ichneumons, are so effective at finding and inserting their eggs into their hosts that many collected caterpillars will be playing host to one of these wasp larvae.

Further Reading: Quicke, D. L. J., Wyeth, P., Fawke, J. D., Basibuyuk, H. H., and Vincent, J. F. V. Manganese and zinc in the ovipositors and mandibles of hymenopterous insects. *Zoological Journal of the Linnaean Society* 124, (1998) 387–96; Spradbery, J. P. Host finding by *Rhyssa persuasoria* (L.), an ichneumonid parasite of siricid woodwasps. *Animal Behaviour* 18, (1970) 103–14.

SACCULINA

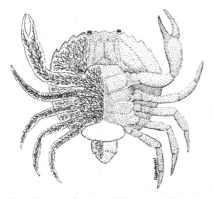

Sacculina—A female sacculina grows like the roots of a plant into every part of a crab's body. (Mike Shanahan)

Sacculina—The brown, globular egg-sac of the parasite can be seen protruding from beneath the crab's abdomen. As the crab cannot shed its exoskeleton, barnacles have taken up residence. (Adam Petrusek)

Scientific name: *Sacculina carcini*
Scientific classification:
   Phylum: Arthropoda
   Class: Cirripedia
   Order: Rhizocephala
   Family: Sacculinidae
What does it look like? These bizarre barnacles are saclike in form; they lack all appendages and don't have any obvious segmentation. The adult female resembles the branching roots of a plant. The adult male has no distinct form as he essentially becomes part of the female once inside the host.
Where does it live? This animal is found wherever the green shore crab is found, normally the intertidal zone and very shallow water.

The Crab's Nemesis

The crustaceans are normally regarded as quite benign animals, familiar to us as crabs, prawns, shrimps, and lobsters, yet among their numbers are some very interesting parasites, which can only be described as remarkable. *Sacculina* is one of these. It is a type of barnacle, and its host is the shore crab. A young female *Sacculina* finds its host and hooks onto it using the feelers at the front of her body. The parasite goes for the base of one the crab's hairs or another part of the host's body where the exoskeleton is thin, such as the thin membrane between the leg segments. This larva undergoes metamorphosis, losing all of its limbs and other crustacean characteristics to become what is essentially a living syringe capable of injecting a cuticle-clad mass of cells into the crab. The clump of cells migrates to one of the crab's nerve cords where it grows like the roots of a plant, eventually branching out through most of the host's body, absorbing nutrients from the crab's blood. Once fully grown inside its host, the parasite can apply itself to reproduction, involving the formation of an external brood chamber resembling the egg mass of the female host. The female parasite, her branches snaking their way into every part of the unlucky crab, then awaits the arrival of a male. As luck would have it, a suitor alights on the crab and enters the brood chamber in the same way the female parasite got inside the host—self injection. The injected cells become a spine-covered larva that migrates to a special chamber in the female before developing into a testis. With the male now in place and ready to produce sperm, the parasite can apply itself to producing the yellowish, spongy brood chamber, which protrudes from beneath the crab's abdomen.

> ## ♀ Go Look!
>
> If you go to the beach, you can try to look for these parasitic barnacles by searching for crabs at low tide. A crab parasitized by *Sacculina* will have lots of encrusting material on its shell as it cannot shed its skin. The brood chamber of the parasite can be seen by turning the crab over. A smooth, yellowish or brown sac may be found poking out from the beneath the abdominal plates of the crab, and you can tell if it is a *Sacculina* egg mass or not as a crab's egg mass is granular rather than smooth. *Sacculina* parasitizes up to 50 percent of crabs, depending on the location, so you have a good chance of finding some of these very interesting crustaceans.

The effects of the parasitism on the crab are severe. For instance, it can no longer molt its exoskeleton, which is how all crustaceans grow. Also, the parasite absorbs so much nourishment from the crab's body that there is not enough for the sex organs to mature sufficiently, and the host is sterilized, although it is possible that substances produced by the parasite may also account for this. Interestingly, the crab host will behave as if the parasite's egg mass is its own, protecting and cleaning it and maintaining a flow of water around it to provide the developing parasite larvae with sufficient oxygen. It is highly likely that the parasite is responsible for this behavior, producing substances that control the crab.

Even if the parasite develops in a male crab, the above behavior is played out as the male is turned into a female due to the constricting effect of *Sacculina* on the male's genitals.

+ Barnacles are very interesting animals. They are all marine and are the only sessile crustaceans. Until around 1830, they were thought to be mollusks, related to animals like limpets and snails. The vast majority of barnacles encrust rocks and the like; however, there are some barnacles (i.e., *Sacculina*) that have forsaken the typical barnacle existence and live in the bodies of other sea creatures.

+ There are many barnacle species that live on or in the bodies of other marine animals; thus making the evolutionary step from commensal to parasitic a relatively small one. Parasitism is just a small part of a continuum of interactions between different organisms, including phoresis, commensalism, symbiosis, and so forth.

+ The host crab is so duped by *Sacculina*'s scheme that it will even perform spawning behavior when the parasite's offspring are ready to emerge. It is thought that the parasite produces chemicals controlling this behavior. The misguided tender loving care shown by the crab for the parasite's young is essential. The parasite brood chamber of crabs that are not able to offer cleaning services soon become diseased.

+ As if the relationship between the crab and *Sacculina* weren't complicated enough, nature heaps another layer of complexity on top: *Sacculina* is also parasitized and by another type of crustacean.

Further Reading: Goddard, J.H.R., Torchin, M. E., Kuris, A. M., and Lafferty, K. D. Host specificity of *Sacculina carcini*, a potential biological control agent of the introduced European green crab *Carcinus maenas* in California. *Biological Invasions* 7, (2005) 895–912.

STREPSIPTERANS

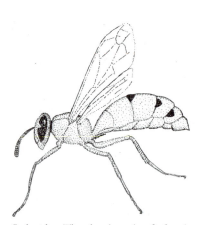

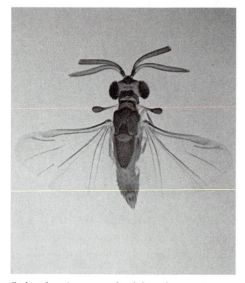

Stylopids—The head end of female strepsipterans emerges from between the abdominal plates of an unlucky wasp. (Mike Shanahan)

Stylopids—A preserved, adult male strepsipteran. Note the elaborate antennae, unusual hind wings and the fore wings, which are reduced to club-like structures called halteres. (Ross Piper)

Scientific name: *Strepsipterans*
Scientific classification:
   Phylum: Arthropoda
   Class: Insecta
   Order: Strepsiptera
   Family: several families

What do they look like? These are small insects; adult males of the largest species are around 5 mm in length. The females are grublike without any discernible features, whereas the short-lived males have large, delicate wings, protruding raspberry-like eyes, and antler-like antennae. The first pair of wings is hugely reduced, forming small, club-shaped structures called halteres.

Where do they live? These insects are found all around the world in all kinds of habitats, wherever there are suitable hosts for their development. They are very difficult to find due to their small size and specialist ways.

A Pocket Parasite

The strepsipterans are a unique group of parasitic insects. The adult males and females are very different in appearance, but both start their life as a small, hyperactive larva, scooting freely around in the body cavity of their mother. When the time comes to leave, they exit the body cavity through a genital pore and move down a narrow brood canal to the outside world. The female produces so many young that the vegetation where they are released becomes a bustle of squirming and jumping larvae, all eager to find a host. A safe bet for a host is a bee or a wasp, but this depends on the species of strepsipteran in question. The jostling larvae make for flowers, which may be visited by bees and wasps foraging for nectar and pollen. When an insect that is the right size, shape, and color for a bee or wasp comes within range of the larva, it uses the long, stiff paired bristles at its hind end to launch itself into the air with the hope of hitting the buzzing insect, which it cling on to for dear life with its clawed feet. The bee or wasp, unperturbed by the presence of its new passenger, heads back to its nest to feed its own larvae with the food it collected on its foraging trip. Once in the nest of the host, the strepsipteran larva disembarks from the ride it hitched and makes for one of the plump wasp grubs in its little cell. The tiny parasite creeps along the body of the host until it reaches a spot it can burrow into. Using enzymes, the parasite larva dissolves the skin of the host and sinks into its body, wriggling furiously as it goes. This frantic squirming separates the various layers of the host's skin, forming a pocket that the parasite slips into. Its place inside the host now secured, the larva goes through its first metamorphosis, turning into a grublike larva. All the nourishment it requires is obtained from the bodily fluids of the host, and the little skin pocket ensures the growing parasite is safe from the dreaded rigors of the host's immune system. Feeding on the fluids of the host, the strepsipteran larva grows, eventually taking up most of the space in the host's abdomen. The effect of this parasitism on the fully grown host is not to be sneezed at. The sexual organs of the adult wasp do not have room to mature due to the space taken up by the young parasite. The host develops into adulthood but it is very damaged; it is sterilized and sexless, and from between the tough plates of its abdomen protrudes the head end of a strepsipteran pupa. Soon after the ravaged host begins its normal adult activities, the cap of the parasite pupa opens and out pops an adult male strepsipteran. His mouthparts are small and useless, and the energy reserves he built up as a larva will not sustain him for long, so he must seek out a mate as quickly as possible. His mate is nothing like him. She found her own host in the same way as the male, but she still looks like a grub and is still to be found in the host insect with her head end projecting from its abdomen. The male is attracted to the scent of his unsavory mate and copulates with her by introducing his sperm into her brood canal, the entrance to which is found just behind her head. The sperm fertilizes the female's eggs, and before long, a new generation of mobile larvae will be ready to begin the complex cycle all over again.

- Around 600 species of strepsipteran have been identified, but their small size and the fact that they are rarely encountered means that there are probably many more species yet to be discovered.
- They parasitize a huge range of insects: 34 families in 7 orders, including bees, wasps, leafhoppers, grasshoppers, crickets, cockroaches, and silverfish. The ability of strepsipterans to parasitize a diverse range of hosts mystified scientists for a long time, as other parasitic insects are restricted to a small range of hosts because of the difficulties of eluding or overcoming the varied immune systems of different insects. The key to the strepsipterans' varied tastes is the pocket of host cuticle it lives in as a larva. This little pocket shields the strepsipteran larva from the substances and cells that make up the host's immune system.
- The complex life cycle of these parasites also took many years to unravel. They are among only a few insects that go through the process of hypermetamorphosis. This is where the egg hatches into a highly active, mobile larva (triungulin) with eyes and legs before undergoing a transformation into a typical, limbless grublike larva. It is this larva that pupates and metamorphoses into the adult insect.

♦ The best way to find strepsipterans is by using some form of trap. In the warm tropics, the adult males are active at night and are attracted to light; therefore, a moth trap should bring in a few specimens, but as they are so small, an alcohol-filled collecting tube may be needed beneath the trap to ensnare the males. The females will only be seen if the host is found, although there is one group of these parasites where both the adult males and females are free living. Under a microscope, the delicate and graceful appearance of the male can be appreciated, from the large, fine wings to the unusual eyes, which are unique in the invertebrate world. Unlike the compound eye of an insect, which forms a mosaic image, a male strepsipteran has a series of eyelets, each of which forms an entire image.

Further Reading: Kathirithamby, J. Review of the order Strepsiptera. *Systematic Entomology* 14, (1989) 41–92; Kathirithamby, J., Ross, L. D., and Johnston, S. J. Masquerading as self? Endoparasitic Strepsiptera enclose themselves in host-derived epidermal "bag." *Proceedings of the National Academy of Sciences* 100, (2003) 7655–59.

WARBLE FLIES

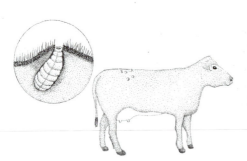

Warble Flies—Warbles on the back of the cow and a large warble fly larva within its warble. (Mike Shanahan)

Warble Flies—A female reindeer warble fly photographed in Norway. Note the similarity to a bumblebee. (Arne C. Nilssen)

Scientific name: *Hypoderma* species
Scientific classification:
　　Phylum: Arthropoda
　　Class: Insecta
　　Order: Diptera
　　Family: Oestridae
What do they look like? An adult warble fly is around 13 mm long and bears a striking resemblance to a small bumblebee. It is a hairy insect with yellowish white fur on its head and thorax, and alternating bands of light and dark hair on its abdomen.
Where do they live? The warble fly is very widely distributed throughout the northern hemisphere, ranging from the United States through Europe and into Asia. Its habitat is grassland and wood pasture where cattle are reared.

Nipping at the Heels

During the long, lazy days of summer, large grazing mammals are a magnet for flies of every description. Some come to drink the cow's sweat, others to lap at wounds, while still others come to make use of the abundant dung produced by these animals. A swish of the tail is normally enough to keep this irritating swarm at bay. Apart from being an annoyance, these flies are not that dangerous and are largely tolerated by the cattle. The warble fly, however, is feared by cattle, and to get close to their quarry, these flies must employ some very cunning means. Before they get within earshot of their target cow, female warble flies will take to the ground and cover the remaining distance in a series of hops before crawling up the animal's leg. Carefully, the female lays numerous 1 mm, pallid eggs that look like miniature grains of rice. Each of these eggs is attached to the hairs of the host by a small stalk. Within a week, tiny larvae have hatched from the eggs. These grubs make straight for the skin of the beast and delve into a hair follicle where they use digestive enzymes and their paired mouth hooks to break through the skin in to the tissue beneath. There, underneath the tough hide of the animal, they embark on a fascinating migration. Using their mouth hooks, they excavate a tunnel in the flesh of the cow, growing as they feed on the nutritious material. They slowly but steadily make their way toward the head of the animal, but when they reach the esophagus in the neck, they a rest for a while and then make an about-turn for no particular reason and head for the rear of their relatively gigantic host. They tunnel back to the rear of the animal through the muscular tissue of the back. They are about 10 mm long when they arrive at the lumbar region of the cattle's back, and it is here that they cut a small hole in the animal's hide and turn around so that their rear end, with its breathing tubes, is facing the small hole. The feeding larvae produce a very obvious, raised lump, and it is these lumps that are known as warbles. With their heads down in the back muscle of the cattle, the larvae continue to feed and grow, held in place by a number of spines on their bodies. When they are around 30 mm in length, the grubs are mature, and they take their leave of the host and fall to the ground. The big wide world is no place for succulent grubs, so they quickly burrow into the soil and undergo metamorphosis in an earthen chamber. Pupation can take 2–8 weeks, and at the end of this time, the perfectly formed adult flies emerge to seek out more hapless mammals.

- Some warble flies make directly for their host on the wing without surreptitiously hopping along the ground. Host mammals are seemingly aware of the approach of these species and become, quite rightly, terrified. They run around in panic trying to get away from the flies and will often injure themselves by running into trees, fences, and water. This panicked behavior is known as *gadding*, and in some parts of the world these warble flies are known as *gadflies*.
- The adult flies live for only around five days. They have no mouthparts and do not feed at all.
- Although the warble flies are fascinating insects perfectly adapted to a parasitic way of life, it is no surprise that farmers the world over would like to see them eradicated. Even as far back as 1965, the U.S. Department of Agriculture reported that these flies were responsible for around $192 million of loss to the cattle industry. Exiting larvae damage the hides of infected animals, and the migrating larva make large cuts of meat worthless

as the tunnels fill with what is known as butcher's jelly. The cattle also lose weight and produce less milk when they are continually alarmed by the presence of these flies.

+ As the United Kingdom is a small island nation, it has almost managed to eradicate the warble fly, which is quite unfortunate. The warble fly may have some very grisly habits, but it is a remarkable little animal that is able to exploit large mammals very effectively.

+ Warble fly larvae can be removed from their feeding nodule by being carefully squeezed out. Care must be taken not to rupture the grub as this can cause infections and severe immune reactions.

+ Interestingly, it is young host animals that are most susceptible to the ravages of the warble fly. It appears that older animals build up an immunity to the larval infections.

+ In those people who are often around grazing animals, it is rare, but not unknown for them to become parasitized by this fly. In cases of human infection, the effects are often gruesome as the larvae will end up in the head or the spinal column, causing the loss of an eye or paralysis of the legs.

Further Reading: Jelinek, T., Dieter, N. H., Rieder, N., and Loscher, T. Cutaneous myiasis: review of 13 cases in travelers returning from tropical countries. *International Journal of Dermatology* 34, (1995) 624–26; Scholl, P. J. Biology and control of cattle grubs. *Annual Review of Entomology* 39, (1993) 53–70; Warburton, M. A. C. The warble flies of cattle, *Hypoderma bovis* and *H. lineatum. Parasitology* 14, (1922) 322–41.

7

THE CONTINUATION OF THE SPECIES: SEX AND REPRODUCTION

BLUE-HEADED WRASSE

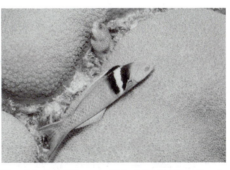

Blue-Headed Wrasse—When the situation arises, a female blue-headed wrasse (back) can make a one way change into a boldly patterned male (front). (Mike Shanahan)

Blue-Headed Wrasse—This fully developed male is pictured swimming among a coral reef in the Caribbean. A young moray eel is clearly visible in the background. (Bart Hazes)

Scientific name: *Thalassoma bifasciatum*

Scientific classification:

  Phylum: Chordata

  Class: Actinopterygii

  Order: Perciformes

  Family: Labridae

What does it look like? The largest males of this fish species are around 80 mm long. They are very brightly colored, the males more so, but the vividness of their colors depends on their degree of development. The body is long and cigar shaped, and the pectoral fins are well developed.

Where does it live? The blue-headed wrasse is a tropical fish of the western Atlantic. They can be found around Bermuda and the waters south of Florida, extending south to northern South America and west into the southeast area of the Gulf of Mexico. It is fond of reefs, although it can sometimes be seen in inshore bays and shallow sea grass beds.

A Sex Change on the Caribbean Reef

In the azure waters of the Caribbean, amid the ancient and elaborate coral reefs, live large schools of blue-headed wrasse. In common with most tropical reef fish, its blue head is strikingly colored, but what makes this little fish stand out from the rest of the reef community is its amazing social behavior. The adults of this fish are divided into three distinct types. There are females and not one, but two types of male: initial-phase males and terminal-phase males. The females and initial-phase males are yellow and white with dark stripes along their bodies. The terminal-phase males are larger and have a striking blue head, black-and-white markings behind the head and a shimmering, dark green body. On a prime bit of reef, it is possible to find a group of females that are ready and willing to breed. Overseeing these females is a terminal-phase male, jealously guarding his harem from the initial-phase males that try with all their guile to sneakily mate with the females. A large male will chase these interlopers, changing color to an intense metallic green as he does so to signal his aggression. When the threat has been dealt with, the male will turn his attentions to his harem, and quick-as-a-flash, his colors change once again, but this time to an opalescent pink-grey with distinctive black circles on his pectoral fins. The initial-phase males will, if they are lucky, grow up to become big, aggressive males with harems of their own. However, it is quite common for an aggressive male to get snapped up by a predator or simply die of old age (guarding a harem is exhausting work!) with no worthy successor to take on his mantle. In this situation, a most extraordinary thing happens. The largest of the females in the harem sees her opportunity and goes through a rapid sex change to fill the vacant lordship. In a little more than a week, the female has grown and developed the colors and markings of the terminal-phase male. Not only does her appearance change, but her behavior changes from that of a meek, supplicant concubine to that of a domineering, aggressive harem owner. The sex change is more than skin deep as her reproductive organs also go through a massive transformation to enable the production of sperm instead of eggs. This sex change is a one way trip, and the newly formed male can never change back into a female.

+ Sex changing is quite common in fish. The sex of the individual is not determined at fertilization, as in mammals, but changes as the individual grows and is controlled by genes *and* environmental conditions. Some species are truly hermaphroditic and can produce eggs and sperm at the same time. Even more bizarrely, there are some fish species that are all-female. They are normally hybrids of hermaphroditic species. Sperm is still required in all except the mangrove killifish (obtained from males of one of the parent species), but only to trigger the development of the embryo. There is absolutely no fusion of the DNA in the sperm with that contained in the egg.

+ The reef habitat may be able to support many of these fish, but suitable spawning sites are in short supply, as they need to be situated in areas where the current is sufficiently strong enough to carry the fertilized eggs safely away from the shore. Because of their rarity, these spawning sites are coveted, and females will not leave them. The remarkable sex-changing behavior of this fish ensures that a female is able to take over a spawning ground and the harem of females, but only when she has had a good stint at reproducing young herself.

+ The sex-changing abilities of this fish were proved by curious scientists. They selectively removed the large, terminal-male fish from a spawning site and were able to show that within a week or two the largest female in the group had successfully changed sex.

+ The terminal-phase males are mating machines and can fertilize the eggs of more than 100 females a day if the spawning site that they guard is sufficiently attractive to members of the opposite sex.

+ Interestingly, the small initial-phase males of the blue-headed wrasse have comparatively larger testes than their larger, more aggressive brethren. This enables them to produce lots of sperm for the snatched opportunities they must take when trying to copulate with the females in the guarded harem.

+ Blue-headed wrasses eat a wide variety of animal food on the coral reef. They are partial to all manner of invertebrates, including worms and crustaceans. Outside of spawning time, packs of females and initial-phase males will stalk the reef searching for the nesting sites of nest-guarding fish. The blue-headed wrasses in their groups distract the nest guards and plunder the eggs before they are driven off.

+ Blue-headed wrasses, especially the initial-phase males are important sanitary species for the other reef fish. They remove parasites from fish that stop at so-called cleaning stations and cleanse injuries, promoting the healing process. Occasionally, the fish receiving the good turn will turn on the little wrasse and gobble it up. The menial task of cleaning for crumbs of food is beneath the terminal-phase males, whose well-developed teeth allow them to eat hard-shelled crustaceans and other reef invertebrates.

Further Reading: Dawkins, M. S., and Guildford, T. Colour and pattern in relation to sexual and aggressive behaviour in the Bluehead wrasse *Thalassoma bifasciatum. Behavioural Processes* 30, (1993) 245–52; Warner, R. R. Mating behavior and hermaphroditism in coral reef fishes. *American Scientist* 72, (1984) 128–36; Warner, R. R., and Swearer, S. E. Social control of sex change in the Blueheaded Wrasse, *Thalassoma bifasciatum* (Pisces: Labridae). *Biological Bulletin* 181, (1991) 199–204.

COCKROACH WASP

Cockroach Wasp—A female cockroach wasp leads her prey around by its antenna. (Mike Shanahan)

Cockroach Wasp—An adult female cockroach wasp, not long emerged from the dead husk of her cockroach host. (Ram Gal)

Scientific name: *Ampulex compressa*
Scientific classification:
    Phylum: Arthropoda
    Class: Insecta
    Order: Hymenoptera
    Family: Sphecidae
What does it look like? The cockroach wasp is around 15–25 mm in length with long curving antennae, a large thorax, and an oval abdomen, which tapers towards the rear. The colors of this insect are very dramatic, as it can be a myriad of metallic blues and greens.
Where does it live? The wasp is native to the tropical forests of Africa and is also found in India and some of the Pacific Islands.

Taming the Quarry

In many species of venomous animals, the poison injected via a sting or a bite has evolved into much more than just a means of killing prey and predators. In some insects, the venom has become so sophisticated that it controls the behavior of the prey, affecting its movement and activity. The cockroach wasp is one such insect. This small flying jewel preys exclusively on the much maligned cockroach. Using its powerful senses, it homes in on one of these unwary pests and administers two stings. When delivering a sting, the wasp faces the cockroach and curves its flexible abdomen around to inject the venom. The first of these stings is directed to the tiny nodes of the central nervous system located in the insect's thorax. These minibrains control the cockroach's legs, and the wasp's venom blocks their activity, paralyzing the victim. This paralysis is only temporary, lasting for about 2–5 minutes. This is more than enough time for the wasp to deliver its second sting, which requires the skill and precision of a brain surgeon. Using its very sensitive sting, it delivers a tiny dose of venom to a region of the cockroach's brain, which affects, among other things, its escape reflex. When it eventually recovers from the paralysis of the first sting, the cockroach does not try to flee for the nearest cover, but grooms itself excessively for around 30 minutes, while the wasp scuttles off to look for a suitable lair. When the wasp returns, it bites off one of the long antennae of the cockroach before lapping at the hemolymph (insect blood) that flows from the severed appendage. Then the wasp grabs the cockroach by one of its antennae stumps and takes it for a walk, like a very docile pet, leading it to the refuge the wasp found earlier. There, the wasp lays a single egg on the stupefied host. The cockroach, essentially incapacitated but still alive is sealed in this hideaway with small stones and other debris, not to prevent it from escaping (it has no urge to!), but to keep it safe from predators. The wasp larva hatches to find itself sitting on a mound of self-cleaning food, which it starts tucking into. After two days, the young larva is big enough to tunnel into the host, and after four or five days, the voracious feeding of the larva takes its toll, and the cockroach dies. After about eight days, the wasp larva is ready to pupate, and it spins itself a silken cocoon inside the drying carcass of the cockroach. The adult hatches after about four weeks and leaves the lifeless husk of its host.

    • There are around 200 species of cockroach wasps, all of which parasitize cockroach species.
    • The cockroach wasp is a parasite of the American cockroach (*Periplaneta americana*), which, confusingly, is actually a native of tropical Africa that has been accidentally

transported around the world. It is one of the cockroaches much maligned by people, and it is an unwelcome guest in dwellings everywhere.

+ Because the jewel wasp preys on the American cockroach, it was introduced to Hawaii in 1941 as a biological control agent of this pesky species. Unfortunately, this scheme was not very successful as the wasp is very territorial and the areas in which they hunt are normally quite small.

+ The stinger of the jewel wasp, like that of other wasps, is a modified egg-laying tube. On its surface, there are a number of microscopic sense organs allowing the wasp to use it like a precision instrument. These receptors are used to good effect to hit a target in the tiny brain of the cockroach.

+ The mind-bending venom of this wasp does not directly control the movements of the cockroach. It just renders the cockroach easily led so the parasitoid can lead it like a mild-mannered hound.

+ Many parasitic wasps tackle hosts that can be carried to a suitable brood chamber. The mind-controlling venom of the wasp allows it to use a host that is much too large to be carried away.

+ If offered other insects similar in size to the American cockroach, the jewel wasp will give them a cursory once over but make no attempt to sting them.

+ Other types of parasitoid wasps inject venom directly into the central nervous system of their host, but the jewel wasp is the only known animal that targets the brain specifically.

+ The blood-brain barrier of animals stops chemicals from entering the brain. The stinging strategy of the wasp overcomes this obstacle.

+ Grooming is fundamentally important to all insects. Using their feet, they can rigorously clean their entire body. This behavior allows them to keep their exoskeleton free of potentially disease-causing microorganisms.

+ It is possible to mimic the effect of the jewel wasp's venom by conducting a small operation on the cockroach's brain, but the result is nowhere near as subtle as the chemicals introduced by the parasitoid.

Further Reading: Gal, R., Rosenberg, L. A., and Libersat, F. Parasitoid wasp uses a venom cocktail injected into the brain to manipulate the behavior and metabolism of its cockroach prey. *Archives of Insect Biochemistry and Physiology* 60, (2005) 198–208; Haspel, G., Rosenberg, L. A., and Libersat, F. Direct injection of venom by a predatory wasp into cockroach brain. *Journal of Neurobiology* 56, (2003) 287–92; Piek, T., Hue, B., Lind, A., Mantel, P., van Marle, J., and Visser, J. H. The venom of *Ampulex compressa*—effects on behaviour and synaptic transmission of cockroaches. *Comparative Biochemistry and Physiology* 92, (1989) 175–83.

DEEP-SEA ANGLER FISH

Scientific name: *Ceratioids*
Scientific classification:
  Phylum: Chordata
  Class: Actinopterygii
  Order: Lophiiformes
  Family: many families
What do they look like? Probably the best word to describe the deep-sea angler fish is grotesque. Many species looks like a swimming head. They are often coal black, and the

Deep-Sea Angler Fish—A fierce looking female deep-sea angler fish with a tiny, withered male attached to her side. (Mike Shanahan)

large mouth bristles with savage looking fangs. On the top of their head there is a thin stalk ending in a quill-like structure. They vary in size from species as big as a baby's fist to larger, football-sized specimens. The males are many times smaller than the females.

Where do they live? These are deep-sea animals, found at depths of at least 1,000 m, the most common habitat in the world's oceans. They are found throughout the world's oceans.

Joined at the Hip

The deep-sea must rank as one of the most exceptional habitats. It is pitch black, and the pressure, due to the great weight of water above, is immense. Also, these waters never feel the warming rays of the sun and are therefore very cold. If the dark, the pressure, and cold were not enough, there is also very little food down in the depths, but as is always the case, life has found a way, and these foreboding places with their exceptional circumstances are inhabited by an array of exceptional animals. The deep-sea angler fish is a perfect example. It is not a looker, but what it lacks in appearance, it more than makes up for in sheer peculiarity. Its small, seemingly malformed body is perfectly adapted to this harsh world. Food is so scarce in the depths that the deep-sea angler fish has come up with a means of attracting what little there is. Lures on its head emit an eerie greenish/blue glow. These beacons attract other, curious animals of the deep. This fish cannot afford to miss a potential meal, and its long, needle-like teeth ensure that any slippery customers investigating its lure are well and truly impaled.

Food is not the only thing difficult to come by in the depths. Mates are also very hard to find, and these fish have evolved to make sure that whenever they find a mate, they never lose them. The tiny males, after they hatch, swim freely in the water, but their gut is useless for feeding; therefore, the race is on to find a mate before they starve. They detect the scent of a female in the water and track it to its source. If they are lucky enough to find a member of the opposite sex, they dispense with the common pleasantries of courtship and grab on to her with their teeth. As they are so small, they hardly interfere with the female's slow progress through the water. The male, latched on to his gigantic mate's flank, releases an enzyme that breaks down the skin of his mouth and that of the female's, all the way down to the blood supply. Slowly but surely, the skin and the blood supply of the male fuses with that of his partner. Nourished by his partner's blood, the male has no need to feed, and his organs eventually begin to degenerate until he is little more than a sperm-producing appendage of the female. He may not be the only male carried by the female, as several more may be latched on to the rear portion of her body, all producing sperm to fertilize her eggs.

- There are many species of angler fish, and a significant number live in deep water. Exceedingly little is known about what these fish do in the wild. They are only ever caught in deep trawl nets and occasionally observed using submersibles. New species are caught and identified regularly, and as exploration of the deep sea continues, more species will undoubtedly be found.
- In all species of deep-sea angler fish, the first spine of the dorsal fin is modified to form the lure used to attract prey. The light from this lure is not produced by the fish, but by marine bacteria, which enter the fish's lure through small vents. Floating about in the sea water, these bacteria never emit light as they are never found in high enough densities, but safely inside the fish, they can multiply rapidly, producing chemicals that eventually reach high enough concentrations to trigger the production of an eerie glow.
- The production of light is actually quite common in nature. Many deep-sea creatures do it, as do many land invertebrates and fungi. Whether this light is produced by a fish or a glowworm, the chemicals involved are quite similar. All hinge on a substance called *luciferin* and an enzyme called *luciferase*. The enzyme breaks down the luciferin, releasing light as a by-product. Normal light bulbs can only turn about 2 percent of the energy they use into light, with the rest being wasted as heat. Light from a biological source is produced much more efficiently than man-made light, as approximately 95 percent of the energy used is turned into light.
- The intense pressure, low temperature, and perpetual dark of the deep sea make it a hard place to live; therefore, animals like the deep-sea angler fish grow very slowly and are normally small. The age at which they reach sexual maturity is comparable to that seen in humans, and they probably live for many decades. The conditions and limited food means they are emaciated animals, their tissues containing little in the way of

protein. The stomach and bones of these fish are also very flexible, enabling them to swallow prey that may be twice their size.

- Deep-sea angler fish and other animals of the depths migrate vertically on a daily basis. During the day, they stay deep, but during the night, they rise up to take advantage of more abundant prey in the near-surface waters. Exactly what cues these animals base their movements on is a mystery. They can't be sensing the onset of twilight, as light does not penetrate very far into water.
- The deep sea, in terms of area, is the commonest habitat on earth, but startlingly, less than 1 percent of it has been explored. It is often said that we know more about the surface of the moon than the deep sea. To put this in context, over the last 50 years or so, countless man-hours and huge sums of money have been thrown at a dead lump of rock in space, while all around us there is a unknown ecosystem inhabited by a dizzying array of bizarre organisms. Only through the cameras of a submersible can we snatch glimpses of this mysterious world.

GREEN SPOON WORM

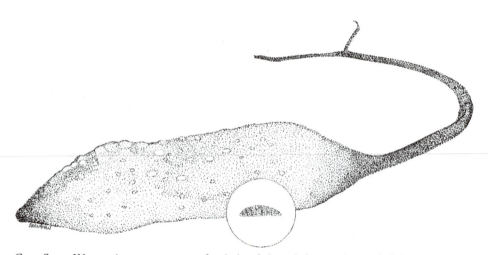

Green Spoon Worm—A green spoon worm female dwarfs the male (inset and magnified). (Mike Shanahan)

Scientific name: *Bonellia viridis*
Scientific classification:
  Phylum: Echiura
  Order: Echiuroidea
  Family: Bonellidae
What does it look like? The body of the female is around 8 cm long, green or blue/green, sausage shaped, and covered in small lumps. The female also has a long proboscis, which, when fully extended, can be up to 2 m long. The male on the other hand is tiny, only 1–2 mm long.
Where does it live? The green spoon worm is a marine invertebrate, a denizen of the shore in northeastern Atlantic and Mediterranean waters. The females are found in areas where the surface is hard, with crevices in which they can hide.

Living inside Your Mate

Sex is one of the biggest conundrums in the animal kingdom, but regardless of its true origins and function, male and female animals are often quite distinct in appearance and lifestyle. Nowhere is this more apparent than in the green spoon worm, a species in which for many years the male was unknown. It was thought the species had no need of males and could reproduce without sex, but after many fruitless searches, the males were discovered—and they were very different from the females. Female green spoon worms are about the same size and shape as a gherkin, but with the addition of a long, thin proboscis. The males on the other hand can only just be seen with the naked eye. They are about 1–2 mm long, slipper shaped, and covered in a coat of hairlike material. Most interestingly of all, though, is the fact they live *inside* the female. Up to 30 tiny males, many hundreds of times smaller than their mate and bearing no similarity to her whatsoever, live in a special chamber called the *androecium* where they absorb all of the food they need from the fluids they are bathed in. The males are concerned solely with the taxing job of producing sperm, and their bodies contain the organs needed to produce this prized substance, but precious little else. The males are in a prime position to fertilize the massive partner's eggs, and they also help out by producing a substance that sticks the eggs together.

The eggs are kept inside the female's body until the larvae hatch, after which they drift in the current near the seabed. After seven days, they begin to settle. Unlike the vast majority of other animals, the gender of a spoon worm is not determined by its genes, but by its environment—more specifically, the proximity of a female. Should one of these young spoon worms settle on a female of the same species, it will become a male. It will then grow a sucker and migrate to the female's androecium where it will be given free room and board in return for fertilizing eggs. Young spoon worms that settle away from others of their species grow into adult females over a period of two years.

Why should the sex of this little sea creature be determined in such a bizarre way? The chances of a male finding a mate are slim, which is also the case for the chances of a female worm finding an empty burrow. The ability to decide on a gender at the last moment gives this species a degree of flexibility, enabling the larval worm to make the best of the situation in which it finds itself.

- There are around 140 known species of spoon worms, although there are probably many more to be identified, especially in the deep ocean.
- Spoon worms are marine, although there are a few Indian species that live in brackish water. Typically, they are inhabitants of the shore, but several species have been collected at great depths. They excavate burrows in the seabed or make use of crevices in rock. They will also inhabit empty shells and the spaces between the roots of mangrove trees. Due to their secretive nature, very little is known about spoon worms. Normally, the only visible part of the animal is the elongated proboscis, which in some species can be extended to 100 times the length of the body trunk.
- These animals feed on particles of edible matter by sweeping their long proboscis across the seafloor. The proboscis can be extended, but not withdrawn into the body of the spoon worm. The food particles are trapped in sticky mucus covering the proboscis, and in a conveyor belt fashion, the mucus is moved by cilia to a furrow that runs the length of this structure. The particles are transported to the mouth of the animal where they are ingested. Some species also use their proboscis to scoop up detritus, while others produce a mucus net from it to trap food.

- Spoon worms in the right conditions can be reasonably common, and as a result, they can be important components of the marine food chain. Although it is known that *Bonellia viridis* does contain a neurotoxin in its skin, spoon worms in general are easy targets for a range of marine predators.
- These animals do not have the segmented bodies of true worms, but the two groups are closely related, indicated by the similarity of their larvae. New evidence obtained from the study of DNA indicates spoon worms and true worms may be much more closely related than currently believed.
- Gender determination in some reptiles and fish is similar to that of the spoon worms.

Further Reading: Berec, L., Schembri, P. J., and Boukal, D. S. Sex determination in *Bonellia viridis* (Echiura: Bonelliidae): Population dynamics and evolution. *Oikos* 108, (2005) 473–84; Stephen, A. C., and Edmonds, S. J. The phyla Sipuncula and Echiura. *Trustee of the British Museum (Natural History),* (1972) London.

NARWHAL

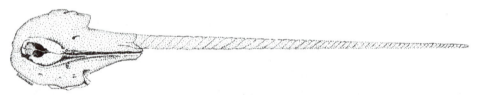

Narwhal—A skull of a narwhal showing its curious tusk. (Mike Shanahan)

Scientific name: *Monodon monoceros*
Scientific classification:
    Phylum: Chordata
    Class: Mammalia
    Order: Cetacea
    Family: Monodontidae

What does it look like? The narwhal is a small whale. The males are around 5 m in length and up to 1,600 kg in weight. The females are slightly smaller. The back of the torpedo-shaped body is mottled black and white, and the underside is white. There is no dorsal fin. The male has a distinctive spiral tusk.

Where does it live? This whale is found in northern temperate and Arctic waters, typically among the drifting and pack ice around northern Russia, the United States, Canada, and Greenland.

A Curious Tooth and the Legend of the Unicorn

The narwhal is one of the most fabled whale species. Pods, numbering more than 2,000 individuals, cruise the northern seas spending a lot of their time beneath the ice. Because they live in inaccessible areas, few people have ever seen a living narwhal, and even fewer can get close enough to study and observe these mysterious beasts. The single most unusual characteristic of the narwhal is the large tusk brandished by the males and, very rarely, the females. Like the tusk of an elephant, the growth is actually a modified tooth, and it is unique in the animal kingdom for being the only straight tusk and the only spiral tooth. The animal has two teeth, but it is normally the left one that becomes a tusk with its counterclockwise spiral. In some large males, the tusk may reach a length of 3 m, its handsome spiral clearly visible. It protrudes from the left upper lip and is sometimes skewed off in awkward directions. A very small proportion of males have a pair of tusks.

Not only does this structure look fantastic, but no one is entirely sure exactly what it is for. There are a number of theories. Some people believe the tusk is actually a tool for finding food, used by the whale to probe the seabed for crustaceans, mollusks, and worms. Others suggest the tusk is a fishing spear, used by the whale to impale fish as it hunts in the cold Arctic waters, or that it is used as a means of making breathing holes in the sea ice. Another common theory is that the tusk is used like a sword in jousts between males during the breeding season. Jousting! Perhaps not, but as the tusk is an adornment almost unique to males, it does lend some credibility to this theory. Many other male animals have hugely exaggerated features, such as the elaborate antlers of male deer, the mandibles and mouthparts of certain insects and spiders, not to mention the ludicrously flamboyant plumage of some male birds, which are used to attract females during the breeding season.

More recently, dentists working in the United States have been intrigued by the narwhal and its mysterious tusk. They have proposed that the tusk is actually a sensory organ. They found over 10 million tiny nerve channels stretching from the core of the tusk to its outer surface. Perhaps the nerve channels are sensing chemicals in the water or the distinctive electrical field of its prey. Whatever the actual function of the bizarre tusk, it is very likely that its sensory function is entwined with some aspect of the whale's breeding behavior.

Further observations will reveal the true function of this unique tooth, but as the narwhal is protected and very difficult to observe, it may be sometime yet before we unravel the mystery completely.

- The narwhal's closest relative is the white beluga whale. The two animals are found in the same locations, and sometimes they may be seen in the same Arctic estuaries.
- The unusual appearance and rarity of the tusk has intrigued people for centuries. Their sporadic appearance in medieval Europe gave rise to an abundance of legends. The legend of the unicorn probably arose from the fertile imagination of a medieval storyteller who had once seen or been told of the narwhal's tusk. Centuries ago, the real animal would have been unknown to all except the Inuit of its native habitat. Stories would have quickly grown up around these peculiar creatures.
- One of the most cherished possessions of Elizabeth I, Queen of England, was said to be narwhal tusk given to her as a gift. In medieval times, these objects were so rare that they were more valuable, pound for pound, than gold. They were thought to have miraculous curative properties.
- Today, the narwhal is a protected species, but the tusks, coveted by collectors and museums the world over, still change hands for several thousand U.S. dollars.
- The name *narwhal* is thought to come from the Scandinavian word *nar*, meaning "corpse." The narwhal's mottled coloration is said to look like that of a decaying human body.
- The narwhal is a deep-diving whale and can reach depths of at least 1,200 m. Underneath its skin, there is thick layer of blubber as much 10 cm thick to provide insulation against the numbingly cold waters. The narwhal can produce a wide range of sounds, some of which are beyond the range of human hearing and are probably used to detect underwater obstacles and food by echolocation.

- As the narwhal ages, the tips of its flippers and tail fluke curl inward.
- There are thought to be around 50,000 narwhal in the cold seas of the north, and about 1,000 are killed every year by Inuit hunters for their meat and skin. The chewy meat of the narwhal, known as *muktuk,* is an important source of vitamin C for the native people and their sled dogs.
- Even though the narwhal can use the back of its head to break through the ice to make breathing holes, it can occasionally become stuck and suffocate. Large numbers of narwhal can also become trapped by encroaching sea ice during the winter.
- The narwhal has few predators to fear, but it is vulnerable to attacks by marauding killer whales, and the young may also be attacked by walruses and polar bears.

Further Reading: Best, R. C. The tusk of the narwhal (L.): Interpretation of its function (Mammalia: Cetacea). *Canadian Journal of Zoology* 59, (1981) 2386–93; Nweeia, M. T., Eidelman, N., Eichmiller, F. C., Giuseppetti, A. A., Jung, Y. G. and Zhang, Y. Hydrodynamic sensor capabilities and structural resilience of the male narwhal tusk. 16th Biennial Conference on the Biology of Marine Mammals. Dec. 13, San Diego, CA 2005.

PALOLO WORMS

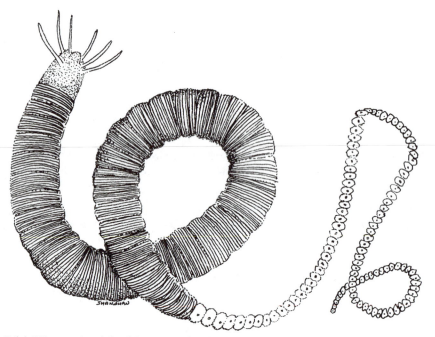

Palolo Worms—An adult palolo worm with its body divided into the atoke and epitoke. (Mike Shanahan)

Scientific name: *Eunice* species
Scientific classification:
　Phylum: Annelida
　Class: Polychaeta
　Order: Aciculata
　Family: Eunicidae

What do they look like? These worms display the segmentation that is common to all the annelids (polychaetes, earthworms, and leeches). When fully grown, they can be approximately 1 m in length. The head end bears sensory tentacles and complex jaws.

Where do they live? Palolo worms are found around the world in tropical and subtropical seas. They normally live below the low-water mark in coral and rocky crevices, although some species build burrows in the mud or sand of the seabed.

Severing the Body for the Species

For much of the year, the Palolo worm, safe in its rocky or coral crevice, looks like any other sedentary polychaete, its body divided into a long series of identical segments. The worm in this stage of its life is called the *atoke*, which is asexual and cannot breed. However, with the changing seasons, the time is right for breeding, and soon the worm's appearance undergoes a radical transformation. From the rear of the worm, new, highly modified segments start to grow, until a long chain of these identical units hangs from the rear of the atoke. This new and completely different section of the worm is called the *epitoke*, and it is this part that is involved in the process of reproduction.

The segments of the epitoke are exact replicas of one another, and each is packed with eggs and sperm. On the surface of the epitoke segment there is a single eyespot; however, this is not an organ for discriminating shapes and detail, but rather for telling night from day. Huge numbers of these worms, with their epitokes fully formed, wait in their refuges for a specific cue to spawn. The cue is the moon, and in October or November, at the beginning of the last lunar quarter, all the worms release their epitokes at the same time. The epitokes, free of the atoke, swim with undulations of their spaghetti-like form to the surface of the sea, and as the sun rises (detected by the eyespot), they burst, releasing their eggs and sperm. The sea, close to shore, will be a soup of gametes. The eggs are fertilized rapidly, and by the next day, tiny larvae have formed. After drifting for two or three days, these offspring begin to settle and find rocky hideaways of their own where they will develop into adult worms in preparation for the next year's mass spawning. The atoke, still in its burrow, will regenerate a new epitoke for the following year's breeding season.

- The polychaete worms, of which the Palolo worms are one type, are very common invertebrates, but as with many smaller animals, their secretive lives result in being overlooked by the casual observer. There are about 9,000 species of these worms, which are mostly marine, although a few species are found in soil and freshwater. Most are rather small animals, but the Palolo worm, *Eunice gigantean*, can be over 3 m in length. Some of the species are beautifully colored in shimmering reds, pinks, and greens. They can be active detritivores, scavengers or hunters (errant polychaetes), or burrow dwelling (sedentary polychaetes). It has been calculated that the burrowing polychaetes living on the seabed turn over more than 4,500 tonnes of mud per hectare of seafloor each year.

- The synchronized release of the epitokes by Palolo worms is greatly anticipated by the people who live in the areas where these animals occur. *Eunice viridis* (the Samoan Palolo worm) is a particular favorite among the people of Samoa, who wait for spawning before wading into the shallows with torches to collect the spaghetti-like epitokes with whatever equipment they can find. Many simply eat the epitokes raw as they are taken from the water, although great quantities are taken ashore where they are boiled, baked, stewed, or fried and eaten in a range of dishes. Epitokes on toast is a particular favorite.

+ A relative of the Samoan Palolo worm, *Eunice fucata* (the Atlantic Palolo worm), synchronizes its breeding for the second or third day before the third quarter of the moon, sometime in June or July.

+ Another species of polychaete, *Odontosyllis enopla*, found in the West Indies and Bermuda, also swarms. However, in this species, there are separate males and females. In the summer, 50 to 60 minutes after sunset and up to 12 days following a full moon, the worms swim from the seabed to the surface where they start to glow. They swim around and around each other, forming small circles of light in the water.

+ Epitoky and swarming is quite common among the polychaetes. This phenomenon makes it possible for a population of worms, which live out their adult lives on the seabed in dispersed populations, to come together briefly for the purposes of reproduction. Swarming is a way of increasing the chances of fertilization.

+ The synchrony of the worm's breeding is controlled by the lunar cycle—the effect the moon's gravitational field has on the earth. Exactly what the worms are detecting is unknown, but many animals, especially those in the seas, synchronize their activities with the waxing and waning of the moon. The influence of the moon on the earth follows a predictable pattern; therefore, it is unsurprising that many organisms have taken to using it as a kind of timer.

Further Reading: Andries, J. C. Endocrine and environmental control of reproduction in Polychaeta. *Canadian Journal of Zoology* 79, (2001) 254–70; Caspers, H. Spawning periodicity and habitat of the Palolo worm *Eunice viridis* (Polychaeta: Eunicidae) in the Samoan Islands. *Marine Biology* 79, (1984) 229–36.

POCKETBOOK MUSSELS

Scientific name: *Lampsilis* species
Scientific classification:
   Phylum: Mollusca
   Class: Bivalvia
   Order: Unionoida
   Family: Unionidae
What do they look like? These mollusks have a pair of shells (valves) that completely envelope the body, linked by a strong, hinge ligament. The head of these mussels is very poorly developed because they are sessile and have little need of sense organs to find food and detect danger. The gills in these animals are normally very large and are involved not only in gas exchange but also assist in feeding. When fully grown, the shells of these mollusks can be 10 cm across.
Where do they live? These animals are typically found in the sediment or gravel of shallow, clean freshwater streams and rivers. They are native primarily to temperate waterways of the United States.

A Fishy Mollusk

Adult pocketbook mussels are sedentary. They wait for food to come to them, filtering edible particles from the water. This is a very low-energy lifestyle, but it presents difficulties when it

Pocketbook Mussels—A curious fish gets doused with larval pocketbook mussels when it investigates the mollusk's intriguing mantle. (Mike Shanahan)

comes to the dispersal of young. The adult female cannot disperse her young far and wide. To counter this, the pocketbook mussels use a very interesting ploy. They let other animals disperse their young for them.

The eggs of these mollusks develop inside the female into small larvae called *glochidia*. These range in size from 0.2 to 0.5 mm, with two simple valves. The larvae remain inside the gill of their mother where they are neatly packed in what look like shelves. The female then goes about attracting a ride for her young. The edge of her body, which protrudes from between the valves of her shell, develops into an astounding mimic of a small fish, complete with markings and false eyes. This decoy moves in the current and attracts the attention of fish. Some fish are attracted to it and get closer because they see the mollusk's adornment as prey, while others approach because they see it is a shoal member or a potential mate. When the fish moves in closer or nips the decoy, the female releases huge numbers of her larvae from her gill, dosing the inquisitive fish with her tiny, parasitic babies. The larvae are drawn in by water currents to the fish's gills, where they attach. An attached larva stimulates the tissue of the fish to produce a small cyst in which it will be protected and nourished. The mantle of the baby mussel contains cells that break down the tissue of the fish and digest it, providing sustenance for the glochidia for 10–30 days, at which time it breaks out of the cyst and sinks to the bottom to begin its sedentary way of life. All of this will probably be a long way from where they were originally released by their mother. This parasitic hitchhiking ensures that a very sedentary species can spread, thus exploiting new areas of habitat.

- There are approximately 30,000 species of bivalves. They are very well represented in fossil deposits, and in some places their fossilized shells can be found in huge numbers. They have been around for at least 500 million years, and today, they are still a successful group of animals. Their bodies have become modified for a sedentary existence. Most species will spend their entire adult lives in the place where they settled as larvae. Some species occur in huge concentrations. Mussels, for example, can completely cover huge areas of rocky shore.

- *Lampsilis* mussels are native to the United States, and many species can be found there. In the United States, these animals have an abundance of interesting common names, including pink mucket, fat mucket, and Higgins' eye. The United States has a rich fauna of freshwater mussels, with some 300 species.

- Apart from using a decoy, pocketbook mussels also trick fish in to dispersing their offspring in other ways. Some species release their glochidia in colored masses that look like tasty worms attached to a gelatinous fishing line. A fish comes along and gobbles up these so-called worms, giving the young mussels easy access to its gills.

- The native mussels of the United States are under threat because of the accidental introduction of the zebra mussel from the Caspian Sea. The zebra mussel has thrived in the Great Lakes and several other waterways in North America. It competes with the native mussels for food and space and even attaches to the shells of these bigger species.

- Bivalves are also economically important animals. Many species are valued as food, such as mussels and scallops. Some are even regarded as delicacies that command high prices (e.g., the oyster). Pearls and mother-of-pearl are obtained from bivalves. Bivalves have also attracted attention from scientists seeking new materials for medical and engineering applications. The byssal threads of mussels are incredibly strong and can anchor the animal to just about any surface, including wave-battered rocks and the hulls of ships.

- The marine pearl oysters are seeded with small spheres taken from the shells of American freshwater mussels. It is reckoned that 95 percent of commercially produced pearls have at their center a bead made from an American mussel shell. Pearls are a defensive reaction in response to an irritant within the shell of a mollusk. Today, they are commercially produced, so a small fragment of shell and mantle tissue from another bivalve are implanted near the pearl oyster's gonads. Over three to four years, or occasionally six, the oyster deposits layer after layer of mother-of-pearl around the irritant, forming a pearl, which is eventually harvested.

SPOTTED HYENA

Scientific name: *Crocuta crocuta*
Scientific classification:
    Phylum: Chordata
    Class: Mammalia
    Order: Carnivora
    Family: Hyaenidae
What does it look like? The spotted hyena has a total body length of up to 1.4 m, a height of up to 90 cm, and a weight of up to 80 kg. Females are heavier than males. The limbs of the

Spotted Hyena—A female spotted hyena sniffs at the pseudopenis and scrotum of another female. (Mike Shanahan)

Spotted Hyena—A fully grown spotted hyena looking with curiosity at the photographer. (Stephanie Dloniak)

hyena are distinctive, as the front pair is longer than those at the rear, giving the animal a sloping back when standing. The muzzle of the spotted hyena is pronounced. The coat of the animal is light brown and shaggy, with a short mane. The tail is relatively short with a brush of long black hairs. As the name implies, dark brown oval spots dapple the coat.

Where does it live? The spotted hyena is found practically everywhere in sub-Saharan Africa, except South Africa and the Congo basin.

Is It a Girl or Is It a Boy?

The spotted hyena is a beautifully adapted predator and scavenger of the African continent living in so-called clans containing up to 80 individuals occupying territories of 10–65 sq. km. The group structure in these clans is complex because, unlike other social, carnivorous mammals, females are the dominant sex. The adult females in a clan are dominant to all the males, and they assume the normal male role in clan protection and territorial disputes. Not only are the females aggressive, but the sex-role reversal is even apparent in their appearance, as the female genitalia are astonishingly malelike to the extent that it is very difficult to tell the difference between a male and female. Should you ever be in the position to look between the rear legs of a female spotted hyena, you will see a perfectly formed false scrotum and false penis. The pseudopenis is actually a hugely modified clitoris, which is erectile just like a real penis. The pseudoscrotum is formed from the exterior skin of the female genitals. This massive modification means the female spotted hyena must urinate, mate, and give birth through the conduit of her elaborate clitoris.

The last of these acts presents a great deal of difficulty as the aperture at the end of the clitoris is small and the young are the largest of any carnivore (in relation to the size of the mother). Giving birth to a baby through the clitoris is a long and probably very painful process. Not only is the birth canal small, but it is also oddly orientated due to the internal anatomy. In a female hyena giving birth for the first time, the false penis may tear as much as 15 cm along its length to accommodate the passage of the baby. Should the young survive the birthing process, they are active almost immediately and have teeth that are put to good use on their siblings, which are attacked as soon as they emerge from the clitoris. The fur around a young hyena's neck is often damp with saliva following an attempted throttling from one of its siblings. These attacks are rarely fatal; it is just a way of establishing a system of dominance in the litter.

- Hyenas are interesting for a number of reasons, not least due to the peculiarities of the female spotted hyena's external genitals. The general appearance of the hyena suggests a close evolutionary link to the dog family; however, hyenas are an offshoot from the cat branch of the carnivores and are more closely related to cats than dogs.

- Today, there are four extant species of hyena: spotted, brown, striped, and the aardwolf. The prehistoric hyenas were very large, and the cave hyenas were at least twice as large as the biggest spotted hyena.

- Anatomically, the spotted hyena is adapted for a scavenging way of life, feeding on the scraps left by other carnivores. The jaws, in particular, are very robust and are equipped with interesting teeth. The premolars are large and adapted for crushing bones, whereas the carnassial teeth are perfectly suited for slicing and shearing. The bite of the spotted hyena is hugely powerful and, relative to its size, is probably the most powerful bite of the carnivorous mammals. The strength of the spotted hyena's bite enables it to splinter and break the bones of carcasses. Not only can it break bones, but it can also swallow and digest them. The stomach acid of the hyena is so powerful that it can digest even large bone fragments. The hyena will not only eat bones, but horns, hooves, ligaments, and hair, much of which is regurgitated later as a pellet. Due to the high proportion of bone in the diet, the feces of the hyena are white and crumbly.

- Hyenas are very efficient predators. A group of 38 hyenas has been observed to consume an adult zebra in 15 minutes, leaving little in the way of scraps. The voracity of the hyena can also be its undoing as sharp fragments of bone and other material will be swallowed, which can sometimes prove difficult to digest even in the harshly acidic environment of the animal's stomach.

- Although excellent scavengers, spotted hyenas are also first-rate opportunistic hunters. Individual hyenas have been observed pursuing an adult wildebeest for 5 km at speeds of up to 60 km/h. Once the prey is captured, the spotted hyenas have no problem dispatching it.

Further Reading: Di Silvestre, I., Novelli, O., and Bogliani, G. Feeding habits of the spotted hyaena in the Niokolo Koba National Park, Senegal. *African Journal of Ecology* 38, (2000) 102–7; Frank, L. G., and Glickman, S. E. Giving birth through a penile clitoris: parturition and dystocia in the spotted hyaena (*Crocuta crocuta*). *Journal of Zoology* 234, (1994) 659–65; Frank, L. G., Glickman, S. E., and Powch, I. Sexual dimorphism in the spotted hyaena (*Crocuta crocuta*). *Journal of the Zoological Society of London* 221, (1990) 308–13; Neaves, W. B., Griffin, J. E., and Wilson, J. D. Sexual dimorphism of the phallus in spotted hyaena (*Crocuta crocuta*). *Journal of Reproduction and Fertilisation* 59, (1980) 509–13.

SURINAM TOAD

Scientific name: *Pipa pipa*
Scientific classification:
  Phylum: Chordata
  Class: Amphibia
  Order: Anura
  Family: Pipidae

What does it look like? The Surinam toad looks as if it has been squashed as it has a very flattened, rectangular body. The head is triangular, and at its front, in what looks like a very

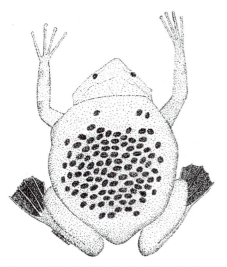

Surinam Toad—The back of a Surinam toad studded with developing young. (Mike Shanahan)

Surinam Toad—Photographed in a shallow pond, this picture shows the flattened appearance of this toad. (Jean-Pierre Vacher)

odd position, are the eyes. The rear legs are long and powerful and have broadly webbed feet. Fully grown specimens may attain a length of 20 cm, although 10.5–17 cm is more typical. The long, thin digits on the forelimbs end in small, starlike structures. Color varies, but they are generally grey brown to black with diffuse mottling.

Where does it live? This is an animal of the Amazon region of South America. It can be found in Peru, Guyana, Surinam, Brazil, and islands such as Trinidad. They prefer muddy, slow moving, or still water with plenty of vegetation in which to hide.

Youngsters that Get Under Your Skin

The Surinam toad is far from being a handsome creature. On appearance alone, the observer would be forgiven for thinking that this amphibian abomination is nothing short of an evolutionary accident. However, dig a little deeper, and you will find this odd little animal is fascinating for a number of reasons, not least of which is the way in which it reproduces and looks after its young. During the breeding season, in the sluggish and still waters of this toad's home, males begin calling to attract a mate. Unlike many other toads, they are unable to croak or emit any of the sounds we associate with these animals; they lack the vocal cords and vocal sacs. Instead, they are able to produce sharp, clicking sounds by snapping the hyoid bones in their throat. These clicks travel far and rapidly in the dense medium of water, and females will appear, investigating the sounds. Exactly what a female Surinam toad looks for in a mate is unknown, but if she finds one that she likes the look or sound of, she allows mating to proceed. The male grips his larger partner by her waist and clings on for dear life as he is carried through the water in a series of daredevil loop-the-loops. On the up part of these somersaults, when the female is above the male she releases, a few eggs. These fall on to the male's belly, and as the pair continue their loop, the eggs fall off the male and onto the female's back, undergoing fertilization as they move through the sperm ejected by the male. With the fertilized eggs now on the female's back, her partner hugs her tighter and presses them into the thick, spongy skin. In preparation for the breeding season, the skin on her back has been getting progressively thicker and softer. After several of these somersaults, the pair stays locked together for as long as 12 hours. This long bout of amplexus keeps the eggs in place until a degree of skin swelling holds them firmly to give an effect like small yellow balls embedded in plasticine. Eventually, the pair part company, and the female wanders off with many growing young lodged in the skin of her back.

Steadily, the skin around the eggs continues to swell, until after about 10 days, when each egg is sealed in its own little chamber. At this stage, the back of the female looks like a section of honeycomb, with numerous small cells, each containing a developing toadlet. In the safety of their own private nursery, the young develop rapidly, nourished by the yolk contained within their egg, and after 12–20 weeks in a scene reminiscent of gory science-fiction films, numerous tiny toadlets break out from the brood chambers. When they hatch, they are fully formed miniature toads. The safety of the pockets on their mother's back saw them through their tadpole and transitional stages. Surinam toadlets are like their parents in every respect, even down to the carnivorous tendencies of the adults. The young will quite happily snatch at any living thing big enough to fit in their capacious mouths, including their brethren.

- The Surinam toads and their relatives, the African clawed frogs, number approximately 30 species. The former are found throughout tropical South America, while the latter are native to sub-Saharan Africa. The body of these amphibians is very well suited to an aquatic existence, and they rarely leave the water, only doing so to locate a new home if space is scarce. They are also unusual in that they have no tongue.
- One of these amphibians, the African clawed toad (*Xenopus laevis*), is the laboratory animal of choice for molecular biologists and geneticists everywhere. Development is quick, and they are easy to maintain in captivity, making them a firm favorite for experimentalists. Research on this animal has paved the way for a huge number of medical and scientific breakthroughs, one of which was the birth control pill.
- Adult Surinam toads are out and out carnivores. They sit motionless on the bed of their muddy pools or rivers for hours, blending in unerringly to the submerged carpet of dead leaves. When the opportunity arises, they strike with explosive speed, gulping the unfortunate victim into their voluminous maw.
- The unusual breeding behavior of the Surinam toad underlines the diversity of reproductive strategies employed by the frogs and toads. From a basic-body-plan point of view, the frogs and toads are all very similar; however, over millions of years, they have evolved a huge number of ways to ensure the maximum survival of their young. We have already seen how tiny tropical frogs nurse their young in small arboreal pools of rainwater. Others whip up a frothy nest for their young from the female's genital secretions. Back breeders also nurture their young in cavities on the female's back, with the added advantage of nourishment supplied by a placenta-like arrangement. In mouthbreeders, the young develop in the male's vocal sacs, in some cases emerging about 50 days later as fully developed froglets. There is, it seems, no limit to the ways in which the amazing amphibians ensure the survival of the next generation.

Further Reading: Rabb, G., and Rabb, M. Additional observations on breeding behavior of the Surinam toad, *Pipa pipa*. Copeia 4, (1963) 636–42; Rabb, G., and Snedigar, R. Observations on breeding and development of the Surinam toad, *Pipa pipa*. Copeia 1, (1960) 40–44.

TAITA HILLS CAECILIAN

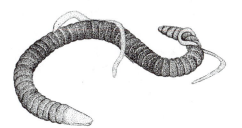

Taita Hills Caecilian—Young Taita Hills caecilians nibble at the nutritious skin of their
mother. (Mike Shanahan)

Taita Hills Caecilian—This photograph of a captive caecilian shows just how odd these amphibians
look. (Alexander Kupfer)

Scientific name: *Boulengerula taitanus*
Scientific classification:
  Phylum: Chordata
  Class: Amphibia
  Order: Gymnophiona
  Family: Caeciliidae
What does it look like? This is a long and thin animal, and superficially it resembles an
  earthworm. Adults are 20–33 cm in length, and along the length of the steel blue body are
  small encircling grooves, which give the animal a segmented appearance. The head and tail
  are bullet shaped. The eyes of this amphibian are much reduced and are concealed beneath
  bone and skin. A pair of small sensory tentacles can be projected from the head.
Where does it live? This species is only known from the Taita hills and their immediate
  vicinity in southeastern Kenya. It is a fossorial animal, and spends almost all of its time
  underground. The caecilians are widespread in soils of forests and agricultural landscapes.

Many Mouths to Feed

Caecilians make excellent parents. They invest a great deal of time and energy in their young
to ensure they have the best start in life. Nowhere is this more apparent than in the Taita hills
caecilian of Kenya. It has been known for a long time that the female would lay her eggs in a subterranean lair and jealously guard them until they hatched by coiling her sinuous body around
them. It was also known that the fetuses of this species had well-developed teeth. The presence
of these teeth in such young animals puzzled scientists, and the full significance of them has only
recently been discovered. It turns out the complete nurturing behavior of this animal is a lot
more complicated and interesting than originally assumed.

The young caecilians stay with their mother for around two months after they hatch. Only
mammals usually show this level of maternal care, and it was unclear exactly what the young
were doing for such a long period of time. Caecilians do not have mammary glands and teats
like mammals; therefore, extending periods of breast feeding are out of the question. The slender, pink young probably survive on yolk stores for their first few days of life, but scientists

videotaping captive litters noticed some odd behavior. The young could be seen nuzzling and butting their mother's body, and it was eventually found that they were peeling off the outermost layers of her skin and eating it. When females of this species are brooding, their skin becomes pale and thickens considerably until it is twice as deep as that of a nonbrooding female. The skin cells change from the normal flat, dead variety to succulent parcels filled with proteins and fats. To enable the young to peel away this nutritious skin they are born with a number of small, hooklike teeth. On this nutritious diet of maternal skin, the offspring thrive, and every week their body length increases by 11 percent. After two months of feeding in this way, the young have grown sufficiently to enable them to leave their mother and go their own way.

+ Around 170 species of caecilians have so far been identified, but due to their very secretive ways and rarely visited habitats, it is very likely that many more species are still to be discovered. They range in size from 10 to 150 cm. They are all found in the tropics.

+ The caecilians first appear in the fossil record more than 150 million years ago. They have become so well adapted to a subterranean existence that they have lost many of the typical outward characteristics of the amphibians.

+ Of all the land-living vertebrates, only the birds have no truly subterranean representatives. The amphibians have the caecilians, the reptiles have the worm-lizards and snakes, and the mammals have the moles and a multitude of other burrowing forms. An underground life promises abundant food and relative safety from surface predators, but it requires major changes to the animal's body. In amphibians and reptiles, these changes have led to animals that look strikingly similar.

+ One of the unique features of the caecilians is the pair of small tentacles found between the eyes and nostrils. These can be projected and withdrawn and depend on many muscles and other features typically associated with the vertebrate eye. Their exact function is unknown, but it is thought they gather small samples of air for the taste organs in the roof of the mouth. In snakes this is achieved by the flicking tongue.

+ Most caecilians are terrestrial burrowers, excavating tunnels by using their rigid, compact head as a battering ram. Some species will venture to the surface during the night or if their tunnels are flooded by heavy rains.

+ Some of the caecilians in the family Typhlonectidae are aquatic as well as being the largest of these limbless amphibians. The Typhlonectidae are only found in South America and lack the little tentacle that all other caecilians have. The aquatic species have a fleshy fin extending along the rear section of their body enabling them to swim through the water in the same way as an eel.

+ All caecilians are carnivorous. They have an under-slung jaw, enabling them to subdue and eat suitably sized prey encountered in their tunnels. Worms will often be taken, as will other soft-bodied invertebrates of the soil and leaf litter.

+ Other than skin feeding, the caecilians show a range of reproductive oddities. Many species give birth to live young, and the fetuses of some species are nourished inside the female on uterine milk and the thick lining of the uterus. The fetuses have elaborate gills. In the terrestrial species, these are very long and branching, extending from just behind the head of the baby amphibian. In the aquatic species, these gills look like small, leaf-shaped sacs. In the adults of all but one species, the job of gas exchange is taken over by lungs.

Further Reading: Kupfer, A., Muller, H., Jared, C., Antoniazzi, M., Nussbaum, R. A., Greven, H., and Wilkinson, M. Parental investment by skin feeding in a caecilian amphibian. *Nature* 440, (2006) 926–29.

TARANTULA HAWKS

Tarantula Hawks—A female tarantula hawk confronts her prey, an intimidating tarantula. (Mike Shanahan)

Tarantula Hawks—An adult female tarantula hawk who will soon begin her search for a tarantula to feed her young. (Gonzalo Useta, Laboratory of Ethology, Ecology and Evolution, Estable Institute, Uruguay)

Scientific name: *Pepsis* species
Scientific classification:
    Phylum: Arthropoda
    Class: Insecta
    Order: Hymenoptera
    Family: Pompilidae

What do they look like? These are very impressive looking insects. They can be large, with some species having a body length of 8 cm and a wingspan of 10 cm. They are handsomely colored, typically being a metallic blue or black with brightly colored wings. They have the typical wasp body plan: a head with large eyes and antennae, a bulbous thorax that bears the long legs and also contains the powerful flight muscles, a thin waist, and a tapering abdomen.

Where do they live? These insects have a worldwide distribution, with species being found in India, Asia, Africa, Australia, Europe, and the Americas. They are often associated with arid habitats but are also found in areas with lush vegetation.

A Mother Not to Be Messed With

Tarantulas are the stuff of nightmares. Their appearance alone is enough to make the flesh crawl. Imagine, then, going into the confines of a tarantula's lair, a narrow, silk-lined burrow. This is exactly what a type of female wasp, known as a tarantula hawk, must do to continue the species. The female tarantula hawk will pick up the scent of a tarantula and trace it back to its source. Occasionally, the spider will be in the open, hunting, but the odor may just as easily be emanating from the burrow of the spider.

In the small number of tarantula hawk species that have been studied, the female wasp is very specific about which species of tarantula she requires, and to confirm this, she needs to get as

🔍 Go Look!

The smaller spider-hunting wasps can be found in a number of habitats, particularly those that are dry and warm. Look out for medium-sized wasps with black bodies and red markings. They are typically seen scurrying along the ground, with characteristic nervous twitching of their wings, applying their antennae to the ground searching for odor clues, which may belie the presence of a suitable spider. If you watch them for long enough, you may see them find a spider, and a fight will normally break out as the wasp attempts to sting its prey. The venom is fast acting and the wasp will then drag and carry the paralyzed spider back to its brood burrow. Should you be lucky enough to find a spider that carries an egg or larva of a pompilid, you can take it home and keep it in a small container where you can watch it develop. Ectoparasitoids, like the pompilid larva are very easy to rear. They already have their food, so all they need is a small, relatively humid cage.

close as possible to the spider. Should the wasp have found the correct species of prey, something odd happens, the exact cause of which is not fully understood, but the spider becomes quite docile and very rarely attacks the female wasp. Perhaps the wasp produces a pheromone, which stupefies the spider. The wasp now crawls around and over the spider, vibrating its antennae furiously, a behavior which is thought to confirm to the wasp that she has the correct species of tarantula. Once sure, the wasp delivers the coup de grâce and uses her formidable sting to inject potent neurotoxic venom through the thin membrane that joins the base segment of the leg to the body. Though the venom is not fatal to the spider, it does cause paralysis. The spider can then be dragged by the wasp to the bottom of its burrow, and once there, she lays a single egg on the arachnid's abdomen. She then seals the spider and her egg in the burrow. Eventually, the egg hatches and the larva thrusts its mouthparts into the abdomen of the paralyzed spider and begins to feed on the animal's juices. The larva grows rapidly, but in completing its development, it switches to a diet of solids, so it feeds on the spider's organs, leaving the essentials, and the spider survives and does not rot. With pupation imminent, the larva consumes the essential organs as well, killing the long-suffering spider before weaving a silken cocoon in which to pupate. With metamorphosis complete, the adult wasp chews itself clear of the cocoon and exits the burrow to search for a mate and begin the process all over again.

- The Pompilidae, the wasp family containing the tarantula hawks, is composed of approximately 5,000 species, and like any invertebrate group, there are probably many more species yet to be identified. Commonly, they are known as spider wasps as the larva develop, typically, on a spider provisioned by the female. Although widely distributed, the Pompilidae are predominantly a tropical and temperate wasp family.

- Experiments conducted with several species of tarantula hawks have demonstrated their very exacting prey preference. A female of one of these species placed in a cage with the wrong species of tarantula will take no interest in the spider, and it may in fact be killed by the large arachnid.

- Some of the pompilids, known as pirate pompilids, have dispensed with finding their own spider and instead wait for another species to do it for them. In one group, the pirate pompilid will open the sealed nest and lay an egg of its own, which will hatch to feed on the original egg/larva and the paralyzed spider. In another group, the female of the pirate species will lie in wait for an unsuspecting female of another species of pompilid, dragging its prey to the nest before rushing out and quickly laying its own egg in the book lung of the paralyzed spider. After the prey has been buried by the original hunter, the pirate's egg hatches first, and the larva immediately searches out and consumes the original hunter's egg before consuming the spider.

+ The tarantula hawks are reputed to have one of the most painful stings of any insect. The Schmidt Sting Pain Index was devised by an entomologist after being stung by most stinging Hymenoptera. The tarantula hawks are second in this scale, with a sting said to be excruciatingly painful. Although painful, the sting of the tarantula hawk is not as dangerous as the sting of the honeybee, which can cause anaphylactic shock in hypersensitive people. The painful sting of the tarantula hawk is an effective defense against vertebrate predators.

Further Reading: Costa, F., Pérez-Miles, F., and Migone, A. Pompilid wasp interactions with burrowing tarantulas: *Pepsis cupripennis* versus *Eupalaestrus weijenberghi* and *Acanthoscurria suina* (Araneae, Theraphosidae). *Studies on Neotropical Fauna and Environment* 39, (2004) 37–43.

TRANSVESTITE ROVE BEETLE

Transvestite Rove Beetle—A transvestite rove beetle attracts flies to the odorous secretion it smeared onto a rock. (Mike Shanahan)

Transvestite Rove Beetle—An adult of one of these beetles perches on a leaf waiting for some suitable prey to come within pouncing distance. (John Alcock)

Scientific name: *Leistotrophus versicolor*
Scientific classification:
   Phylum: Arthropoda
   Class: Insecta
   Order: Coleoptera
   Family: Staphylinidae

What does it look like? The beautiful transvestite rove beetle has a covering of short hairs, giving it a furry appearance. It is between 18 and 25 mm in length, and the head carries a large pair of eyes and enormous sickle-shaped mandibles. Like all rove beetles, the wing cases are short and do not cover the abdomen. The abdomen is often upturned in this species.

Where does it live? Found in central and northern South America in forested areas. It is often seen on forest tracks or perching on low vegetation and is usually found near freshwater.

Looking Feminine to Find a Mate

In terms of appearance, the males of this species are divided into two types. There are the big, bullying so-called normal males, who are large and powerful, and then there are the sneaky males, the transvestites, who mimic the appearance of the females and are therefore considerably smaller than the normal males. Both types of males defend territories centered on a resource that will attract females, such as a small pile of dung. Dung is important to these beetles as they depend on it to lure their favorite food—flies. The big males attempt to defend these territories by repelling the attentions of other males who are looking for a dung pile of their own. Putting their appearance to good use, the transvestite males can get the girl if dung resources are thin on the ground. A big male guarding a good pile of dung will attract lots of females, and this is just what the transvestite male needs. He looks and probably smells like a female and can therefore scamper undetected amongst the normal male's harem and, more importantly, sneakily mate with the females under the nose of the normal male. Sometimes the illusion is a little too good, and the smaller male has to bow to the amorous attentions of the big male or run the risk of being exposed as a fake and attacked. His disguise is so complete that instead of begrudgingly giving in to the larger male, he actively encourages the larger male to copulate by presenting the tip of his abdomen.

Duplicity is a way of life to these beetles. Not only do they lie about their gender, but they also use a cunning trick to dupe their prey. Normally, the beetles will congregate around some feces, the corpse of an animal, or rotting fruit and wait for flies to descend, attracted by the scent of decay. When the fly is busy feeding, the beetle will pounce and dispatch the victim with its lethal mandibles. Sometimes, however, the beetle will adopt a quite different strategy, a strategy that has no need of rotting matter. The beetle sits on a leaf, stone, or similarly exposed spot, often near a stream, and with its abdomen pointing upward, everts a pair of glands. The beetle will even rub the tip of its abdomen on the surface where it stands. Small flies find the odor and secretion from these glands very alluring. They are fixated by this smell and will not fly away even if the beetle accidentally brushes against them. They will edge ever close to the beetle to find the source of the odor, until they are within range of the beetle's jaws.

- Sneak copulations are common in animals where males guard a resource (i.e., food or nesting sites) that attracts females. In many mammal species, there are individual males who dominate mating. The other, lowly males in the group will also want to mate, but will have to do so covertly, without the dominant male catching them.
- Rove beetle species in all parts of the world are attracted to decaying matter because of the other animals that are drawn to it. Rove beetles associated with dung are normally ambush predators, but they will also stalk their quarry over short distances. The mandibles of these species are always large and sickle-shaped, but they are useless

for breaking down food, so, like many predatory beetles, they regurgitate digestive juices onto their prey, turning it into a liquid mush that can be sucked up.

- In the tropics and subtropics, dung and other rotting matter does not last long before it is broken down by a range of animals, bacteria, and fungi. As soon as it appears in the habitat, it will be colonized and utilized, representing for a short time, a very rich habitat for scavengers and their predators. The clever strategy of using an odor to attract small flies is important when their main resource is not available.

- As with the Stenus rove beetle, the substance produced by the anal glands of the transvestite rove beetle was probably originally a deterrent to predators, but over time, it has evolved to fulfill another function. It is not yet known what odor it is mimicking to attract the flies.

⌕ Go Look!

Animal excrement, particularly the large amounts produced by grazing animals, such as cattle and horses, is used by a myriad of insect life. If you see a field of horses or cattle in the summertime, go and take a look at these hotbeds of insect activity. Many of the insects attracted to this resource are fast moving and flighty; therefore, approach with stealth. There will be flies on the surface of the matter, feeding, mating, and laying eggs, and at its edges, there may be rove beetles waiting to pounce on the flies. If the feces are a few days old, scarabid beetles will have arrived and will be burrowing through it to make their dung-stocked nests in the soil below. Underneath a cowpat there will be lots of animal life. There will be fly and beetle larvae and animals that specialize in feeding on these, such as small rove beetles and histerid beetles, which are normally very shiny and black and can retract their head, legs, and antennae into grooves on their body. Mites are also very common in dung, and often the animals that live on it or in it have a few mite passengers sticking to their undersides. Today, it is common practice for farmers to feed their cattle and horses drugs called *ivermectins* (avermectins), which kill worms living in the gut of the animal. These chemicals are not only toxic to parasitic worms, but they also end up in the animal's dung where they kill the many insects that depend on this resource. This has huge repercussions for the populations of birds, bats, and small mammals that feed on the dung insects.

Further Reading: Alcock, J., and Forsyth, A. Post-copulatory aggression toward their mates by males of the rove beetle *Leistotrophus versicolor* (Coleoptera: Staphylinidae). *Behavioral Ecology and Sociobiology* 22, (1988) 303–8; Forsyth, A., and Alcock, J. Female mimicry and resource defense polygyny by males of a tropical rove beetle, *Leistotrophus versicolor* (Coleoptera: Staphylinidae). *Behavioral Ecology and Sociobiology* 26, (1990) 325–30; Forsyth, A., and Alcock, J. Prey luring as alternative foraging tactics of the fly catching rove beetle *Leistotrophus versicolor* (Coleoptera: Staphylinidae). *Journal of Insect Behavior* 3, (1990) 703–18.

8

PUSHING THE BOUNDARIES: SURVIVING EXTREMES

ANTARCTIC TOOTHFISH

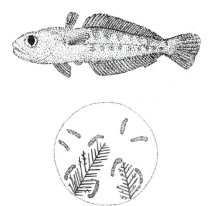

Antarctic Toothfish—Antifreeze proteins block the formation of ice crystals in the blood of the Antarctic toothfish. (Mike Shanahan)

Antarctic Toothfish—A large specimen of this supremely cold adapted fish looms from the dark chilly waters of the Antarctic. (Paul Cziko)

Scientific name: *Dissostichus mawsoni*
Scientific classification:
    Phylum: Chordata
    Class: Actinopterygii
    Order: Percifomes
    Family: Nototheniidae
What does it look like? A fully grown Antarctic toothfish may be as much as 140 kg in weight and 2 m long, although the average size is considerably less. They have a long, cigar-shaped

body with a broad head bearing big rubbery lips and large eyes. Long fins run along the back and the underside, and behind the gills are the large fanlike pectoral fins. Their skin has quite a somber color and ranges from grey to black or olive brown. The underside is typically paler.

Where does it live? This fish lives in the deep, cold waters around Antarctica. It is often caught in the Ross Sea and has been hooked at depths of more than 2,000 m.

Cool but Not Freezing

The frigid waters of the Southern Ocean are home to a surprising diversity of animals. The water is rich in nutrients, supporting huge densities of plankton. This plankton is munched by bigger planktonic organisms, which are in turn caught and eaten by small fish. Bigger fish eat these small fish, and so the food chain goes until you get to the top predators. In these cold waters, fully grown toothfish, cruising through the chilly waters, sculling their pectoral fins, are amongst the top predators. They cruise through the cold, dark waters on the look out for suitable prey. To enable them to live in these waters, the toothfish have some remarkable adaptations. The waters in the Southern Ocean are on the cusp of freezing, and any animal that swims through them must be suitably protected against the cold. Freezing is fatal to most animals as the tiny ice crystals that form in their tissues will rupture cells, killing the animals. Put a lettuce or a strawberry in the freezer for a while to get an idea of how damaging ice crystals can be to cells.

The blood of the toothfish contains some substances to counteract the lethal effects of ice crystal formation. As the temperature drops toward $-2°C$ (the temperature at which seawater freezes), water molecules will begin latching together firmly, forming tiny ice crystals that coalesce into the ice we are so familiar with. Circulating in the blood of the Antarctic ice are lots of large molecules called antifreeze proteins. These proteins are very strongly attracted to the surface of an ice crystal, and when they bump into a tiny sliver of ice, they stick fast to its surface. With this molecule on its surface, the ice crystal can't grow any further. Without these tiny antifreeze molecules, the blood and the tissues of the Antarctic toothfish would quickly freeze, and the animal would die if it ingested some ice or if an ice crystal penetrated a wound. Not only is the toothfish equipped with antifreeze, but it also has other adaptations to enable it to thrive in the frigid waters of the Southern Ocean. As a direct result of the chilly conditions, the fish has a very slow metabolic rate. Its heart beats about once every six seconds, which is astonishingly slow for such a large animal. With its blood coursing with antifreeze and its metabolism barely ticking over, the toothfish is hardly a hyperactive fish. To snaffle prey, it can use a short but powerful burst of speed, but generally it skulks in open water, mostly near the bottom without expending too much energy. Its skeleton is composed mostly of cartilage, making it lightweight, and around its body are various pockets of fatty tissue. Both of these adaptations contribute to the fish's ability to remain neutrally buoyant in deep water without expending muscular effort. Although there is very little light at the depths at which this fish lives, its eyes are very sensitive to even the faintest scattering of light from above, enabling it pick out the ghostly shadows of other fish in the gloomy depths. The Antarctic toothfish is an excellent example of how animal life can thrive in even the most inhospitable of conditions.

  + The Antarctic toothfish belongs to a family of fishes known as the cod icefishes. Although they can resemble cod, they are a completely different type of fish. Worldwide there are around 50 species of cod icefish. They are mostly found in the Southern Ocean and around the coast of Antarctica.

+ The larger species of cod icefish, such as the toothfishes, are among the dominant predators in the cold waters of the far south. They are thought to fulfill the same ecological role as sharks do in more balmy waters.

+ The antifreeze proteins in the blood and tissues of these fish are very important in their ability to tolerate cold conditions, but it is also thought the spleen plays a crucial role. It has been suggested that the spleen removes the tiny, thwarted ice crystals from the circulating blood.

+ As they have a very slow pace of life, the toothfish can reach a considerable age. Large specimens may be at least 50 years old and are probably far older.

+ In recent years, stocks of other types of fish have dwindled, forcing commercial fishing boats to cast their nets farther afield. The fishing fleets have turned their attention to the waters around the Antarctic, the home of the toothfish. The Antarctic toothfish, and to a greater extent the Patagonian toothfish, are finding themselves the targets of intensive fishing efforts. The flesh of these fish is good to eat, comparable to the northern true cod, and large numbers of them are being hauled from the depths every year. A single high-quality specimen, good enough for sushi, can be sold for as much $1,000. One of the largest fisheries operates from South Georgia, and it is allowed a catch of 3,000 tonnes per year. There is scant information on the populations of these remarkable fish, but it is known that they take a long time to mature and can reach grand old ages. Any fish with these characteristics will be very vulnerable to the effects of overfishing, and populations may take decades or longer to recover from overexploitation.

Further Reading: Di prisco, G. Life style and biochemical adaptation in Antarctic fishes. *Journal of Marine Systems* 27, (2000) 253–65; Eastman, J.T. *Antarctic Fish Biology: Evolution in a Unique Environment.* Academic Press, New York 1993; Kock, K.H. *Antarctic Fish and Fisheries.* Cambridge University Press, Cambridge, UK 1992; Somero, G.N., and DeVries, A.L. Temperature tolerance of some Antarctic fishes. *Science* 156, (1967) 257–58.

BEARD WORMS

Scientific name: Siboglinids
Scientific classification:
Phylum: Annelida
Class: Polychaeta
Order: Sabellida
Family: Siboglinidae

What do they look like? The beard worm can be a large animal, as much as 2.5 m long and 4 cm wide. Its long body is divided into a frontal crown bearing a spray of tentacles. This fore-part is attached to a long trunk containing most of the animal's organs. Attached to the trunk is a segmented, knoblike structure that anchors the worm in its tube home. Remarkably, the adult worms have no sign of a mouth, gut, or anus.

Where do they live? The beard worms are found throughout the world's oceans. They are often encountered on continental slopes and areas of seafloor that are spreading due to the

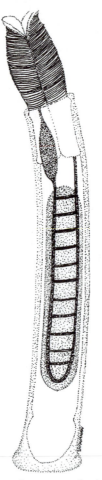

Beard Worms—A cutaway of a beard worm showing its trophosome which is packed with symbiotic bacteria. (Mike Shanahan)

presence of a rift in the earth's crust. All of the large beard worms are found at depths exceeding 100 m, in the ooze of the seafloor.

Gas Guzzlers

Between 1977 and 1979, discoveries were made in the deep Pacific Ocean off the Galapagos Islands that revolutionized our understanding of some of the most fundamental biological principles. Before this time, it was thought that the sun was the ultimate source of energy for all living things. Plants, algae, and an abundance of smaller life-forms use the power of the sun's rays to produce organic molecules via the process of photosynthesis. These organisms are the producers, and they are the foundation of all life on Earth, or so it was thought. Over 2,600 m down in the Pacific, it is pitch black. The life-giving energy transmitted in the rays of the sun is all absorbed by 125 m, yet in these dark places, dense communities of animals thrive, dominated by the beard worms, which are found in dense aggregations of more than 200 per square meter.

How are these animals living and thriving at such great depths without the power of the sun? It took several years to learn the secret of these worms, but eventually, the puzzle was pieced together, and the result is astonishing.

These deep-water worms are all found around what are known as *hydrothermal vents*, essentially underwater geysers that belch out warm water and huge quantities of gases such as hydrogen sulfide. The worms, most of their body safely in the confines of a tube, bathe their splendid crown of red tentacles in this noxious, heated water. The tentacles are red because blood containing hemoglobin is pumped through them, absorbing oxygen and sulfide-containing chemicals from the water. The hemoglobin molecules carry their cargo to a special part of the worm called the *trophosome*. In an adult worm, a large part of the trunk is taken up by the dense, brown tissue of this organ. The trophosome is not an organ in the normal sense. It is actually a part of the body that is dedicated to a symbiotic relationship with huge numbers of bacteria. These bacteria absorb the gases dissolved in the worm's blood and convert them into organic molecules, some of which are secreted and used as nourishment by the worm. This amazing symbiotic relationship allows the beard worms to generate their own nutrients from the chemical energy contained in the warm water of the deep-sea vents, in closed ecosystems far from the life giving rays of the sun.

+ There are approximately 120 species of beard worm, and because they dwell at such great depths, it very likely that there are many more species yet to identified. The first specimens were dredged from the seabed in 1900 off the coast of Indonesia. As the worms are rather fragile and easily broken by dredging, it was not until the 1960s that scientists got a look at a complete specimen.

+ There are two types of beard worm, perviate and obturate. The perviate beard worms are much smaller than their relatives, measuring between 5 and 85 cm, and hardly ever more than 1 cm in diameter. The perviate beard worms have a knot of tentacles at their head end, instead of a tentacular crown. These threadlike tentacles are used to absorb oxygen and sulfide-containing chemicals from the water, but it is thought that the nourishment provided by the symbiotic bacteria is supplemented by the absorption of edible particles in the seabed detritus.

+ The beard worms are the only group of animals where all the representative species completely lack a mouth, gut, and anus. These features are important in understanding the relationships of different groups of animals; therefore, for many years, it was difficult to know what the beard worms were related to. In 1988, it was found that the larvae of beard worms possessed a complete gut, which bore a resemblance to that of juvenile annelid worms. So, it seems the beard worms are related to the annelid worms, but their paths probably diverged over 500 million years ago, as fossils of beard-worm-like animals have been found in North American, north European, and Greenlandic Cambrian rocks.

+ The beard worms construct tubes made from chitin, the same material that makes an insect's exoskeleton. They secrete this tube on the seabed, among shells or on decaying wood.

+ The trophosome of the beard worms is very tightly packed with symbiotic bacteria. One gram of the tissue from this organ can contain 1 billion bacteria. It is still not clear how a young beard worm strikes up a relationship with bacteria. It is thought that when the worm is very small, a tiny vent on its body allows marine bacteria to enter and set up residence, eventually giving rise to the trophosome. Although the worms primarily rely on the secretions of their bacteria for sustenance, it is also possible that their helpful passengers are digested now and again.

- The bacteria living inside the worm are chemosynthetic. Unlike photosynthetic organisms (trees, alga, etc.), they can produce living matter from chemical energy instead of solar energy.
- Every symbiosis is a two-way relationship. It may appear that the worms are taking advantage of the industrious bacteria, absorbing the extra nutrients they produce without any anything in return. In fact, the bacteria are provided with a safe place to live and reproduce. It is also possible that the full extent of their symbiotic relationship is yet to be understood.
- The fact that these animals dwell in places far from the limits of normal, sun-dependent life is made even more remarkable by the fact that they lay down bulk by using hydrogen sulfide gas, which is toxic to all other forms of life. The unique makeup of their blood allows them to transport oxygen and this normally toxic gas through their tissues without any ill effects.

COCONUT CRAB

Scientific name: *Birgus latro*
Scientific classification:
  Phylum: Arthropoda
  Class: Malacostraca
  Order: Decapoda
  Family: Coenobitidae

What does it look like? The coconut crab is a very large crustacean. Reports of the full size this animal can reach vary wildly, but a body length of at least 40 cm and a weight of 4 kg are realistic. The first pair of legs is very well developed, and they bear huge pincers. The second, third, and fourth pairs of legs are slimmer and used for walking, while the last pair are very small and thin. The body is divided into two major parts, the carapace (cephalothorax) and the abdomen. All parts of the body have a very tough exoskeleton. Color also varies greatly, but these crabs are normally blue or brownish with pale patterning.

Where does it live? The coconut crab is found in a large swathe of the Indo-Pacific region, stretching from the Andaman Islands in the Indian Ocean to the Pitcairn and Easter Islands in the Pacific. It prefers coastal habitat, but has been found up to 6 km from the ocean. In some areas they live in rocky crevices, while in others they construct burrows in sandy ground.

Life on Land Is a Tough Nut to Crack

The crabs, regardless of their success in the world's oceans, never really made much of an impact on dry land. However, one exception to this is the coconut crab. As an adult, this animal is so adapted to a terrestrial way of life that if it is submerged in water for any length of time, it will drown. Not only have they turned their back on a marine way of life, but they have also developed into by far the largest land-dwelling invertebrate. Although the normally accepted limit is around 4 kg and 4 cm in length, there are reports of specimens more than double this length and four times the weight.

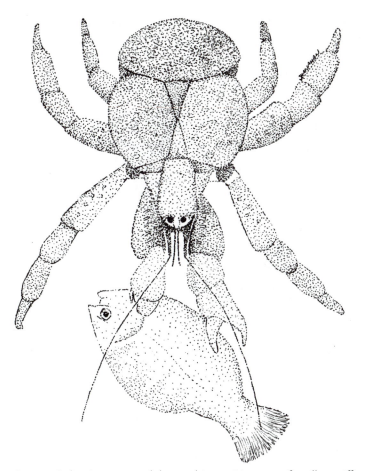

Coconut Crab—A coconut crab has used its sensitive sense of smell to sniff out a dead fish. (Mike Shanahan)

These gigantic crabs start life as eggs attached to the underside of their mother, who must venture down to the shore to release her brood into the water. The larvae that hatch from these eggs are marine and spend the first month or so of their lives floating around in the plankton. After this aquatic phase, the young, which have managed to survive the rigors of an aquatic exist-ence, drop to the seabed where they seek the empty shells of marine snails. The young crab backs into a suitably sized shell and uses it as a mobile home, disappearing into it at the first sign of danger. The crab's soft abdomen is particularly vulnerable, so it is concealed in the empty shell at all times. During this time, the young coconut crab may leave the water occasionally to explore dry land. This amphibious stage lasts another month, during which time the crab may have changed its shell to accommodate its growing body. With dry land beckoning, the crab leaves the sea forever, but it will be another two or three years before it can ditch its protective shell. Its ability to survive in water is lost, but its modified gills enable it to breathe air. These gills are situated in the carapace of the crab and are surrounded by a spongy tissue, which must be kept moist at all times to ensure sufficient gas exchange. This is where the frail, rear legs come in handy as they are dipped into water and brushed against the gill tissue.

With these lungs, the coconut crab can explore its terrestrial habitat. They haul themselves over the sandy ground of their islands and can even scale coconut palms. They may take to a palm to get out of the sun or to evade a predator. It is doubtful that they actually cut developing coconuts from their stalks, but they can use their fearsome pincers to open coconuts that have already fallen. Anyone who has tried to open a coconut will know this is no mean feat, but the crab's pincers are strong enough to lift a weight of 29 kg, and with patience the crab can whittle down the tough covering of these huge seeds. Although tough nuts can be dealt with by the crab, it eats a wide variety of other fruits and decaying animal matter. Small animals, like newly hatched turtles, which are too slow to escape the crab's clutches, are also on the menu.

All this food leads to increased bulk, and the only way the coconut crab can grow is by shedding its skin. In a fully grown specimen, this is an elaborate process, which takes place in its lair over a period of 30 days. The new skin is soft and flexible, and the burgeoning bulk quickly takes up the slack. Not wanting to be wasteful, the crab eats its shed skin after it has recovered from the considerable exertions of changing its armor.

+ The coconut crab is a type of hermit crab, of which there is 500 known species. All spend at least part of their life in the discarded shells of marine snails.

+ Mating in the coconut crab is a brusque affair and involves the larger male wrestling the female onto her back where he precedes to copulate. Shortly after, the female produces her fertilized eggs and adheres them to her abdomen where she keeps them until they are ready to hatch.

+ The coconut crab is also known as the robber crab, as it apparently fond of stealing pots and pans from houses and tents. It may mistake these objects for food and will drag them back to its burrow, which is what it does with real food. Competition for food can be fierce; therefore, tasty morsels are normally taken back to the animal's lair before being consumed.

+ During the larval stage, the vast majority of coconut crabs will fall foul of predators, but an adult specimen has nothing to fear except humans. The flesh of this animal is a delicacy in many areas, and so hunting pressure can be quite intense. Fortunately, in several areas the crab is protected, or collecting is limited to sustainable levels.

+ With the present conditions on Earth, the size of the coconut crab is probably the greatest that can be attained by a terrestrial invertebrate. It has been proposed that the concentration of oxygen was far higher in the atmosphere many millions of years ago. This allowed many types of invertebrate to reach gigantic proportions.

Further Reading: Greenaway, P., Taylor, H.H., and Morris, S. Adaptations to a terrestrial existence by the robber crab *Birgus latro*. VI. The role of the excretory system in fluid balance. *Journal of Experimental Biology* 152, (1990) 505; Morris, S., Taylor, H.H., and Greenaway, P. Adaptations to a terrestrial existence in the robber crab *Birgus latro* L. VII. The branchial chamber and its role in urine reprocessing. *Journal of Experimental Biology* 161, (1991) 315; Stensmyr, M.C., Erland S., Hallberg E., Wallén R., Greenaway P., and Hansson B.S. Insect-like olfactory adaptations in the terrestrial giant robber crab. *Current Biology* 15, (2005) 116–21; Taylor, H.H., Greenaway, P., and Morris, S. Adaptations to a terrestrial existence in the robber crab *Birgus latro* L. VIII. Osmotic and ionic regulation on freshwater and saline drinking regimens. *Journal of Experimental Biology* 179, (1993) 93–113.

COELACANTH

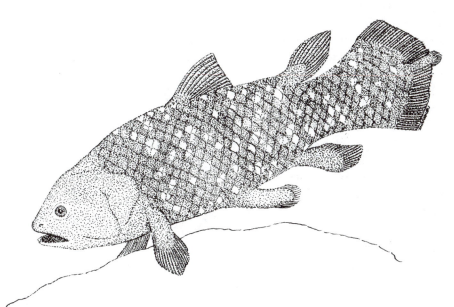

Coelacanth—A coelacanth uses its flexible fins to steady itself on the rocky sea bottom. (Mike Shanahan)

Scientific name: *Latimeria chalumnae*
Scientific classification:
  Phylum: Chordata
  Class: Sarcopterygii
  Order: Coelacanthiformes
  Family: Latimeriidae
What does it look like? Coelacanths can grow to be more than 2 m in length and 80 kg in weight. They are a steely blue gray color and are dappled with irregular white spots. Their eyes have an ethereal golden reflection due to a layer of tissue in their retina.
Where does it live? These fish have been found around the Comoros Islands, Sulawesi (Indonesia), Kenya, Tanzania, Mozambique, Madagascar, and the Greater St. Lucia Wetland Park in South Africa. All of Comoros Island coelacanths have been captured on long lines in 260–300 m of water, about 1.5 km offshore.

If It's Not Broke, Why Fix It?

Countless times throughout Earth's history, cataclysmic events, the likes of which we can only make educated guesses at, have wiped out much of the planet's life. Yet some animals survived these dark times and spawned lineages that have persisted into the present, more or less unchanged. A perfect example of an animal representing a lineage that has scarcely changed in millions of years was discovered in 1938, when Courtney Latimer, a curator of the museum of East London (South Africa), was in the Comoros islands looking for interesting specimens among the catch of the local fishermen. Amid a pile of fish, she spied an unusual, steel blue fin.

She exposed the owner of this fin and found it to be a species of fish she had never seen before. Convinced it was something special, she bought it from the fishermen and took it with her. In 1930s Africa, refrigeration was rare indeed, so to preserve the fish, she took it to her local taxidermist who promptly stuffed and mounted it. Unsure of what she had found, she contacted James Smith of Rhodes University in South Africa who was intrigued by the possibility of a discovery of a species new to science. Smith was not to be disappointed. He visited Latimer and her stuffed fish and was bowled over by what he saw. There in front of him was a coelacanth, a bit dog-eared, but a coelacanth nonetheless. The coelacanths were known only from ancient fossils and were thought to have become extinct at least 70 million years before Latimer's discovery. The observant curator had inadvertently discovered what would become the most famous so-called living-fossil of all time. When Smith announced the discovery, the scientific community was in disbelief, and many thought the living fossil was an imaginative hoax. A fresh specimen was needed to prove to the naysayers that the coelacanth was a surviving relic of a bygone age. A bounty of £100 was offered to anyone who could present a coelacanth. In those days, £100 was a large sum of money, so the search began in earnest. It was 14 long years until another specimen was captured. This proved beyond any reasonable doubt that a descendent of a fish from the mists of time had survived the mass extinction at the end of the Cretaceous period and was alive and well and living in the Indian Ocean, almost indiscernible from long dead animals only known from fossils. Since 1952, more than 150 specimens of this fish have been caught in the waters around the Comoros Islands.

It was not until the late 1980s that living coelacanths were filmed in their natural environment. Hans Fricke along with some of his colleagues used a submersible to descend into the waters where coelacanths had been caught. Their expedition was a great success, and they saw at depths of almost 200 m, six coelacanths going about their everyday business. All were seen in the middle of the night on or near the seabed, and all moved their fins in the same way as a four legged animal moves its legs.

Another chapter in the saga of the coelacanth began in 1997, when a British couple on their honeymoon in Indonesia spotted what looked like a brown coelacanth being brought into the market. The fish was collected, and it was indeed a coelacanth, but it was a different species from the one found around the Comoros Islands. This species was dubbed *Latimeria menadoensis*. To the locals, this fish was known as *rajah laut* (king of the sea). In 2000, three deep-water divers in the St. Lucia marine park off South Africa unexpectedly spotted a coelacanth while at a depth of 104 m. Further searches revealed the coelacanth, a relic from a long ago, was thriving in many areas.

+ The coelacanth is so important to zoology because it represents a group of animals from which all land-living vertebrates evolved. All the amphibians, the reptiles, and the abundance of mammals and birds evolved from an ancestor very similar to this fish. The fins of these fish are what developed into legs, enabling large animals to carry their weight on dry land.

+ The coelacanths and their relatives first appeared about 380 million years ago, or so the fossil record suggests. They were numerous and diverse up until the Cretaceous, but whatever cataclysmic event occurred at the end of this era spelled the end of these animals, or so it was thought. The ancestors of the modern day coelacanth obviously managed to hang on during these torrid times. Their deep-water habitat and ability to scavenge would have given them an edge, but it is likely they too teetered on the brink of extinction.

+ Apart from being an ancient relic, the coelacanth is peculiar for a number of reasons. The locals in the Comoros Islands had known about the coelacanth for many years before Courtney Latimer discovered it. Once hooked on a line, they make powerful and aggressive opponents and will struggle for a long time before being reeled in. The local name for it was *gombessa*, which translates as "worthless." The tissue of the coelacanth exudes oils even when dead, making its flesh taste foul. Salting it can make it semiedible, but it is far from coveted by the Comoros fishermen.

+ Live coelacanths swim slowly near the seabed with languid strokes of their fleshy fins. Their body is held at an angle with the head pointing toward the bottom. Sometimes they flip over and swim upside down. A gold reflective layer in the coelacanth's retina enables it to see in the dim world of the deep sea.

+ Female coelacanths give birth to live young. One captured specimen was dissected to reveal five young, all of which were about 30 cm long. It is a mystery how they mate, as the males have no special structure to introduce sperm into the female.

+ There is a great deal still to learn about these ancient fish. The fact that they were unknown to the world of science before 1938 shows how little is known about the oceans in general.

Further Reading: Erdmann, M. V., Caldwell, R. L., and Moosa, M. K. Indonesian "King of the Sea" discovered. *Nature* 395, (1998) 335; Fricke, H. Coelacanth. *National Geographic* 173, (1988) 824–38; Fricke, H., and Hissmann, K. Natural habitat of the coelacanth. *Nature* 346, (1990) 323–24.

GIANT MUDSKIPPER

Scientific name: *Periophthalmodon schlosseri*
Scientific classification:
   Phylum: Chordata
   Class: Actinopterygii
   Order: Perciformes
   Family: Gobiidae
What does it look like? This fish looks a lot like a gargantuan tadpole with flippers. It is one of the biggest mudskippers, reaching lengths of almost 30 cm. Its long body is speckled brown with a broad black stripe running from the eye all the way to the tail. The pectoral fins are well developed, and the gill covers are large, giving it a rather fat-cheeked look. The first spine of the dorsal fin is large and can be erected to hoist up the fin. The mudskippers eyes are large, goggling, and positioned on top of its head.
Where does it live? The giant mudskipper is native to Southeast Asia and can be found from Indonesia to Borneo. They are typically found in mudflat-type habitats, which represent the boundary between land and sea.

A Fish Out of Water

Think of a fish, and your mind's eye automatically conjures up an animal beneath the waves, perfectly adapted to an aquatic lifestyle. Yet, in some parts of the world where the sea is calm, some fish have left their watery world for a moist life on the margins of the land. These fish

Giant Mudskipper—A giant mudskipper hauls out on the exposed root of a mangrove tree. (Mike Shanahan)

are the mudskippers, and the giant mudskipper is one of the largest species. In the sheltered lagoons and bays of the Indo-Pacific, silt, sand, and organic matter is deposited by rivers or by the sea itself to form large mudflats. These habitats are very rich in animal life, and in burrows within the mud dwell the giant mudskippers. Although the giant mudskipper spends most of its time with its body hauled out of the sea on its powerful pectoral fins, it has not completely turned its back on the water. Unlike true terrestrial vertebrates, the mudskipper does not have lungs. Instead it must obtain its oxygen and get rid of carbon dioxide by using its gills and so-called skin breathing. The membranes in its mouth and throat are packed with blood vessels enabling gas exchange to take place as long as they are kept moist. Unlike other fish, the gill filaments of mudskippers are short and sturdy and do not collapse in on themselves when removed from the support of the water. These too, need to be kept moist to work effectively, so the gill chambers are actually quite large, allowing a small reservoir of water to bathe the gills. A rudimentary respiratory system is one thing, but how does a fish with fins move around when it is on land? Fins are perfectly adapted to propelling an animal through the dense medium of water, but on land, they are essentially useless. Evolution has equipped the mudskipper with modified fins that have a sort of elbow joint. These fins support the body of the fish and enable it to move forward in a series of short hops, or skips. The fins cannot be moved alternately to allow the animal to walk; nonetheless, the fish makes rapid, albeit ungainly, progress across the mud. When skipping just isn't fast enough, the fish can flip its muscular body and catapult itself up to 60 cm into the air.

Although the giant mudskipper is at home on land, it will take to the water where it can still swim well. The large, protruding eyes are all that can be seen of the fish when it is in the water.

When it skips from the water, the big eyes must be periodically moistened and so are retracted into the sockets. By sucking in its eyes now and again, the fish also swirls the water in its gill reservoirs, further improving gas exchange. All in all, the giant mudskipper is a very successful animal on the cusp of an amphibious existence. It gives us a fascinating, modern-day take on how backboned animals took their first tentative steps, or skips, on land.

- Worldwide there are around 35 species of mudskipper, most of which are found around the coasts of the Pacific and Indian oceans. A few species are found around Africa; yet none are known from the New World.
- Most mudskippers are carnivorous animals—catching and eating a wide range of prey, although there are some species that graze on algae.
- The giant mudskipper builds a nest in the mud using its mouth. In this hole, the mudskipper can seek refuge from the tropical midday sun and predators. The mouth of the burrow is surrounded by a low wall of mud, which traps a small pool of water around the entrance when the tide recedes. The burrows can sometimes be very elaborate, reaching a depth of more than 1 m and with many entrances.
- It is in these burrows that the mudskipper lay its eggs. Although there is water in the burrow, it is very low in oxygen, so the proud parents bring mouthfuls of air into the burrow to aerate the water. The eggs are actually laid on the roof of a chamber at the bottom of the burrow, and the larvae that hatch stay in the safety of this refuge until they transform into miniature mudskippers. They then swim out of the burrow and loiter in the small pool of water that surrounds the burrow entrance.
- Mudskippers are very territorial animals. They defend small areas of mud around their burrows and behave very aggressively toward trespassers, especially around breeding time. They gesture to one another by erecting their dorsal fins, and if these visible warnings are insufficient, they will chase each other. These chases may end in the pair joining battle where they wrestle with their mouths.
- The erectable fins in many species are brightly colored, and the male uses these to impress females. He flips into the air and flashes his dorsal fin at the peak of the jump.
- The mudskipper's view of the world is an unusual composite: half color and half black and white. The retina has color sensitive rods in the top half and monochrome cones in the lower half. The significance of this is not yet understood.
- There are even mudskippers that can climb trees. Using their powerful fins, they edge themselves along the exposed roots and lower branches of mangrove trees. The fins not only propel them but also provide a certain degree of suction, giving the fish a good grip during these arboreal explorations.

Further Reading: Graham, J. B. *Air-Breathing Fishes. Evolution, Diversity and Adaptation.* Academic Press, San Diego, CA 1997; Ishimatsu, A., Hishida, Y., Takita, T., Kanda, T., Oikawa, S., and Khoo, K. H. Mudskippers store air in their burrows. *Nature* 391, (1998) 237–38; Ishimatsu, A., Takeda, T., Kanda, T., Oikawa, S., and Khoo, K. H. Burrow environment of mudskippers in Malaysia. *Journal of Bioscience* 11, (2000) 17–28.

GIANT SQUID

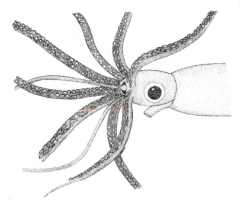

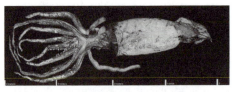

Giant Squid—A giant squid has large eyes and a mass of tentacles to hunt other animals in the ocean depths. (Mike Shanahan)

Giant Squid—A giant squid recovered from a trawl net. The long feeding tentacles have broken off, but the scale bar shows the size of this creature. (Gene Carl Feldman)

Scientific name: *Architeuthis dux*
Scientific classification:
    Phylum: Mollusca
    Class: Cephalopoda
    Order: Teuthida
    Family: Architeuthidae

What does it look like? The squid is a peculiar-looking creature. At the head end, there are eight muscular arms and two long tentacles, all of which bear suckers. At the center of the rosette of tentacles is the mouth with its powerful jaws, which bear a remarkable similarity to the beak of a giant parrot. Behind the tentacles is the head with a large pair of eyes, and further back still is the mantle—the housing for the animal's organs. The mantle is hydro-dynamically tapered, sporting a pair of fins for steerage.

Where does it live? The deep, uncharted oceans are the habitat of the giant squid. It is known from deep water in the South and North Atlantic Ocean and the North Pacific Ocean. Its specific habitat requirements are poorly understood.

Calamari as Big as Car Tires

The support provided by water allows the animals to attain great size. The giant squid is a perfect example of this as it is an invertebrate; yet when we think of animals without backbones we normally think of worms, insects, and the like. Imagine, then, a giant squid washed up on a Newfoundland beach in 1878 with a body length of almost 17 m. Of course, most of this length was due to the tentacles, but the animal still tipped the scales at 2,200 kg, millions of times bigger than a typical invertebrate.

Not only are giant squid large, but they are also active hunters, which whip their tentacles around with great speed when attempting to seize prey. Suction cups, surmounted by a rim of horny, toothed chitin are found on the underside of the squid's arms and on a palm-like pad at the end of the tentacles. These cups, ranging in size from 2 to 5 cm, are mounted on a stumpy stalk and latch onto the prey, bringing it toward the gnashing beak where the shorter arms maneuver it. The beak can tear the prey into pieces before it is taken into the mouth by the rasping radula, a structure common to all mollusks, which functions like a scraping tongue. The few giant squid that have been found show that this animal feeds on fish, other squid, and octopuses, and it was thought for a long time that they were little more than scavengers, eating whatever food they came across. However, recent footage from a Japanese vessel in the Pacific Ocean showed a giant squid taking bait from the end of a 900 m line. The sequence of pictures shows the squid is a fast-moving predator, using its numerous limbs to explore the bait before lunging at it. The animal in the pictures was in such a rush to take the bait that it got itself snagged on the line and struggled for four hours before eventually freeing itself by losing a 5 m section of tentacle.

Although the giant squid is one of the biggest invertebrates and an effective predator, it is not invulnerable. The sperm whale appears to a specialist predator of this massive mollusk. The big rectangular head of the sperm whale is often covered with scars made by the suction cups of struggling squid. Some of these scars are so large that it points to the possibility of some monstrous squid, as yet undiscovered, stalking the ocean depths.

Clearly, sperm whales relish giant squid, and it would be reasonable to assume you could make the mantle of the giant squid into massive rings of delicious calamari. Unfortunately, you would be very much mistaken. Giant squid calamari would taste disgusting, at least to us. In the tissues of the squid there are high levels of ammonium chloride, a substance that acts like a buoyancy aid but imparts the flesh with a foul taste to which the sperm whale must be oblivious.

- Cephalopods are represented by approximately 600 living species, but as with any ocean creature, it is more than likely that many more species are yet to be discovered. As a group they first appear in fossil record during the Cambrian period, around 500 million years ago.
- The giant squid is probably the inspiration for the legends of sea monsters, such as the Norwegian kraken, the Caribbean lusca, and the Mediterranean Scylla. Rarely seen sea creatures were often made out to be terrible monsters, responsible for the disappearances of ships and their crews. The giant squid is just a fascinating animal and represents no danger to humans whatsoever.
- The eye of the giant squid is one of the largest in the animal kingdom, with a diameter of at least 25 cm. Not only are the eyes of the creature big, but they are sensitive enough to allow the brain to form images comparable to those of higher mammals, such as humans. In some ways, the eyes of squid and octopuses are superior to those found in vertebrates. The light-collecting cells in the cephalopod eye face the incoming light instead of facing away from it (the vertebrate eye). Also, the cephalopod eye does not have a blind spot like that of the vertebrates. Cephalopods probably cannot see in color, but they can probably discern small differences in tone.
- The giant squid has a very interesting way of mating. The male has a long, prehensile tube at least 90 cm long, loaded with small packets of sperm. This tube is used like a hypodermic syringe, and when the male encounters a female, he injects the female's arms with his sperm parcels. As giant squid live in the blackness of the ocean depths, no one has seen the prelude leading up to what appears to be a brutal means of reproduction.

+ The giant squid, like all squid and octopuses, uses a form of jet propulsion to move short distances. Water in the body cavity can be shot out from a short, mobile siphon near the animal's head by powerful contractions of the mantle muscles. As the siphon is mobile, the squid can squirt itself forward just as well as backward with surprising speed. This is a very inefficient way of swimming, so the larger species normally use their fins and arms when cruising.

+ In 1925, another species of squid, now termed the *colossal squid*, was identified from fragments of two tentacles found in a sperm whale's stomach. Recent findings suggest this species is even larger than the giant squid. It is a heavy-bodied animal, with bigger eyes and a bigger beak than the giant squid. Its tentacle clubs brandish some impressive, swiveling hooks that are used to catch prey. It is found in the Southern Ocean, especially around Antarctica and is probably another favorite snack of the sperm whale.

Further Reading: Ellis, R. *The Search for the Giant Squid.* Lyons Press, London 1998; Kubodera, T., and Mori, K. First-ever observations of a live giant squid in the wild. *Proceedings of the Royal Society B: Biological Sciences,* 272(1581) (2005) 2583–86.

HAGFISH

Scientific name: *Mxyine* species and others
Scientific classification:
   Phylum: Chordata
   Class: Myxini
   Order: Myxiniformes
   Family: Myxinidae
What do they look like? Hagfish are elongate, wormlike animals ranging in length from 18 to 116 cm. The skin is smooth without scales and can be pinkish, grey, or black. They lack eyes and have sensory barbels around the mouth. They have no fins, but the tail is flattened from side to side.
Where do they live? Hagfish are found around the world, in cold temperate waters. They are found on or near the sea bottom, down to depths of 1000 m and probably more, and will often makes burrows in the muddy ooze of the seabed.

It's Not All about Looks

The first time someone sees a hagfish, the typical reaction is abject disgust. Although these are not the most attractive animals, hagfish are nonetheless very interesting. They are scavengers, and they swim slowly, sniffing for the odor of death and decay in the water, which may indicate the presence of a dead or dying fish nearby. As the hagfish has no jaws, it cannot bite chunks from the carcasses it finds. Instead, it fixes two pairs of horny rasps carried on a tongue-like structure to the carcass and then ties a knot in its very flexible body. The knot is forced down toward the head end of the animal, and when it reaches the carcass, it provides leverage to the rasping apparatus, which is retracted, pinching some of the flesh from the carcass. This technique is excellent when the carcass in question is small; however, occasionally the hagfish may chance upon the bounteous feast of a whale carcass. In this situation, the hagfish will be joined by huge numbers

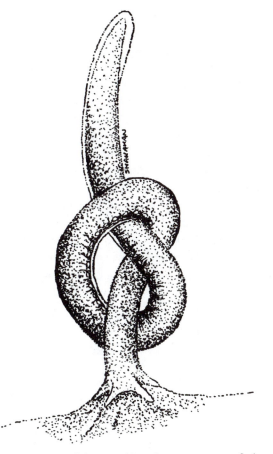

Hagfish—A hagfish ties itself in a knot to tear some flesh
from a fish. (Mike Shanahan)

of its kin that search for an easy way into the dead animal. Of course, the most obvious route
into the body is through a natural aperture, such as the anus, eyes, ears, or blowhole. Once inside
the carcass they can devour the soft organs.

The rasps of the hagfish can also be used to catch marine invertebrates, such as polychaete
worms, which appear to make up the bulk of its diet. Although the hagfish is a very efficient
scavenger and predator of small animals, its metabolic rate is very slow, so it can go for as much
as seven months without feeding.

Not only is the hagfish's feeding strategy rather peculiar, but the way in which it defends
itself is even more extraordinary. Along the flanks of the hagfish are a line of slime glands, which
during periods of danger or inactivity pump out huge quantities of gelatinous slime. The slime
is used to deter potential predators, heal wounds, and produce a lining for the burrows in which
these animals sometimes seek refuge. Slime production is so prolific that a single hagfish can turn
a whole bucket of seawater to slime in a little over 30 minutes. So much slime is produced that
the animal will often block its single nostril and have to essentially sneeze to clear the passage of
the thick mucus.

- There are 64 known species of hagfish in 5 genera (*Eptatretus, Myxine, Nemamyxine, Neomyxine*, and *Notomyxine*). Although they are thought to represent an ancient lineage, fossil evidence supporting this theory was lacking, until recently. A single fossil discovered at the end of the 1980s shows an animal so strikingly similar to today's hagfish that it seems they have not really changed for at least 300 million years. This is remarkable but is testimony to how well these animals are suited to their environment.

- Unlike the vast majority of other chordates, the hagfish lacks a cerebrum, cerebellum, jaws, and stomach. Instead of one heart, the hagfish has a systemic heart, plus three accessory hearts. The skeleton of the hagfish is composed of cartilage and not bone. The digestive tract of the hagfish is unusual because a lump of food is enclosed in a permeable matrix, and the only other animals known to produce this peritrophic membrane are insects. Also, unlike other fish, hagfish do not have a larval stage but develop directly into miniature adults. The eggs of the hagfish are also very large, approximately 2.5 cm long, and only small numbers are produced. But although the fecundity of the hagfish is not high, they are often found at very high densities—up to 15,000 in a very small area. This suggests that the hagfish have very low levels of mortality, but then again, what animal would be stupid enough to eat one of these slimy creatures?

- The slime produced by the hagfish is a complex and interesting substance. When it emerges from the slime glands, it absorbs water, swelling enormously. Unlike the slime produced by other animals, hagfish slime contains numerous threadlike fibers, which probably give it mechanical strength. The slime produced by hagfish is currently being studied for its potential in medical applications. One interesting application is in the promotion of wound healing, as it has been observed that hagfish wounds heal very quickly and cleanly.

- For many years, hagfish were of no commercial importance, but relatively recently a thriving hagfish fishery has developed. In Korea, the flesh of these animals is eaten, and the skin is made into a fine leather called eelskin. The Oregon fishery for the Pacific hagfish began in 1988 and peaked in 1992, with 15 vessels catching over 330 tonnes. The fish are caught with either lines baited with dead fish or baited plastic cylinders. As the reproductive rate of the hagfish is low, overfishing could have drastic effects on the populations of these very interesting animals, even though they have existed on Earth, unchanged, for many millions of years.

Further Reading: Crystall, B. Monstrous mucus. *New Scientist* 2229, (2000) 38–41; Jørgensen, J.M., Lomholt, J.P., Weber, R.E., and Malte, H.E. *The Biology of Hagfishes.* Chapman and Hall, London 1997; Martini, F.H. Secrets of the slime hag. *Scientific American*, (1998) October, 70–75.

HUMAN

Scientific name: *Homo sapiens*
Scientific classification:
  Phylum: Chordata
  Class: Mammalia
  Order: Primates
  Family: Hominidae

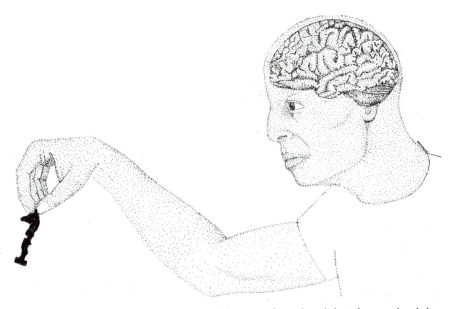

Human—Humans have a very large brain and their use of speech and their dextrous hands have helped them change the world. (Mike Shanahan)

What does it look like? Humans vary greatly in size and appearance. In North America, the average height of the male is 1.75 m with a weight of around 78 kg. Females are smaller, measuring 1.6 m on average and weighing in at 62 kg. Humans walk on their rear limbs, and unlike the rest of the primates, their body hair is fine. A human's eyes are large, and both are directed forward to give good binocular vision.

Where does it live? Humans are probably native to the continent of Africa, but in a geologically short period of time, they have spread to all parts of the world. The only environments where humans have not established communities are the harshest ones, such as high mountain tops, the sea, the driest deserts, and the polar regions.

We Have Come a Long Way

Some people may disagree with the inclusion of humans in this book. They might argue that we aren't animals. Let me assure you that from a zoological point of view, we are simply a type of primate, albeit a very intelligent one. In a geologically short period of time, the human species has spread over the globe and has become the most successful animal the world has known—a true survivor. Other animal species are constrained by the environment and have evolved adaptations to cope with what their environment throws at them, but the human species is equipped with three unique tools—the brain, the hands, and language—and it has used these to turn the tables and mould the environment to suit itself.

The inner workings of our brain are still poorly understood, yet it is the ultimate instrument for all of our successes. What the brain thinks up, our voice utters and our hands make real, and it is the combination of the three that has made our meteoric rise possible.

The secret of the human brain is thought to lie in its complexity. In an average human brain there are 100 billion separate cells, known as neurons, and each one of these is connected to a multitude of other cells, to the extent that there may be somewhere in the region of a million billion neuronal connections. This complexity comes at a price. The brain is very energy hungry, consuming one-fifth of the whole body's energy requirements, and the only fuel it can use is glucose. Such prolific energy consumption in a small space produces huge amounts of heat, which must be dissipated if the brain is not to overheat and suffer irreparable damage.

With the great complexity of this organ comes many complex functions. The brain gives us consciousness, complex problem-solving abilities, abstract thought, and many emotions. These have led to the evolution of technology, culture, complex societies, and religion. The brain controls the muscles that moves our limbs and directs the fingers in deft tasks. When you stop to consider the achievements of the human race, they are astonishing. We are equipped with a tool that allows us to peer into the mysteries of nature. We can grasp abstract concepts that are at the very core of the universe and gaze with curiosity at our fellow animals, explaining how and why they do certain things. We have invented computers and a multitude of other machines and have domesticated countless animals and plants. All of these things essentially come down to the brain. Even though this organ makes us what we are, it is still shrouded in mystery. However, it is becoming increasingly clear that the secret of its abilities lies in the connections between the individual cells.

Thanks to our origins in the trees, we are equipped with very sensitive hands that make even the most delicate manual task possible. The hands instructed by the inventiveness of the brain enabled humans to manufacture tools and clothes—the first technologies. The last part of the puzzle, speech, allowed us to share ideas and learn from one another. Long before the days of writing, speech ensured that ideas and new ways of doing things trickled down from one generation to the next.

+ There is only one living human species, although three types are identifiable (European, Asian, and African), all of which have characteristic features—a result of the environment where each evolved.

+ The origin of humans is a very contentious subject, but most authorities agree that modern humans, anatomically identical to you or I, appeared in the fossil record in Africa about 130,000 years ago. Scientists studying human evolution by looking at DNA from people all over the world have found that there is very little variation in our DNA. This means that at some point in its history, the human race was likely reduced to a very small population (1,000–10,000 individuals). A huge volcanic eruption from the Toba caldera in Indonesia has been proposed as the cause of this near extinction. As the earth shook off the effects of this massive volcanic upheaval, it has been proposed that modern humans began their colonization of the planet.

+ For the majority of human history, we existed as hunter gatherers, foraging for meat, tubers, and fruits. Then, at least 10,000 years B.C., a massive leap forward was made. We learned how to domesticate animals and plants. First to be tamed were the dog and bee, followed by several other animals and many plant species. Our forebears obviously found the wild relatives of these species useful and bred them to amplify the desired characteristics. This single advance was to change the course of human evolution. It meant that humans no longer needed to be nomadic. Settlements could form, and work could be divided among the individuals in the community. With less time being devoted to foraging, more time was available for the learning and sharing of ideas. Later still, these budding settlements in the most fertile areas developed into the first cities.

+ Although the human species is remarkable beyond doubt, it is exerting an ever-greater pressure on the planet's resources and other inhabitants. We are already starting to see the consequences of too many humans consuming the earth's resources at an ever-increasing pace.

Further Reading: Adas, M. *Machines as the Measure of Men: Science, Technology, and Ideologies of Western Dominance.* Cornell University Press, Ithaca, NY 1989; Bloomfield, M. *Mankind in Transition; A View of the Distant Past, the Present and the Far Future.* Masefield Books, New York 1993; Fairservis, W.A., Jr. *The Threshold of Civilization: An Experiment in Prehistory.* Scribner, New York 1975; Greenfield, S. *The Private Life of the Brain: Emotions, Consciousness, and the Secret of the Self.* Penguin Books, London 2002; Seymour, S. *The Brain.* HarperTrophy, London 1999.

LAKE TITICACA FROG

Lake Titicaca Frog—The wrinkled appearance of the Lake Titicaca frog enables it to survive in the oxygen deficient waters of its home. (Mike Shanahan)

Scientific name: *Telmatobius culeus*
Scientific classification:
    Phylum: Chordata
    Class: Amphibia
    Order: Anura
    Family: Leptodactylidae

What does it look like? Adult Lake Titicaca frogs can be up to 50 cm long and 1 kg in weight. The head is broad, with large bugeyes. The color varies considerably. Some are olive green with a pearl-colored stomach, while others are very dark, almost black, with white undersides. The most distinctive feature is the very loose skin, which hangs in large pleats and folds around the animal's neck, legs, and stomach.

Where does it live? This frog is found in Lake Titicaca, a huge lake high in the Andes on the border between Peru and Bolivia. The frog is found in the shallower parts of the lake, amid the reed beds and the other sheltered areas near the shore.

Wrinkly for a Reason

Lake Titicaca is the highest navigable lake in the world, and with an area of more than 8,300 sq. km, it is also very large. Due to the great altitude at which this lake is found, the oxygen levels in its waters are very low, so low in fact that some of the animals living in this lake have evolved some remarkable adaptations. Most famous of all the lake's inhabitants is the Lake Titicaca frog, an amphibian that is found only in this lake and only in its shallower reaches. As frogs go, the Lake Titicaca variety is quite a specimen. There are reports of adult frogs measuring more than 50 cm long, tipping the scales at more than 1 kg. In many respects, the Lake Titicaca frog is much like any other frog. It has a squat body, a wide head, and well-developed hind limbs with webbed digits. What is unusual about this animal is its skin. Its covering is a loose, baggy affair, and around the abdomen, legs, stomach, and neck, the flaccid skin falls into folds giving the animal a rather unsavory appearance. When it was first discovered by two adventurers during an 1876 expedition, it was labeled with the Latin name *Telmatobius coleus*. This translates as "aquatic scrotum" and is a tongue-in-cheek description of its appearance. As unsightly as it may be, the Lake Titicaca frog is perfectly suited to life in the oxygen-deficient waters of the lake. It rarely breathes using its lungs but gets most of the oxygen it needs via diffusion across the moist surface of its skin. The wrinkling and creasing of its covering greatly increases the surface area available for diffusion, allowing the animal to make the very most of the limited quantity of life-giving oxygen dissolved in the water. With the oxygen diffusing into its body and the carbon dioxide heading out into the water, the frog also requires an efficient means for the transport of these gases. The red blood cells tumbling through the amphibians vessels are the smallest of all the amphibians, and they contain the most hemoglobin. As these smaller cells with their oxygen gripping hemoglobin can be more tightly packed within the veins and arteries, sufficient oxygen can be shunted around the body to supply the animal's demands.

These physiological adaptations to surviving in Lake Titicaca are impressive, but occasionally they are not enough, and the frog has to resort to behavioral means of obtaining more oxygen. Resting on the bottom in shallow parts of the lake, the frog engages in what can only be described as push-ups. These aren't amphibian aerobics, but the frog's way of making small water currents around its saggy skinned frame. The push-up-powered flow of water increases the rate at which oxygen diffuses into the animal's blood stream and is another ingenious way in which these frogs have adapted to the tough environment of Lake Titicaca.

- The Lake Titicaca frog is one of 40 related species, all of which are found in South America.
- The Lake Titicaca frog is the largest truly aquatic frog. Its lungs are very much reduced and are of little use in gas exchange, although they can be used when the frog surfaces.
- To the native people who live on the shores of the lake and the artificial reed islands, the frog has always been a source of wonder. As the frogs often come to the surface when it rains, they have generally been revered as rain makers with divine powers. During times of drought, an adult frog is captured and placed inside a ceramic pot.

The pot is taken to a hilltop and abandoned. Its plaintive calls are said to bring rain, and if a deluge does arrive, the urn fills with water, releasing the captive amphibian.

- The frog's large size also makes it an attractive source of protein, and it is a popular food in Bolivia and Peru. Recently, a worrying trend has developed involving a drink called frog juice. This is a simple, but grisly beverage involving a Lake Titicaca frog stripped of its skin, some honey, a local tuber, water, and a blender. Allegedly, the pureed frog drink is a potent aphrodisiac. The frog is also thought to be something of a cure-all. Small ones are swallowed whole to cure a fever, and larger ones are tied to a fractured limb to act as living poultice.
- Many relatives of the Lake Titicaca frog are also vulnerable or endangered due to overhunting for their supposed aphrodisiac qualities.
- At these lofty altitudes, the ultraviolet shielding provided by the earth's atmosphere is much reduced. The skin on the back of a Lake Titicaca frog is normally dark, which provides a certain amount of protection from this potentially damaging radiation.
- Although the frog is a protected species, exploitation has had a drastic effect on its numbers. Very large specimens are hardly ever seen today, and what populations remain have shrunken considerably.
- Depending on where they are found in Lake Titicaca, the frogs vary tremendously in color, leading some scientists to believe that the lake was actually home to several sub-species of frog. In actual fact, comparison of DNA has shown all of these color types are the same species.

LUNGFISH

Scientific name: *Dipnoans*
Scientific classification:
    Phylum: Chordata
    Class: Sarcopterygii
    Order: Ceratodontiformes
    Family: Ceratodontidae, Lepidosirenidae, and Protopteridae
What do they look like? Lungfish are long, stout-bodied fish that are between 1.5–2 m in length. All species have a large, symmetrical tail. The Australian species has the most-well-developed fins, but in the other species, these are greatly reduced.
Where do they live? Found in Australia, South America, and Africa, they are freshwater fish that occur in small pools that are prone to bouts of flooding and drying.

Buried, but Not Dead

The lungfish are among the most unusual of fish and are remnants of a bygone age. All the living species swim with undulating movements of their body through the water of the lakes and pools where they live, skulking around on the bed looking for tasty morsels to gulp into their capacious mouths. When their pools are filled with water, they will feed ravenously on invertebrates and other fish and grow rapidly in preparation for the lean times ahead. Some of the African lungfish species inhabit pools that disappear in the dry season as the relentless heat evaporates the water. Soon, the pool is reduced to a fraction of its original size, and the fish must seek shelter from the worsening conditions. The fish finds refuge in the mud of the pool bed. It excavates a burrow

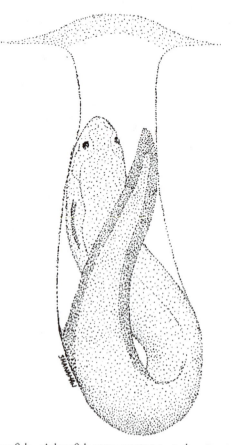

Lungfish—A lungfish prepares to see out the worse
of the dry season in its mucus lined, earthen retreat.
(Mike Shanahan)

in the mud, forming itself a large chamber, which can be as much as 1 m underground. In this
chamber, the fish holds its body in a loose coil with its tail over its eyes and secretes thick mucus,
which eventually encapsulates its whole body. A small hole in the mucus cocoon links the fish to
the surface, allowing air to reach the chamber. As their name suggests, lungfish have lungs and
therefore do not need water to obtain oxygen. Indeed the South American and African lungfish
will drown if they cannot use their lungs, such is there dependence on oxygen in the air. Deep into
the dry season, the pool bed may be reduced to an arid, cracked, apparently lifeless habitat; however, safe in their chambers, the lungfish are encased in hardened cocoons of mucus, envelopes
preventing moisture loss from their skin. To survive these dry conditions, the fish must reduce its
metabolic rate to a bare minimum. In this aestivating state, the fish depends on the muscle bulk
it developed when it was feeding in the pool. Proteins in the muscles and other parts of the body
are broken down to supply the fish's minimal energy requirements. After about six months, the
rains arrive, replenishing the pool and percolating down to the fish's lair, dissolving the mucus and
awakening the fish. Withered and shrunken from its ordeal, the lungfish breaks free of its chamber, with finding food as its highest priority. Although these animals can remain in the aestivating

state for six months, they have been shown to survive four years of entombment. Within one month and after some voracious feeding, the fish will have regained its original size.

+ There are six living lungfish species. Four of these are found in Africa (African, East African, marbled, and slender), and Australia and South America have one species each. The African and South American lungfish represent one lungfish group, while the Australian species represents a second group.

+ Lungfish are living fossils. The remains of similar animals have been found in deposits over 400 million years old. These fossils are the remains of animals that look very much like the living Australian lungfish. The fact these fish have lungs is also intriguing as it from an animal very similar to the modern Australian lungfish that all land-living vertebrates are thought to have evolved. All the amphibians, reptiles, mammals, and birds evolved from lungfish-type creatures, which took their air breathing abilities one step further to spend more time out of the water. The basic body plan of the land-living vertebrates can be seen in the Australian lungfish. There are two pairs of fins, which correspond to the front and hind limbs of terrestrial vertebrates. The Australian lungfish even uses its fins to walk along the bottom of its pool, moving them in the same way a walking, land-living vertebrate moves its limbs.

+ The global distribution of the lungfish also hints at its great antiquity. Like the velvet worm, they are found only in the Southern Hemisphere, and although today they are found in widely separated locations, the populations would have been much closer to one another before the process of continental drift slowly wrenched the ancient landmass of Gondwanaland apart to form the continents we see today in the Southern Hemisphere.

+ Only some of the African lungfish species bury themselves in the mud. The Australian species resorts to using its lung only when the there is insufficient oxygen in the water. Normally this species uses its gills to breathe. Although not all of the African and South American species resort to burying themselves, all are dependent on their lungs as their gills are very small.

+ During the breeding season, the South American lungfish develops a pair of feathery appendages, which are highly modified pelvic fins. These fins are used to increase the levels of oxygen and remove carbon dioxide from around the eggs in the fish's nest. The pectoral and pelvic fins of the African lungfish are modified into long, tapering appendages, the purposes of which are not understood.

Further Reading: Allen, G.R. *Freshwater Fishes of Australia.* T.F.H. Publications, 1989; Nelson, J.S. *Fishes of the World,* 3rd ed. John Wiley & Sons, New York 1994.

MARINE IGUANA

Scientific name: *Amblyrhynchus cristatus*
Scientific classification:
    Phylum: Chordata
    Class: Reptilia
    Order: Squamata
    Family: Iguanidae

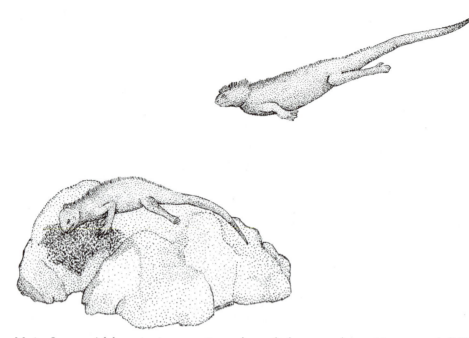

Marine Iguana—Adult marine iguanas swim to the sea bed to graze the nutritious seaweed. (Mike Shanahan)

What does it look like? The marine iguana is a large, disheveled looking lizard. Fully grown males can be 1.3 m long and weigh as much as 1.5 kg. They are normally black or grey. Color can also vary depending on the island on which they are found. In some locations, the adults are brick red and black, while in other places they are red and dark green, but mainly during the breeding season. The tail is very long, and the limbs are well developed with long, curved claws. Along their back, the lizards also have a crest of spines giving them a fierce, dragon-like appearance.

Where does it live? The marine iguana is only found on some of the islands that make up the Galapagos archipelago. They are normally found on the rocky coastline but occasionally can be seen in areas of marshland and mangrove swamp.

A Reptilian Beach Bum

During his expedition aboard the *Beagle*, Charles Darwin saw and studied many creatures unknown to Europeans. The Galapagos were a rich hunting ground for his collecting forays, and on the wave-beaten shores of some of the more barren islands, he took note of a peculiar lizard basking on the black volcanic rock. In Darwin's eyes, this lizard—the marine iguana—was far from being a looker. The young naturalist described it as a "disgusting, clumsy lizard." It's fair to say the marine iguana is not the prettiest animal, but what it lacks in looks it more than makes up for in character.

Large groups of these lizards bask on the wave lashed shore, their spiky bodies almost blending into with rugged volcanic rock. Like all lizards, marine iguanas depend on the sun for their warmth. Before they can search for food, they must bask in the sun's rays until their body temperature has risen sufficiently. With their muscles nicely warmed, they can set about foraging.

This is where the similarity to other lizards ends, for this is the only lizard that can live and search for food in the sea. Clumsily, they plunge from the rocks into the heavy swell. Once in the water, they are graceful animals, and with legs folded along their sides, they use their powerful tails to swim through the water. The big males can quite easily dive 15 m to reach the succulent algae carpeting the rocky seabed. They cling to the rocks with their sharp claws and graze the low-growing seaweed. Although the Galapagos Islands straddle the equator, the water flowing around them is cool—the result of upwellings from the ocean depths. It doesn't take too long until the marine iguana starts feeling the chill, and in a state of cold fatigue, it must surface and haul itself out onto the rocks. This is no mean feat for a cold-blooded animal. Out on the rocks, the iguana returns sluggishly to its territory and sprawls out beneath the warming sun. During its subaquatic forays, the iguana swallows quite a lot of salt water with its food, and to prevent any ill effects, it must get rid of this excess salt. It does this with the use of special glands, the products of which are sneezed from the nostrils. Following a good stint in the sun, the iguana may make another foray into the water to get at its favorite seaweed. These daily forages for food and long periods of sunbathing are how the marine iguana spends its time. This surely makes it one of the most dedicated beach bums in the animal kingdom.

- Iguanas are among the largest lizards and can be found around the world in tropical and subtropical habitats. The marine iguana shares the Galapagos Islands with a close relative, the Galapagos land iguana.
- The existence of these lizards in the Galapagos archipelago is something of a zoological conundrum. These islands are around 1,000 km from the west coast of Ecuador, but since they formed, 5–10 million years ago they have been colonized by various animals, most of which have evolved into unique species. How these animals and plants reached these remote volcanic islands is a bone of contention. One popular theory is that rafts of vegetation torn from river banks by flood waters drifted out to sea, eventually finding landfall on the virgin Galapagos Islands. Among the floating vegetation, and somehow surviving the hardships of a very long and arduous sea crossing, there were land animals. Some of these would have been the ancestors of the marine iguanas we marvel at today.
- The fact that these lizards are essentially marine demonstrates the power of adaptation. Their ancestors would have been suited to eating terrestrial plants, but over the eons, they gradually developed the ability to take to the sea where they could feast on the abundant seaweeds growing in the nutrient-rich waters surrounding these islands.
- The unique assemblage of animals and plants inhabiting the Galapagos Islands and the way in which they vary subtlety from island to island were major inspirations that enabled Charles Darwin to formulate his theory of evolution.
- Today, the Galapagos Islands have been recognized for their ecological importance. Most of the land and a vast swathe of surrounding ocean has been designated as a national park, world heritage site, and biosphere reserve. The human population of the islands continues to expand massively, putting pressure on the flora and fauna, especially as nonnative animals have been introduced.

Further Reading: Berry, R. J. *A Natural History of the Galapagos.* Academic Press, London 1984; Wikelski, M., and Thom, C. Marine iguanas shrink to survive El Niño. *Nature* 403, (2000) 37–38.

OLM

Olm—The eyeless olm is adept at hunting prey in caves. (Mike Shanahan)

Olm—A captive olm, showing the long, thin body of this amphibian, the small limbs, the gills and the complete lack of eyes. (Helmut Presser)

Scientific name: *Proteus anguinus*
Scientific classification:
    Phylum: Chordata
    Class: Lissamphibia
    Order: Caudata
    Family: Proteidae
What does it look like? A fully grown olm is around 30 cm with a sinuous body and long tail. There are two pairs of stumpy legs and three pairs of feathery gills behind the head. In its natural environment, the olm is pink with semitranslucent skin.
Where does it live? The known range of the olm is tiny. It is only found in the Dinaric Alps, which form part of the Adriatic coast in Europe, stretching through Italy, Slovenia, Croatia, Bosnia and Herzegovina, and Montenegro. It inhabits the many elaborate cave systems of this region.

Thriving in the Darkest Corners

The peaks and crags of the Dinaric Alps provide a mountainous backdrop to the Adriatic Coast. Composed of limestone-containing rocks, these mountains have long been at the mercy of the elements. Over millions of years, rainwater, with its slightly acidic bite, has eaten into them, forming a honeycomb of tunnels and galleries. It is in the perpetually dark, subterranean roots of these mountains that you find the olm. Just how long these anomalous amphibians have evolved in the dark, chilly waters of these caves is unknown, but they certainly represent one of the more ancient groups of salamanders, and over the eons they have become superbly adapted to a troglodyte existence.

    The body of the olm has been molded by its environment. Its body is long and snakelike, enabling it to swim through the water with graceful undulations. The short forelimbs and hind limbs of the animal look as though they are in the wrong place. They are so far apart that they are of little use for walking. A troglodyte has no need for the skin pigments, such as melanin, which protect the delicate, outer covering of an animal from the unforgiving ultraviolet rays emitted by the sun. The olm is entirely pink, and beneath the thin skin, the capillaries and the outline of the organs can be seen. Where there is no light, there is also no need for eyes, and the olm has almost lost these organs completely. The eyes are very small and poorly developed and sit beneath a layer of skin. It is only in the fetal olm that the eyes can be clearly seen.

Food in these subterranean habitats is scarce, and with no light or indeed eyes with which to find its prey, the carnivorous olm must slip through the waters of its flooded caves in the hope of detecting the water-borne scent of small crustaceans or other invertebrates that make up the bulk of its diet. As food is so rarely encountered, the olm is able to go for long periods of time without eating a single morsel. The temperature in these caves is cool and constant, allowing the animal's metabolism to tick over at a very slow rate. There is a documented case of a captive olm being kept in a container in a refrigerator for 12 years without a single item of food passing its lips. Needless to say, this long period of enforced starvation took its toll on the animal. It had lost a lot of weight, and to keep itself alive, it had digested some of its own internal organs, including its gut. Even in the wild, it is very likely that these animals can go for several years without food.

- The family to which the olm belongs, the Proteidae, is represented by seven species. The other species, commonly known as mudpuppies and waterdogs are all found in North America. None of these American species are cave dwellers, but they are all aquatic and prefer shallow lakes and streams.

- People have known of the olm for centuries, as every so often, floods wash them out of their caves into surface streams. Local people called this animal the human fish because of its pink skin and limbs, a name that has stuck to this day in its native range. Other people, undoubtedly inspired by tales of fabulous beasts, thought olms were baby dragons, a belief that persisted for some time. In Slovenia, the animal has the name *Mocheril*. This translates as "the one that burrows into wetness."

- Even today in the twenty-first century, countless olms have been caught and studied in captivity, but very little is known about what they actually do in the wild. For example, nothing is really known of their breeding habits in the dark caves of the Dinarics.

- Cave systems, both dry and wet can be found all over the world. Apart from the animals normally associated with these habitats, such as bats, there are in fact huge numbers of animals supremely adapted to a life in the darkness. Apart from salamanders, there are insects, spiders, crustaceans, and fish. All have lost their eyes, either partially or completely; all are very pallid, normally pinkish or white, and many have long, sensory appendages with which to find their prey in the dark. Day length and other rhythms only have a limited effect on true troglodyte animals, as deep down in a cave system, the conditions are pretty much constant through the year.

- Sealed cave systems are discovered by explorers on a regular basis. In some cases, these underground time capsules have been sealed for thousands, perhaps millions of years. The animals within evolve in isolation and are occasionally found to be descendents of primitive creatures only previously known from fossils. Recently, a sealed cave system in Romania was found to contain at least 30 species of organisms new to science. Like some deep-sea ecosystems never touched by the suns rays, the foundation of life in these sealed caves is the gases dissolved in hot water springs. Bacteria thrive on these gases, providing a source of food for other more complex organisms.

SPERM WHALE

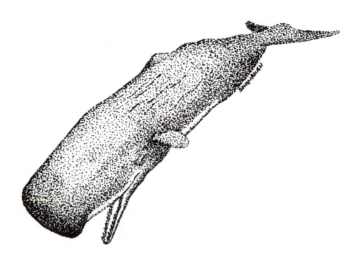

Sperm Whale—A sperm whale dives to catch other marine animals, with giant squid being a particular favorite. (Mike Shanahan)

Scientific name: *Physeter macrocephalus*
Scientific classification:
  Phylum: Chordata
  Class: Mammalia
  Order: Cetacea
  Family: Physerteridae
What does it look like? The sperm whale is the largest of the toothed whales. Males (bulls) can be as long as 20 m and weigh 70 tonnes. Females are much smaller. The whale has a huge block head, small pectoral fins, a large triangular tail fluke and a long, narrow lower-jaw studded with large conical teeth.
Where does it live? Sperm whales are found throughout the world's oceans and in the Mediterranean Sea. Populations are most dense in regions above underwater continental shelves and canyons.

Plumbing the Depths

The head of the sperm whale is immense, accounting for 40 percent of its total body length. If this capacious space was filled with brain, the sperm whale would have more grey matter than it knew what to do with, alas, the rectangular head is filled mostly with wax. This wax is whitish and rather gloopy, giving the whale its common name as whalers in the eighteenth and nineteenth centuries likened it to semen.

Why should this animal have a head brimming with wax? The blunt answer is that no one is really sure. There are several theories. One of these relates to the whale's feeding behavior. The favored prey of the sperm whale are those animals that dwell on or near the seabed, such as squid, fish, and small sharks. To reach these tasty morsels, the whale must dive to great depths, making it the deepest-diving animal on the planet. It is not known exactly how deep they can dive, but it is at least 1,200 m and possibly as much as 3,200 m (about 10 times the height of the Eiffel tower). These dives can last for two hours. The whales don't have an aqualung; however, their muscles can store a lot of oxygen, enabling them to stay underwater for long periods. Even so, swimming down to the seabed would quickly drain their oxygen stores, so it is has been proposed that the wax aids them in their search for food.

The wax organ in the head, the spermaceti organ, has a series of passages and chambers and a blood supply. When the sperm whale is ready to dive, some scientists have suggested that seawater is allowed into these passages, circulating around the wax and cooling it. Cooling results in shrinkage, and the wax increases in density, surpassing that of the surrounding seawater. With a heavy head, the whale sinks to the bottom at a rate of around 170 m per minute. Once at the bottom, the whale can begin to feed. After a good forage, hunting for nameless creatures in inky blackness, the whale's oxygen stores begin to run low. The whale must return to the surface or drown. It has been theorized that blood flow is increased to the wax organ, and as it warms, it expands, becoming lighter and acting like a buoyancy aid to carry the exhausted whale back to the surface. Once at the surface, the animal rests for approximately five minutes before diving again. After several long dives, the whale is thoroughly spent and has to float around on the surface for many minutes.

The buoyancy aid theory is an interesting one, but it has also been suggested the spermaceti organ helps the whale catch its prey in the pitch dark of the ocean depths. It is possible they just reach the seabed and scoop up the bottom mud and anything edible. Tin cans and stones in the stomach of captured whales support this notion, but more interesting is the fact that, like bats, sperm whales can echolocate, using sound to build up a picture of their environment. Echolocation may be used not only to find their prey but also as a weapon. It is thought the sound pulses produced by the whale are focused by the wax in the spermaceti organ, in much the same way as a glass lens focuses light. The focused bursts of sound produce pressure waves in the water that stun prey.

The large wax-filled head of the whale may also be important during the breeding season. Some scientists think the head full of wax acts as a shock absorber to prevent damage to the delicate parts of the head when the animals ram into each other during disputes, or even into ships when defending themselves. In the whaling days, three whaling ships are known to have been sunk after being rammed by large sperm whales. This may explain the wax, but the female also has a wax-filled head, and fighting is normally the pastime of male mammals. Also, some other species of cetaceans have wax in their heads, such as the bottlenosed whales. They are not known to ram into one another, but they do dive to great depths to capture squid, supporting the theory that the wax is somehow involved in feeding.

- + The sperm whale holds other records in the animal kingdom. In addition to being the deepest diving animal, it also has the largest brain, weighing in at 9 kg. In comparison, a human brain weighs 1–1.5 kg. It is also thought to be the largest toothed animal that has ever lived. There is evidence of giant bulls killed in the early days of whaling, which were probably 28 m long and 150 tonnes in weight. If these accolades weren't enough, the sperm whale also has the thickest skin of any animal—more than 30 cm!

* The title role of Herman Melville's classic, *Moby Dick*, was a gigantic white sperm whale.

* This species was once the favored target for whalers, as the whales rest at the surface for a long time, completely exhausted after finishing a sequence of long dives and are therefore essentially defenseless. The whalers used hand harpoons and later ship-mounted ones to kill the whales, before dragging them back to the ship. Its social behavior also made it attractive to whalers. Pods of this cetacean consist of females and young males (adult males are solitary), and if one member of the family is injured, the whales will rally around and support the injured animal (the marguerite formation) to prevent it from drowning. This behavior made it easy for whalers to kill many individuals quickly, as other group members would soon surround an injured animal.

* Once dead, the whales were not only processed for their meat and blubber, but also for the spermaceti wax in the head, which was used as a high quality lubricant. Up until recently it was used as a lubricant in some very high-tech applications, such as the space industry, where synthetic alternatives could not match its specific qualities. Another substance, found in the intestine of the sperm whale, ambergris, is also highly prized. It is thought this substance, also known as so-called floating gold binds hard, indigestible bits of food so they can be safely passed from the whale's digestive system. Fresh ambergris is black and sticky and is far from sweet smelling. Once passed by the whale it can drift for many years, all the time being weathered by the sun and salt water. Eventually, when it gets washed up, its smell has become a lot more alluring. It is this weathered ambergris that is prized by purveyors of fine fragrances, commanding prices of $20 per gram. A 15 kg lump recently found washed up in southern Australia would be worth $295,000, which is a good reason to take up beachcombing.

* The sperm whale was hunted ruthlessly in the 1850s to the point where the population collapsed in 1860. The population recovered, and in the 1960s, when factory ships scoured the oceans vacuuming up the large whales, at least 25,000 sperm whales were killed every year. The peak was in the 1963–64 season when 29,300 animals were taken. This scale of exploitation is not sustainable, and the commercial hunting of these whales is banned; fortunately only Norway, Japan, and some indigenous peoples are allowed to catch a small quota.

Further Reading: Evans, K., and Hindell, M.A. The diet of sperm whales (*Physeter macrocephalus*) in southern Australian waters. *ICES Journal of Marine Science* 61, (2004) 1313–29.

SUN SPIDERS

Scientific name: Solifugids
Scientific classification:
   Phylum: Arthropoda
   Class: Arachnida
   Order: Solifugae
   Family: many

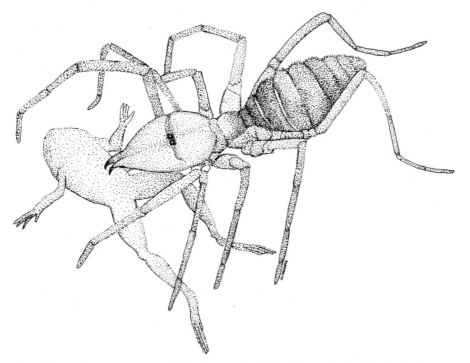

Sun Spiders—A sun spider eats a small amphibian which it has just caught. (Mike Shanahan)

What do they look like? The sun spiders are medium to large arachnids. The largest species can have a body length of more than 7 cm, although the very long legs make them look much larger. The body of the animal is divided into two parts, the clearly segmented abdomen and the head and thorax which are fused into one unit. The jaws are huge and pincer shaped. Color varies according to species, but many species are brown/yellowish, while a minority are dark, and some are even patterned.

Where do they live? Sun spiders can be found in Africa and east into India and Indonesia. Many species are also known from the New World. They can be found in a variety of habitats, including forests and grasslands; however, they seemed to have developed an affinity for the drier parts of the world.

Red Romans and Beard Trimmers?

Few people would know what a sun spider is, let alone have seen a living specimen, and it is to the deserts of the world you must go if you want to see what surely rates as one the oddest members in that band of animal oddities, the arachnids. *Sun spider* is an ironic choice of name for an animal that is no great sun worshipper. Even the diurnal species try to keep out of the scorching, unforgiving rays of the tropical sun as much as possible. The appearance of the sun spider is nothing short of disconcerting. They are reminiscent of the scuttling, malevolent creatures beloved by science-fiction film makers. Add the grotesqueness of their appearance to the speed at which they can move, and you have an animal that makes a tarantula seem benign as a guinea pig.

Although they look mean, the sun spiders are harmless, well, to humans at least. Their single most impressive characteristic and that which attracts the most attention from arachnid aficionados is their huge chelicerae. They are massive, and size for size are probably the largest jaws in the animal kingdom. They sprout from the head like an elaborate pair of nut crackers. Where they join the head, they bulge outward, swelled by the enormous muscles within. These magnificent mouthparts give the animal a rather head-heavy look. Either side of these huge chelicerae are long appendages that resemble thin legs. These are in fact the pedipalps, important sensory organs used in a variety of situations. Behind these pedipalps are the arachnid complement of four pairs of walking legs. These legs enable the sun spider to move with grace and poise under normal circumstances, but when threatened, they can propel the beast to more than 16 km/h, an impressive speed for such a small animal. Scaled up, this would be equivalent to a running human reaching a speed of almost 500 km/h. With such a turn of pace and those jaws, the sun spider is a capable predator, able to get the better of just about any smaller animal. The eyes can detect movement, probably form rudimentary images, but the numerous long, stiff hairs all over the animal's legs alert it to tiny fluctuations in air pressure, indicating the proximity of potential prey. Once the prey has been subdued, the jaws work like an effective pair of shears, slicing it up and allowing digestive fluids to permeate and dissolve the soft inner parts. The resultant soup is then taken into the mouth.

On this diet of invertebrates, small reptiles and other creatures, the sun spider grows quickly and may shed its skin nine times before it reaches maturity. Apart from food, a mature sun spider must also seek out a mate. This is left to the male, who must sniff out a receptive partner and engage in copulation, which, in the tradition of the arachnids, is a perilous affair. To the female, the male is little more than a fat snack, so to stand any chance of perpetuating the species, the male approaches his intended mate very gingerly. Any false move will see him trapped in the formidable jaws. He uses his long, sensitive pedipalps to stroke the female, which seems to put her in some form of trance. With a chivalrous air, he carefully picks her up in his jaws and carries her a short distance before placing her carefully on her side. From his genital opening he produces a small packet of sperm that he delicately picks up in his chelicerae. He introduces this spermatophore into the female's genital aperture and quickly hops away before the inseminated female regains her senses and turns on him.

+ Approximately 900 species of sun spider are known from around the world. In the United States, there are a few species in Florida and more than 100 in the Southwest. Some species are even found as far north as Canada. Unlike some other large arachnids, they tend to be seasonal species, living out their whole life in a single year.

+ They are known by a variety of names other than sun spider; these include, wind scorpion, camel spider, red roman, haarskeerders, and baardskeerders. The latter two are the most interesting and refer to the belief that a female sun spider carrying eggs will use hair clipped from humans and animals to line her subterranean nest. A sun spider has never been witnessed snipping at a sleeping man's beard with its jaws, but nests have certainly been found to contain fur.

+ It is also the mistaken belief that a sun spider will actively pursue a human it chances upon in its habitat. The reason for this is that although the sun spider is a creature of warm, arid places, it keeps out of the direct sun light as much as possible. The shadow cast by a human is a convenient source of shade, and it is only this shade that the arachnid is pursuing.

- Another unique feature of the sun spider is the characteristic racquet or malleoli organs, which can be found along the underside of the back legs of certain species. The function of these, like the pectines of scorpions, is unknown, although they are probably sensory.
- As the sun spiders actually shun solar rays, the Latin name *solifuge* is much more apt as this translates as "fleeing from the sun."

Further Reading: Harvey, M.S. The neglected cousins: What do we know about the smaller arachnid orders? *Journal of Arachnology* 30, (2002) 357–72; Punzo, F. *The Biology of Camel-Spiders (Aachnida, Solifugae).* Kluwer Academic Publishers, Boston 1998.

SYMBION

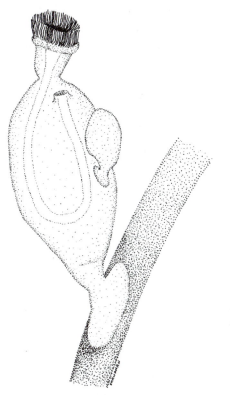

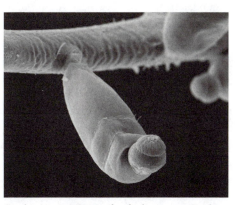

Symbion—The sessile stage of the Symbion with its u-shaped gut clearly visible through its thin skin. (Mike Shanahan)

Symbion—An SEM of a feeding stage Symbion showing the crown of cilia at its head end and the disc that it uses to grip onto the lobster's bristle. (Matthias Obst)

Scientific name: *Symbion pandora*
Scientific classification:
    Phylum: Cycliophora
    Class: Eucycliophora
    Order: Symbiida

What do they look like? These are microscopic animals. The females are slightly less than double the width of a human hair, whereas the males are only one quarter this size. In shape, the adult resembles a miniscule urn, with a short stalk at its back end and a crown of short hairs at the other.

Where do they live? Symbion has so far been around the coast of Norway and in the Mediterranean. It lives on Norwegian lobster and has been recorded at depths of 20–40 m.

The Strangest of the Strange

Just before Christmas in 1995, two zoologists from Denmark announced the discovery of a new animal from the Kattegats Straits. Their discovery was heralded as the "zoological highlight of the decade." The new animal was so odd that it didn't fit into any of the accepted categories of animal life (phyla). A new phylum had to be created to accommodate it.

The animal had actually first been observed in the 1960s, but it had gone undetected for so long because of its small size and strange habits. Symbion, for much of its life, lives on the bristles that grow on the mouthparts of the Norwegian lobster, commonly known to many people as scampi, Dublin bay prawns, or langoustines. Apart from this most unusual habitat, Symbion also has a bizarre life cycle, composed of a variety of forms, the most obvious of which is the feeding stage. At this point, Symbion is asexual, so there are no males or females. This asexual feeding stage is found attached to the bristles around the lobster's mouth. It fixes itself to these hairs with a small adhesive disk at the back end of its body. At its head end there is a coronet of small hairs (cilia), which it uses to engulf edible particles that have come adrift from the lobster's mouth. This food is funneled down into the mouth and into the loop of the gut. Undigested food passes from the anus, which opens near the thin stalk of the feeding crown.

To continue the species, the feeding stage reproduces asexually by a process known as budding. Small offshoots develop on the body of the feeding stage and develop in to males and females, which represent the sexual stage of Symbion. The males are very simple. They have no mouth or anus and lack even the rudiments of an alimentary canal. What they do have, however, is well-developed genitalia. A male detaches from the asexual, feeding stage and drifts through the water, hopefully to find himself a feeding stage specimen containing a developing female, which he will impregnate with his penis. The inseminated female leaves the body of the asexual stage, which has nurtured her, to find a place of her own on the lobster's bristles. The life of the female is short, and before long, her digestive organs deteriorate and are reordered as a larva. This larva turns into a young asexual feeding stage, and it breaks its way from the shell of the female and swims off to find a suitable place to fix itself on the lobster, where it will grow to begin the complex process all over again.

- Since Symbion was discovered, other similar species have been found in distant locations from the original spot off the coast of Denmark. New species have been found on the mouth bristles of American lobsters and European lobsters, extending the distribution of these unusual animals to include the West Atlantic and the Mediterranean. Close examination of other marine crustaceans may yield even more species.
- Although Symbion was first observed in the 1960s, it took many years for microscopy to become sufficiently advanced to be able to see the features which make this animal unique.

+ One of the scientists who discovered this species, Reinhardt Kristensen, has also discovered two other, completely new groups of animals. In 1983, he discovered the animals known as loriciferans and in 1994 the micrognathozoans. Both of these are very small marine animals that had simply been overlooked. The discovery of these animals is important as they add another branch to the already busy tree of life. They also show how animal life has evolved into a multitude of different forms to exploit different habitats. The animal life on earth is typically classified into 35 major groups (phyla) and more are being classified as our understanding of how animals are related increases. Our ability to decipher the genetic code of a species has shed an incredible amount of light on how different groups of animals are related. Reading the genetic code has shown us that animals previously thought to be only distantly related are in fact quite similar and vice versa.

+ The marine environment continues to throw up all sorts of biological surprises. One very rich habitat for new types of animal life is the intricate network of microscopic, water-filled channels and passages that exist between the sand grains of the seabed. The animals that dwell here are known as *meiofauna*. They squirm among the sand grains, eking out an existence in this strange, miniature world.

+ The sexual reproductive cycle of Symbion is triggered when the host lobster molts its skin in order to grow. In one way or another, its little passengers sense this and begin the production of males and females.

+ There is still much we don't understand about the life of Symbion. They are so small that it is very difficult to study live specimens without exposing them to the bright lights and high temperatures found beneath the gaze of a powerful microscope.

Further Reading: Funch, P., and Christensen, R.M. Cycliophora is a new phylum with affinities to Entoprocta and Ectoprocta. *Nature* 378, (1995) 711–14; Obst, M., Funch, P., and Giribet, G. Hidden diversity and host specificity in cycliophorans: A phylogeographic analysis along the North Atlantic and Mediterranean Sea. *Molecular Ecology* 14, (2005) 4427–40.

WATER BEARS

Scientific name: *Tardigrades*
Scientific classification:
  Phylum: Tardigrada
  Class: Heterotardigrada, Mesotardigrada and Eutardigrada
  Order: many
  Family: many
What do they look like? Water bears are tiny, compact creatures with four pairs of stumpy legs ending in claws. The majority are no longer than 0.3–0.5 mm, although some reach 1.2 mm. There are four body segments and the head ends in a snoutlike projection. Long filaments project from the body of some species. They can be red, purple, blue, olive, yellow, brown, or orange.
Where do they live? Tardigrades have been identified from a huge variety of habitats, although many species have been found inhabiting the water films surrounding terrestrial mosses and lichens. They are found all over the world.

Water Bears—When its habitat dries, the tardigrade becomes almost lifeless and takes on a barrel-like shape. (Mike Shanahan)

Clinging on to Life

Dormancy, in its many forms, is a common phenomenon in the animal world. Many creatures have evolved reproductive strategies, such as resistant eggs and so forth, to survive extreme conditions, but the water bears, tiny and overlooked though they are, must be the toughest animals on the planet. They inhabit ephemeral pools or the water film around mosses, lichens, or soil particles, and when their habitat dries up, they can enter a state of deathlike, suspended animation, known as *cryptobiosis* (Greek for "hidden life"). In this state, water bears have been shown to survive amazingly harsh conditions. They can tolerate temperatures ranging from close to absolute zero (much colder than liquid nitrogen) up to 250°C (hotter than an oven). They can also survive X-ray bombardment, the conditions inside a vacuum, and pressures equivalent to 6,000 atmospheres. Entering the state of cryptobiosis is no mean feat and is not something the water bear does lightly. The process takes several days, and the animal's body changes from that of a charming, stubby-legged critter to little more than a featureless speck that you would be hard pressed to identify as an animal. The changes include gradual dehydration where water molecules are replaced by glycerol and a simple sugar—trehalose. These take over the role of water in maintaining the structure of large molecules and cellular features (i.e., DNA, proteins and cellular membranes). It is fair to say that a water bear in a state of cryptobiosis is essentially a dried out husk in which the spark of life is vanishingly faint—the basic processes of metabolism fall to almost immeasurable levels (0.01 percent of normal). Although the water bear is on the cusp of death, cryptobiosis ceases as soon as life giving water returns. The animal slowly resumes its normal shape and continues with its life, apparently none the worse for wear. This ability not only makes the water bears incredibly tough, but it also makes them very long lived. No one is exactly sure how long these animals can remain dormant for, but it could be centuries or even millennia. The fact is that the ability to become cryptobiotic is a huge advantage to any animal living in a habitat that dries out periodically, as they can survive the lean periods, barely ticking over, and bounce back as soon as conditions become favorable again.

> ### ♀ Go Look!
>
> You can look for these animals yourself, especially if you have a microscope. A handful of moss or lichen can be soaked in distilled water for a few days and then squeezed or shaken to release the water bears, which can be examined under a microscope. Notice the four pairs of legs and the slow, sedate way in which they move. Your chances of seeing a water bear can be improved by removing a small quantity of moss or lichen from a clean habitat, as they are very sensitive to pollutants.

+ There are approximately 700 species of water bear, but due to their small size and habitat they are often overlooked, and it is highly likely that there are many more species as yet unidentified, especially in the deep oceans.
+ Due to the slow, deliberate motion and endearing appearance of these animals, Thomas Huxley, the English naturalist, gave them the name *water bears*, which seems to have stuck.
+ The water bear forms what is known as a *tun* during cryptobiosis. Their body contracts and any recognizable features disappear. The term *tun* comes from the similarity of these animals, in their deathlike state, to the large kegs once used to hold wine and beer.
+ With repeated bouts of cryptobiosis, a water bear's 30-month life span can be increased to at least 70 years, and possibly far longer.
+ Water bears in the cryptobiotic state can not only survive extremes of temperature and pressure but also huge doses of radiation. The dose of radiation required to kill a water bear in cryptobiosis is at least 1,000 times greater than the fatal dose for a human.
+ Relatives of the water bears, the rotifers and nematodes, are also captains of cryptobiosis. Even certain amphibians can tolerate periods of intense cold, falling into a dormant state with characteristics similar to cryptobiosis, but they lack the sheer hardiness of the water bears. The size and complexity of the general vertebrate body makes true cryptobiosis impossible; however, the survival of removed vertebrate organs can be greatly extended by cooling them and treating them with trehalose.
+ Some aquatic water bears also have an interesting approach to reproduction. These animals like many other invertebrates must shed their skin in order to grow, and it is at the point of molting that the male mates with the female and deposits his sperm into the space between the new and old skin. The female lays her eggs into the shed skin where they are fertilized. The young can then develop in the safety of this ready-made egg case.
+ Water bears feed by sucking up liquid. Some pierce plant cells with stylets that can be projected from the mouth, while some of the smaller tardigrades are predatory, piercing the bodies of other small animals with their stylets and sucking them dry.

Further Reading: Copley, J. Indestructible. *New Scientist* 23, (October 1999) 44–46; Copley, J. Putting life on hold. *New Scientist* 7, (November 1998) 7; Kinchin, I. M. *Biology of Tardigrades.* Portland Press, London 1994; McInnes, S. J., and Norman, D. B. *Tardigrade biology. Zoological Journal of the Linnaean Society.* Linnaean Society, London 1996.

WATER SPIDER

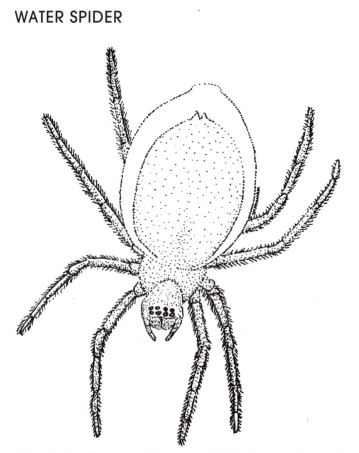

Water Spider—The water spider carries a bubble of air to its diving bell beneath the hairs on its abdomen. (Mike Shanahan)

Scientific name: *Argyroneta aquatica*
Scientific classification:
  Phylum: Arthropoda
  Class: Arachnida
  Order: Araneae
  Family: Agyronetidae
What does it look like? The water spider out of water is a rather drab specimen. It has a mousy, dark grey abdomen. The body of a fully grown female is between 8 and 15 mm long, although specimens 20 mm long are known. Males are between 9 and 12 mm long.
Where does it live? This spider is widely distributed throughout Europe and northern Asia, and where it occurs it can be often be found in quite large numbers. It prefers pools of water rather than streams and rivers, and it can be found in lakes, ponds, and ditches.

A Silken Diving Bell

Spiders, with their eight legs, tough little bodies, and ability to make webs and other silken structures are almost archetypal land lubbers—or so you would think. There is one spider, the water spider, which has moved from the land and back to the ancestral home of all life: the water. This amazing and rarely seen animal has retained the ability to produce silk from its back end, and in the aquatic environment, it has put this skill to very good use. In the sheltered confines of a pond or perhaps a large ditch, the spider spins what looks like a silken balloon anchored to stems of aquatic vegetation. To the underwater observer this structure is mysteriously silver, like an imperfect sphere of mercury suspended in the water. The silvery effect is actually due to the presence of air. The little, tightly woven silk balloon is actually a neat diving bell, and it is in this silvered chamber the spider resides. Like a normal, land-dwelling, web-building spider, this species spends much of it time in the confines of its retreat, drawing on its supply of air and waiting for a tasty snack to swim or drift by. The water spider is a fairly large spider, and with its potent venom, it can subdue animals as large as tadpoles—dashing out from its retreat to deliver a deadly bite.

A diving bell underwater is all well and good, but how on earth does this animal replenish its supply of air? It's not as though it has a support ship on the surface piping oxygen down to it. Each time one of these spiders builds a diving bell, it must ascend to the surface and collect air under the hairs that clothe its abdomen and legs. Enshrouded it what appears to be a suit of quicksilver, the spider descends to its bell and wipes the air from its body into the bell. As soon as the bell is full, it is a self-replenishing air supply. Oxygen from the surrounding water percolates in to the bell giving the arachnid a self sustaining, underwater breathing apparatus.

Even more fascinating is the way these animals go about mating. An ardent male comes along and, without a care for personal space, builds a diving bell right next to that of a mature female. If this were not presuming enough, he then proceeds to build a small tunnel connecting the two bells. Seemingly, the female is impressed by the male's behavior, and he breaks through the wall into her bell where the two mate. The female lays between 30 and 70 eggs in her diving bell, and the spiderlings that eventually hatch will leave this refuge and search for habitat of their own, traversing land where they have to.

- Of all the spiders, the water spider is the only species that has taken to an aquatic existence. There are other spiders that are semiaquatic, living on or around water. One such example is the raft spider. These big, impressive looking spiders can move around on the meniscus of small bodies of water, especially where there is abundant vegetation to act as a base for their rafting activities. They use the surface film of the water as a sort of web, detecting faint ripples of prey beneath the surface. They plunge through the meniscus and grab the prey. They are large enough to subdue small fish and amphibians.
- Although the water spider spends its life in an aquatic environment, it must still breathe air. Truly aquatic animals can extract the oxygen in water through the structures known as gills. Technically, the diving bell of the spider acts like a gill as oxygen diffuses into it from the surrounding water.
- The water spider can use its long legs to swim freely in the water while on its hunting forays.
- Although the water spider is quite big and has powerful fangs, robust enough to pierce human skin, they are eaten by fish and frogs. The usual prey of this spider is the larvae and nymphs of various aquatic insects and crustaceans. The captured prey of the spider is taken back to the diving bell where it is consumed.

+ Uncharacteristic for a spider, male diving spiders are the same size or larger than females. This reverses the normal situation of a male spider tentatively approaching a female when it comes to mating. In water spiders it is the female who must be wary of the male, for he is big enough to overpower and kill her.
+ The water spider is one of the only spiders in Europe that has a bite that can be considered to be even remotely dangerous. The bite is painful and the symptoms of itching, shivering, vomiting and fever can last for as long as three days.

Further Reading: Schütz, D., and Taborsky, M. Adaptations to an aquatic life may be responsible for the reversed sexual size dimorphism in the water spider, *Argyoneta aquatica. Evolutionary Ecology Research* 5, (2003) 105–117.

ZOMBIE WORM

Zombie Worm—The zombie worm's root-like projections house bacteria that break down the marrow in a dead whale's bones. (Mike Shanahan)

Zombie Worms—A zombie worm under a microscope. The trunk and feathery gills can be clearly seen. The root-like mass is the part of the worm that grows inside the whale bone. (Greg Rouse and MBARI)

Scientific name: *Osedax frankpressi*
Scientific classification:
    Phylum: Annelida
    Class: Polychaeta
    Order: Sabellida
    Family: Siboglinidae
What does it look like? These are small animals, the largest yet found are about as long as an adult's index finger and the thickness of a pencil. They lack a mouth, stomach, eyes, and limbs. There is a muscular, extendable reddish-colored trunk that bears a crown of red tentacles or plumes. Beneath the trunk there is a greenish, knotted, root-like mass.
Where does it live? These are animals of the seabed where they make their living on the carcasses of whales and other large, dead sea animals. Individuals have been found in very deep

water (2,800 m) off the Californian coast, while others have been found off the west coast of Sweden in water only 120 m deep.

Marine, Marrow Munchers

In February 2002, a remotely operated submersible discovered the decomposing corpse of a young grey whale 2,800 m under the waves at the bottom of the Monterey Canyon. The powerful lights of the submersible illuminated the remains, revealing numerous small, red tufts that were retracted as the manipulator arm investigated the bones. The operators back on board the mother ship could see they were a worm of sorts, but a type that had never been seen before. The submersible collected a section of bone and its covering of worms for examination at the surface.

The worms, upon inspection, turned out to be very odd indeed. The red tufts and muscular trunk was only a portion of the animal as the rest was concealed inside the bone of the deal whale. Careful dissection of the bone revealed the rest of the worm extended out in the marrow of the bone cavity in the same way as the roots of plant snake out through the soil. These so-called roots were green in color, and further examination showed they were packed with bacteria. The bacteria in these roots were digesting the fats and oils in the whale marrow, not only to sustain themselves but also to nourish the worm which housed them. This was the first time symbiotic bacteria had been found that thrived on fats and oils, and it enabled the worms to exploit the bonanza of a dead whale on the nutritionally poor seabed. The tufts of the worm are gills and are red because they have hemoglobin running through them.

The unique feeding behavior of these worms was not the only peculiar thing about them as it was soon realized that all the specimens they were seeing were females laden with eggs; there was no sign of any males. Eventually the males were found, but they were absolutely tiny and living inside the trunk of the female. One female worm was found to have 111 miniscule males living inside her. These males were scarcely more than larvae, but they were packed with sperm to fertilize the eggs of their mighty mate.

How the males find their partners in the vastness of the ocean is not fully understood, but it is probably in a way very similar to the spoon worms featured earlier in the book. The female worms release huge numbers of eggs. These eggs hatch into larvae that drift around the oceans sustained by a store of yolk. If they are exceptionally lucky, they may alight on the decomposing body of a whale and develop into a female, which sends her roots down into the nutritious marrow of the whale. If they are luckier still the larvae may settle on a female, in which case it undergoes the very minor transformation needed to turn it into a male before heading straight for the security of the female's innards. Those larvae that are unfortunate enough to miss a whale corpse or female of their species simply perish, a fate which befalls the vast majority of them.

+ The zombie worms are new to science, and there is a great deal still to learn about them. Currently, three species are known, *Osedax frankpressi*, *O. rubiplumus*, and *O. mucofloris*. The first two species were found on the grey whale corpse off California, discovered by Bob Vrijenhoek, a marine biologist at the Monterey Bay aquarium research institute, while the third was found on a minke whale that was purposefully deposited in 120 m of water off the coast of Sweden. The original two species led marine biologists to assume that zombie worms were creatures of very deep water, but the Swedish species showed that they are probably to be found in all regions of the seas as long as there are dead whales. Since then, as many as 11 new species of *Osedax* have been discovered.

+ It has been suggested that before hunting brought the populations of the great whales to their knees, many more species of zombie worm existed. The species we find today are maybe just a fraction of those that once sapped the marrow of dead whales.

+ Dead whales represent a feast for deep-sea animals, and soon after one hits the sea bottom, swimming, floating, and crawling creatures will converge from far and wide to take advantage of this bounteous supply of food. A dead whale, represents a 30–150 ton chunk of edible matter large enough to sustain huge numbers of deep-sea animals for many years. Normally, the only source of nourishment in the impoverished environment of the deep seabed is the matter which falls from shallow waters. This is known as marine snow, and a single whale fall is equivalent to approximately 1,000 years of this slow nutrient influx.

+ The zombie worms are only a few of the animals that take advantage of whale falls. Firstly, the flesh of the dead whale will be scavenged by sharks, fish, hagfish, and polychaete worms. Only when the blubber and muscle has been cleared can the zombie worms take up their positions on the exposed bones. Of course, many animals are drawn to the carcass to prey on the great density of scavengers and as result the huge, dead mammal supports an entire oasis of life for several years, even decades, until nothing but its bones lie amid the ooze of the seafloor.

Further Reading: Glover, A.G., Källström, B., Smith, C.R., and Dahlgren, T.G. World-wide whale worms? A new species of *Osedax* from the shallow north Atlantic. *Proceedings of the Royal Society: Biological Sciences* 272, (2005) 2587–92; Rouse, G.W., Goffredi, S.K., and Vrijenhoek, R.C. *Osedax*: Bone-eating marine worms with dwarf males. *Science* 305, (2004) 668–71.

GLOSSARY

Abdomen—the posterior segment of the body in arthropods.

Aestivation—cessation or slowing of activity during the summer; especially slowing of metabolism in some animals during a hot or dry period.

Alate—the winged form of an insect when both winged and wingless forms occur in a species.

Altruistic—of or pertaining to behavior by an animal that may be to its disadvantage but that benefits others of its kind, often its close relatives.

Amoeba—a type of protozoa that moves by means of temporary projections of the cell membrane and is well-known as a representative unicellular organism.

Amplexus—the clasping posture of fertilization in frogs and toads.

Arboreal—adapted for living and moving about in trees.

Arthropod—a phylum of animals that includes the insects, arachnids, crustaceans, millipedes, and centipedes.

Barbel—a slender, external process on the jaw or other part of the head of certain fishes.

Biomass—the amount of living matter in a given habitat, expressed either as the weight of organisms per unit area or as the volume of organisms per unit volume of habitat.

Brackish—slightly salty water.

Caste—one of the distinct forms among social insects performing a specialized function in the colony, as a queen, worker, or soldier.

Cellulose—a complex carbohydrate composed of glucose units; the main constituent of the cell wall in most plants.

Cephalothorax—the anterior part of the body in arachnids and certain crustaceans, consisting of the fused head and thorax.

Cercaria—juvenile digenean trematodes produced by asexual reproduction within a sporocyst or redia.

Cerebellum—the part of the vertebrate brain that is located below the cerebrum at the rear of the skull and that coordinates balance and muscle activity.

Cerebrum—the largest part of the vertebrate brain, filling most of the skull and consisting of two cerebral hemispheres divided by a deep groove; processes complex sensory information and controls voluntary muscle activity.

Chitin—a nitrogen-containing carbohydrate, related chemically to cellulose, that forms a semi-transparent horny substance and is a principal constituent of the exoskeleton, or outer covering, of insects, crustaceans, and arachnids.

Cilia—minute hairlike structures that line the surfaces of certain cells and beat in rhythmic waves, providing locomotion and movement of liquids and solids on inner and outer surfaces.

Class—the usual major subdivision of a phylum or division in the classification of organisms, usually consisting of several orders.

Coelom—the body cavity of higher multicelled animals, located between the body wall and intestine.

Collagen—any of various tough, fibrous proteins found in bone, cartilage, skin, and other connective tissue; collagen has great tensile strength and provides these body structures with the ability to withstand forces that stretch them.

Commensal—creatures living with, on, or in another organism, without causing injury to either organism.

Convergent evolution—in evolutionary biology, the process whereby organisms not closely related independently evolve similar traits as a result of having to adapt to similar environments or ecological niches.

Detritivore—an organism that uses organic waste as a food source.

Diurnal—an animal that is active by day; the opposite of nocturnal.

Echolocation—the sound-based system used by dolphins, bats, and other animals to detect and locate objects by emitting usually high-pitched sounds that reflect off the object and return to the animal's ears.

Ecosystem—a system formed by the interaction of a community of organisms with their environment.

Ectoparasitoid—a parasitoid that lives *on* another animal.

Endoparasitoid—a parasitoid that lives *in* another animal.

Equatorial—refers to any regions on or near the equator.

Exoskeleton—the rigid outer skeleton of arthropods, such as insects, arachnids, and crustaceans.

Family—the usual major subdivision of an order or suborder in the classification of organisms, usually consisting of several genera.

Fecundity—the quality of being fecund; the capacity, especially in female animals, of producing young in great numbers.

Food chain—a series of organisms interrelated in their feeding habits, the smallest being fed upon by a larger one, which in turn feeds a still larger one.

Food web—a series of interconnected food chains; the entirety of food chains in an ecosystem.

Gamete—a mature sexual reproductive cell, a sperm or egg, that unites with another cell to form a new organism.

Ganglia—a mass of nerve tissue existing outside the central nervous system (singular = ganglion).

Genus—the usual major subdivision of a family or subfamily in the classification of organisms, usually consisting of more than one species.

Global warming—the increase in global temperatures that results due to natural events and/or human activities.

Gonad—a sex gland in which gametes are produced; an ovary or testis.

Gondwanaland—a probable landmass in the Southern Hemisphere that separated many millions of years ago to form South America, Africa, Antarctica, and Australia.

Haltere—the vestigial fore or hind wings of certain insects (i.e., flies) that are important in the fine control of stability.

Hermaphrodite—an individual in which reproductive organs of both sexes are present.

Hyoid—a bone or group of bones at the base of the tongue that supports the tongue muscles.

Integument—the outer covering of an animal.

Invertebrate—any animal that lacks a vertebral column.

Keystone species—any species that is fundamentally important to the functioning of an ecosystem. The extinction of a keystone species can have huge repercussions for other species in the same ecosystem, including further extinctions.

Maw—the mouth of an animal.

Metacercaria—stage between the cercaria and adult in the life cycle of most digenean trematodes; usually encysted and usually quiescent.

Miracidium—first larval stage of digenean trematodes; often ciliated and often free swimming.

Molt—the process whereby an invertebrate, especially an arthropod, sheds its skin in order to grow.

Mother of pearl—a hard, iridescent substance that forms the inner layer of certain mollusk shells.

Mygalomorph—the primitive group of spiders to which the tarantulas belong.

Neurotoxin—a toxin that damages or destroys nerve tissue.

New World—the Western Hemisphere, which includes the Americas.

Niche—the position or function of an organism in a community of plants and animals.

Old World—Europe, Africa, and Asia.

Order—the usual major subdivision of a class or subclass in the classification of organisms, consisting of several families.

Palps—sensory appendages of arthropod mouthparts used to taste and manipulate food.

Parasite—an organism that lives on or in an organism of another species, known as the host, from which it obtains nutriment.

Parasitoid—any of various insects whose larvae are parasites that eventually kill their hosts.

Parthenogenetic—development of an egg without fertilization.

Pedipalps—in arachnids, the longer pair of appendages immediately behind the mouthparts.

Pelage—the coat of a mammal, consisting of hair, fur, wool, or other soft covering, as distinct from bare skin.

Peristaltic—of or relating to peristalsis, the progressive wave of contraction and relaxation of a tubular muscular system.

Pharynx—the tube or cavity, with its surrounding membrane and muscles, that connects the mouth and nasal passages with the esophagus.

Pheromone—any chemical substance released by an animal that serves to influence the physiology or behavior of other animals that it interacts with.

Phloem—the vessels in plants through which nutrients are transported.

Phoresis—a type of commensalism, whereby one organism uses a second organism for transportation.

Phylum—the taxonomic category that lies below the kingdom level and above the class level. A phylum groups together all classes of organisms that have the same body plan (i.e., the arthropods).

Plankton—the aggregate of passively floating, drifting, or somewhat motile organisms occurring in a body of water, primarily comprising microscopic algae, protozoa, and the larvae of marine invertebrates and fish.

Prehensile—adapted for seizing, grasping, or taking hold of something (i.e., the tail of a chameleon).

Proboscis—any of various elongate, flexible feeding, defensive, or sensory organs of the oral region.

Redia—larval, digenean trematodes produced by asexual reproduction within a miracidium, sporocyst or mother redia.

Refracted—pertaining to refraction; the change of direction of a ray of light, sound, heat, or the like in passing obliquely from one medium into another in which its wave velocity is different (i.e., from air to water).

Retina—a delicate, multilayered, light-sensitive membrane lining the inner eyeball and connected by the optic nerve to the brain.

Sedentary—animals that move about little or are permanently attached to something.

Sessile—permanently attached; not freely moving.

Sexual dimorphism—the differences in appearance that sometimes exist between male and female animals.

Sinus—any small cavity within the tissues of an organism.

Spawn—the synchronous release of eggs and sperm or larvae by aquatic animals.

Species—the major subdivision of a genus or subgenus, regarded as the basic category of biological classification, composed of related individuals that resemble one another, are able to breed among themselves, but are not able to breed with members of another species. (The definition of the term *species* has divided scientists for years. The definition used here is the most general and suits the purpose of this book.)

Spinneret—the small nozzles at the back end of a spider through which silk is extruded.

Spiracle—the closable pores on an arthropods body through which gases are exchanged.

Sporocyst—an asexual stage in the development of some trematodes.

Symbiosis—the living together of two (sometimes more) dissimilar organisms, usually referring to a relationship where both species benefit.

Thorax—the middle segment of an arthropod between the head and abdomen that bears the limbs and wings.

Tissue—an aggregate of similar cells and cell products forming a definite kind of structural material with a specific function, in a multicellular organism.

Torpor—a state of regulated hypothermia in a warm-blooded animal lasting for periods ranging from just a few hours to several months.

Troglodyte—an organism that is adapted to a cave-dwelling existence.

Vertebrate—any animal with a vertebral column.

SELECTED
BIBLIOGRAPHY

References that have an asterisk (*) beside them are heartily recommended as they are a treasure trove of information for anyone interested in the natural world.

Arnett, R. H., Jr. *American Insects: A Handbook of the Insects of America North of Mexico.* Van Nostrand Reinhold Co., New York 1985.

Barnes, R.S.K. *The Invertebrates: A Synthesis.* Blackwell Science, London 2006.

Bond, C. E. *Biology of Fishes,* 2nd ed. Saunders College Publishing, Fort Worth, TX 1996.

Brusca, R. C. and Brusca, G. J. *Invertebrates,* 2nd ed. Sinauer Associates, Sunderland, MA 2003.

Carrier, J. C., Musick, J. A., and Heithaus, M.R.E. *Biology of Sharks and Their Relatives.* CRC Press, Boca Raton, FL 2004.

*Foelix, R. F. *The Biology of Spiders,* 2nd ed. Oxford University Press, Oxford, UK and New York 1996.

Gauld, I. D. and Bolton, B. E. *The Hymenoptera by Gauld, Ian D. and Barry Bolton (Eds.) 1988. British Museum (Natural History).* Natural History Museum, Oxford University Press, Oxford, UK and New York 2006.

*Gullan, P. J. and Cranston, P. S. *The Insects: An Outline of Entomology,* 2nd ed. Blackwell Science, London 2000.

Hamlett, W. C. *Sharks, Skates and Rays: The Biology of Elasmobranch Fishes.* Johns Hopkins University Press, Baltimore and London 1999.

Hanson, P. E. and Gauld, I.D.E. *The Hymenoptera of Costa Rica.* Oxford University Press, Oxford 1995.

Herring, P. *The Biology of the Deep Ocean.* Oxford University Press, Oxford 2002.

*Hickman, C., Jr., Roberts, L. S., and Larson, A.. *Integrated Principles of Zoology,* 13th ed. WCB Publishing, Dubuque, Iowa 2006.

Hickman, C., Jr., Roberts, L., and Larson, A. *Animal Diversity.* McGraw-Hill, New York 2003.

Howell, W. M. and Jenkins, R. L. *Spiders of the Eastern United States: A photographic guide.* Pearson Education, Boston, MA 2004.

Kavanagh, J. *Pond Life: An Introduction to Familiar Plants and Animals Living in or near Ponds, Lakes and Wetlands,* 2nd ed. Waterford Press, Phoenix, AZ 2003.

Macdonald, D. *The Velvet Claw: Natural History of the Carnivores.* BBC Books, London 1992.

*Macdonald, D. *The New Encyclopaedia of Mammals.* Oxford University Press, Oxford 2001.

*Margulis, L. and Schwartz, W. H. *Five Kingdoms,* 3rd ed. Freeman and Company, New York 1998.

Nowak, R. *Walker's Mammals of the World*, 6th ed. The Johns Hopkins University Press, Baltimore and London 1999.

Perrin, W., Wursig, B., and Thewissen, J.G.M.E. *Encyclopedia of Marine Mammals*. Academic Press, Burlington, MA 2006.

*Pough, F. H., Janis, C. M., and Heiser, J. B. *Vertebrate Life*, 7th ed. Prentice Hall, Upper Saddle River, NJ 2005.

Quicke, D.L.J. *Parasitic Wasps*. Chapman and Hall, London 1997.

Reeves, R. R. *National Audubon Society Guide to Marine Mammals of the World (National Audubon Society Field Guide Series)*. Distributed by Random House, New York 2002.

Reid, G. K. *Pond Life: Revised and Updated*. St Martin's Press, New York 2001.

*Roberts, L. and Janovy, J., Jr. *Foundations of Parasitology*, 7th ed. McGraw-Hill Science/Engineering/Math, Columbus, OH 2004.

Robison, B. and Conner, J. *The Deep Sea*. Monterey Bay Aquarium Press, Monterey, CA 1999.

*Ruppert, E. E. and Barnes, R. D. *Invertebrate Zoology*, 6th ed. Saunders College Publishing, Fort Worth, TX 1994.

Service, M. W. *Medical Entomology for Students*, 3rd ed, Cambridge University Press, Cambridge, UK 2004.

Tricas, T. C., Deacon, K., Last, P., McCosker, J. E., Walker, T. I. and Leighton, T. *Collins Sharks and Rays*. Ultimate guide series. Collins, London, UK 1997.

Ubick, D., Paquin, P., Cushing, P. E., and Roth, V. *Spiders of North America: An Identification Manual*. American Arachnological Society, Poughkeepsie, NY 2005.

White, R. E. *Beetles—A Field Guide to the Beetles of North America*. Houghton Mifflin Company (Peterson Field Guide), Boston, MA 1998.

Zug, G. R., Vitt, L. J., and Caldwell, J. P. *Herpetology: An Introductory Biology of Amphibians and Reptiles*. Academic Press, San Diego, CA 2006.

INDEX

About the Author and Illustrator

ROSS PIPER has a Ph.D. in entomology from the University of Leeds. He lives in Hertford-shire, UK.

MIKE SHANAHAN has a doctorate in tropical rainforest ecology from the University of Leeds.